S0-DVE-651

THE HOME VIDEO SOURCEBOOK

by Jim McNitt

Collier Books
A Division of
Macmillan Publishing Co., Inc.
New York
Collier Macmillan Publishers
London

"Television programs, films, videotapes, and other materials may be copyrighted. The recording of such materials on videotapes without the permission of the copyright owner may be contrary to provisions of the United States Copyright Law."

Copyright ©1982 by Jim McNitt

All rights reserved. No part of this book may be reproduced or transmitted in any form or by any means, electronic or mechanical, including photocopying, recording or by any information storage and retrieval system, without permission in writing from the Publisher.

Macmillan Publishing Co., Inc.
866 Third Avenue, New York, N.Y. 10022
Collier Macmillan Canada, Ltd.

Library of Congress Cataloging in Publication Data
Main entry under title:

The Home Video Sourcebook.

1. Television—Apparatus and supplies—Catalogs.
2. Video recordings—Catalogs. I. McNitt, Jim.
TK6650.H65 621.388'33 81-15603
ISBN0=02=080790=2 AACR2

First Printing 1982

Printed in the United States of America

COVER DESIGN: BILL GRAEF
FRONT COVER PHOTO: RALPH HEIGL
BACK COVER ART: COURTESY OF *NEWSWEEK INTERNATIONAL* AND SSK&F ADVERTISING, INC.

About the Staff

Jim McNitt

Over the past decade THE SOURCEBOOK editor, photojournalist Jim McNitt, has carried out hundreds of assignments in more than 40 countries in Europe, Asia, Africa, South America, and The West Indies for publications such as Time-Life Books, Time Magazine, Newsweek, People, United Press International, Sail, Yachting, Boating, Sea Magazine, *and others. In 1980 McNitt served as a staff screenwriter on ABC's "Fantasy Island" series at the Burbank Studios in California—where the idea of THE HOME VIDEO SOURCEBOOK was hatched. More recently, Jim McNitt has written about home video for a number of business and general interest publications, including* Newsweek International. *He is also the screenwriter of a home video series—*Popular Photography Magazine's *"Photo School"—which is being aired on the Warner Amex cable TV system.*

Bill Graef

Design director Bill Graef is a 1979 graduate of Pratt Institute and is currently operating his own Manhattan design studio. Bill Graef's recent projects, in addition to THE SOURCEBOOK, include work for E.F. Hutton, The New York Stock Exchange, McGraw-Hill, St. Martin's Press, Penthouse Magazine, *and numerous New York-area advertising agencies.*

Jim Wright

THE SOURCEBOOK video reviewer, Jim Wright, is a graduate of Syracuse University's Newhouse School of Journalism. Since 1976 Jim Wright has been the movie critic for The Record, *New Jersey's largest evening paper. In 1981 he launched "On Video," the nation's first weekly newspaper column dedicated exclusively to reviews of video programming. In addition to his regular "On Video" and "Fine Tuning" columns, Jim Wright is the author of two children's books and scores of general interest articles that have appeared in publications such as* The Washington Post *and* Los Angeles Times.

Dave Delong

Boston-based writer and business consultant Dave Delong is a contributor to publications such as Inc. Magazine, The Boston Globe, *and* The Harvard Business Review *on subjects pertaining to business management and high technology. Dave Delong's contributions to THE SOURCEBOOK include the sections on home computers and video games.*

THE HOME VIDEO SOURCEBOOK

by Jim McNitt

Editorial Staff

Jim McNitt
Editor

Bill Graef
Design Director

Jim Wright
Video Reviews

Dave Delong
Computers, Video Games

Dina Boogaard, Don Hinkle
Editorial Research

Susan Graef
Art Editor

Betty Marsh
Illustrator

Rose Shigo
Type

Lou Lumenick
Editorial Assistance

Reg Bragonier, David Fisher
Editorial Consultants

Typography: The Summit Group, Madison, N.J.

CONTENTS

PART 1 That's Entertainment

The Best In Video

page 10

From Academy Award winners to X-rated movies, here's a connoisseur's guide to the very best in programming available on video tapes and discs.

PART 2 The Big Time Players

Selecting Your Videocassette Recorder

page 74

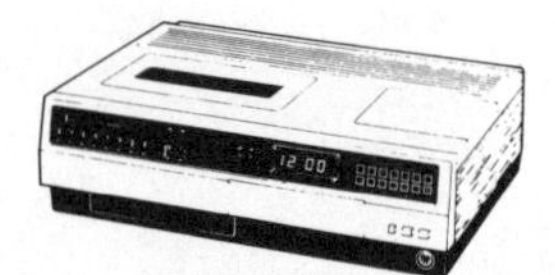

At last you can watch what you want, when you want! Here's how to decide which VCR format and features are best for you.

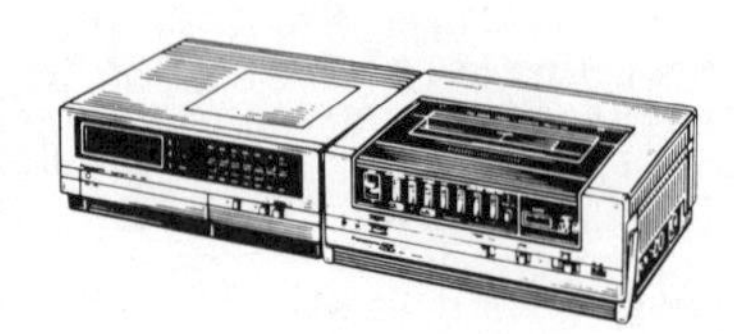

The VCR Buyer's Guide

page 84

The inside story on over 60 videocassette recorders with a model-by-model rundown on their best features—and potential drawbacks! You'll also find our nominees for Home Video's Top 20 VCRs.

Selecting Your Videodisc Player

page 118

Buyers of the early VCRs were left out in the cold when prices plunged and features changed overnight, making their machines obsolete. Could the same thing happen with the videodisc?

The Videodisc Buyer's Guide

page 124

A model-by-model review of what the discplayer manufacturers have to offer—including the brand new VHD format machines.

Videocassette Recorder Vs. Videodisc Player

page 128

What you can—and can't—do with each, along with some thoughts on which home video system may be best for you.

PART 3
The Supporting Cast

Here Come The Video Games
page 130

An amusement arcade in your living room? Here's the score on America's hottest-selling video games.

The Beginner's Guide to Home Computers
page 134

A close-up look at the 7 most popular personal computers and how to shop for them.

The Big Picture —Projection TV
page 140

These big screens are sharper, brighter, and easier to live with than ever before!

Video Camera Roundup
page 144

So long, Super 8! Here come the video cameras.

Video Software & Accessories
page 149

Everything you ever wanted to know about software and accessories—from cassette storage to duplicating prerecorded films. Cables, connectors, matching transformers, installation RF splitters, up-converters, video enchancers, TV commercial eliminators, copyguard stabilizers, video faders, spotting faulty tapes, and more.

Video Furniture
page 156

An illustrated guide to furniture classics for your space-age electronics.

70-Channel Television— A Buyer's Guide to Home Satellite Receivers
page 160

A shopper's directory to the brave new world of home satellite receivers.

The Best of Cable & Pay-TV
page 164

From blue movies to "The Blue Danube," cable TV is exploding across America. Here's what programming to expect from your local system—and how much you should be paying for it!

The Rental Alternative
page 169

Why spend $59.95 for a hot new Hollywood film when you can rent it for $5.99? In "The Rental Alternative" you'll discover why you needn't pay through the nose for top quality video programming.

The Do's & Don'ts of Discount Shopping
page 172

You can save a bundle through a discount house—but beware the pitfalls!

Videospeak
page 173

Behind the high-powered words—some very simple ideas.

INTRODUCTION

On the Threshold of the Video Frontier

High Technology is adding something new to the traditional Big Three constants of American life—birth, death, and taxes. That something is home video.

Business analysts predict by the end of the 1980s almost half of America's households will be equipped with some kind of a home video unit. The new generation of video inventions—cassette tape recorders, laser disc players, cable hookups, video action games, portable TV cameras, home computers, and even backyard satellite stations—offer an almost unlimited potential for personal creativity, entertainment, and education. They have also left those of us without a PhD. in electronics a bit bewildered and confused. And why not? It was only eight years ago that home video meant broadcast TV. Period.

To help you sort the facts from the fiction about the new world of home video equipment and programs is why THE SOURCEBOOK was born.

As on any new frontier, the Videoworld has its share of saints and sinners—the visionaries and pioneers as well as the con men and gyp artists. We have tried to steer you toward those firms whose products and services are known for their reliability and integrity. If you do have a serious problem with any of the products or companies listed in THE SOURCEBOOK, we would like to hear about it. Our address is: THE SOURCEBOOK, Box 2201, Green Village, NJ 07935.

For the first-time buyer of home video equipment, the biggest barrier is not so much the price (which now begins at about $450 for a discounted, no-frills videodisc player) or even the complexity of the equipment—most of which is designed to be incredibly easy to install and operate. The biggest barrier is understanding the language.

Video buffs and video experts have a jargon all their own. We call it Videospeak. Many of them seem to assume that everyone knows VCR stands for Videocassette Recorder and that only an ignoramus isn't aware that VHF means Very High Frequency—which happens to refer to the broadcast frequency range the rest of us know as channels 2 through 13. (UHF, another commonly used—and misunderstood—video abbreviation stands for Ultra High Frequency. You probably are familiar with UHF as channels 14 through 83.)

In THE SOURCEBOOK we've tried to keep the language simple and the jargon to a minimum. But if you don't understand a technical term or an abbreviation, just turn to the "Videospeak" section for a definition.

Home Video Programs (Software)

Most people get into home video because of the control a videocassette recorder or videodisc player gives you over *what* you watch, and *when* you watch it. But contrary to the impression sometimes created by an overzealous video salesperson, not every TV show, feature film, and educational program ever produced is available on pre-recorded cassette tapes and discs. In *videodiscs,* where the mastering and production facilities are still limited, the *combined* disc catalogs of MCA DiscoVision and RCA SelectaVision contain only a few hundred titles —although the level of technical and artistic quality is high indeed.

On the other hand, in pre-recorded *videocassette tapes* there is an apparent embarrassment of riches. In the category of movies and entertainment alone, THE VIDEO TAPE/DISC GUIDE (published by The National Video Clearinghouse, Inc., 100 Lafayette Drive, Syosset, NY 11791; price $12.95) lists over 4,000 pre-recorded programs. There's just one catch. For every gem you find available on videocassette, there are at least a dozen real dogs.

In their rush to cash in on the video boom, many program distributors have padded their catalogs with low-grade films and shows that aren't worth wasting the time it takes to read the screen credits, let alone the price of a rental or—heaven forbid—an outright purchase.

Another problem we've seen with pre-recorded videocassettes is their technical quality. Most distributors try to maintain an acceptable level of visual and audio quality on their tapes. But there are also firms that charge first-class prices for fourth-generation copies.

In "That's Entertainment—The Best of Video," TV and film critic Jim Wright reviews what he considers the best programs available in 26 different categories from dramas to horror films. He tells you why they are worth buying or renting, and where you are most likely to find a quality copy.

If you're thinking about buying a videodisc player, you'll want to pay special attention to "The Disc List" beginning on page 64. Videodisc units can only *play back* pre-recorded programs; they cannot record. "The Disc List" will tell you what shows are available for a videodisc system before you buy it.

Home Video Equipment (Hardware)

Selecting a home video unit can be about as simple as a graduate course in Quantum Mechanics.

If you have friends who are into home video, you should certainly seek their advice. The neighborhood video dealer is another source of information. But with two basic *mediums* (videocassette recorder or videodisc player) and five incompatible *formats* (called Beta and VHS in videocassete recorders and CED, Laser Optical, and VHD in videodisc players) to choose from—not to mention the

70-odd different models sold under 21 brand names—you have every reason to feel a bit confused by it all.

In "The Big Time Players" (starting on page 74) THE SOURCEBOOK introduces you to the world of videocassette recorders and videodisc players. Here you'll find the basic *facts* you need to decide which *medium* and which *format* is best for you. You'll also find a summary of prices, specifications, and features of every major brand and model of videocassette recorder and videodisc player—along with our recommendations on specific models we think offer you the most for your money.

Once you've invested in a home video unit, you may want to take a look at "The Supporting Cast" (beginning on page 130). Here you'll find answers to some of the most often asked questions about video software. Like, *"How should I store my cassette tapes? Or, should I try to splice a broken tape?"* You'll also find "Buyer's Guide" information on the whole potpourri of peripheral video goodies and accessories—video action games, home computers, cable and pay-TV, video furniture, big-picture projection television, video cameras, home satellite receivers, and video exchange and rental clubs.

One last word. The Videoworld is the leading edge of modern electronics technology. It is a new frontier which is not without its risks for both the manufacturer and consumer. Products, concepts, and even whole technologies are changing overnight.

Today's miraculous new video device may become tomorrow's electronic Edsel. It happened with some of the early brands of videocassette recorders—the Cartivison and V-Cord machines are now dinosaurs for which replacement parts and cassette tapes are virtually extinct.

It's possible that one of the new videodisc systems could go the same route. On the Videoworld frontier many waters remain uncharted. In THE SOURCEBOOK we've tried our best to steer you clear of the obvious rocks and shoals. Where we've seen a stormy horizon, we've tried to give you warning. Enter home video and you are crossing the threshold of a new frontier. It is a world of new adventure. But like all adventures, the Videoworld is not without its risks.

Home Video Market Saturation and Average VCR Retail Price

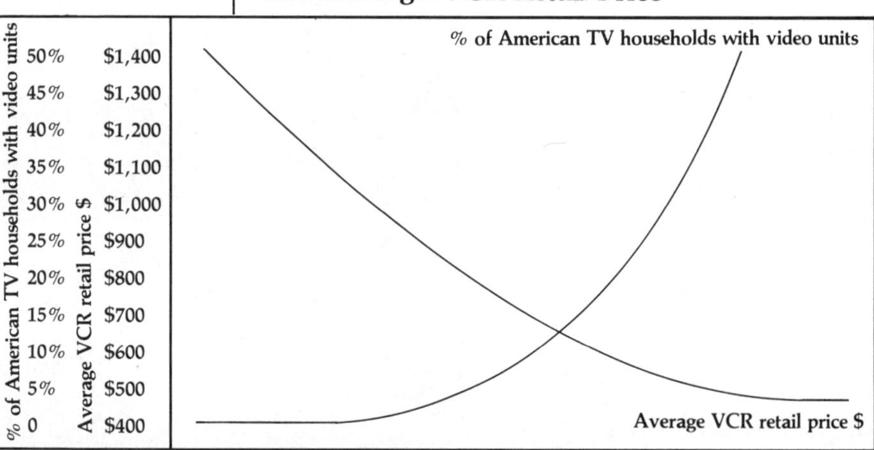

The Home Video Outlook—Production Up, Prices Down

The average VCR costs only $800 today, compared to $1,200 in 1977 and over $2,000 in 1975. Prices should continue to decline throughout the 1980s despite inflation. Mass volume production, more economical technology, and increased competition from the videodisc will all help contribute to the gradual retail price drop.

Even the most conservative business analysts expect the number of American households with video units to grow almost ten-fold by the end of the decade. By 1989 between 40 and 50% of American homes will own some kind of a video system.

©1982 THE HOME VIDEO SOURCEBOOK

THIS IS THE VIDEO WORLD

Your Video Display. Almost any TV in use today—b&w or color—is compatible with the current generation of video devices from VCRs to home satellite receivers.

The Videocassette Recorder. It can tape off your TV, or play back any of the thousands of pre-recorded cassettes. Record one show while you watch another, or up to six hours of programs while you're away. Pre-recorded cassettes of movies are $50 and up. List prices for VCRs start at $695 for the stripped-down Sanyo VTC 9100A.

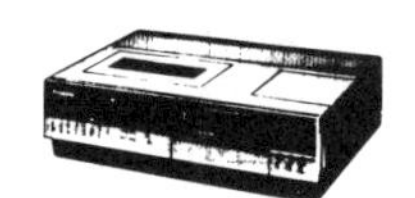

The Videodisc Player. The newest Videoworld "spin-off." You can play back, but not record. Audio and visual quality of some models is incredible. Programs cost from $6.95 to $29.95, but only a few hundred are available. RCA's SelectaVision videodisc player sells for under $500. Pioneer and Magnavox market laserdisc machines equipped with features you won't even find on Han Solo's *Millennium Falcon.* Both units list for about $700.

Cable & Pay-TV. It started as a way to get TV signals to fringe-area viewers, and has become a way of life for over 20,000,000 subscribers. Round-the-clock films and news, as well as expanded live sports coverage, are part of cable's appeal. Added cable TV services like Showtime, HBO, and the Z Channel offer uncut, uninterrupted feature films for a modest monthly charge. Another wrinkle in the pay-TV field is exemplified by Wometco Home Theater, a movie network that goes out over the airwaves instead of via cable.

Video Games & Home Computers. Your TV set can become an amusement arcade or a home computer. Atari's MP-1000 costs under $200 and can be programmed to play almost anything from "Combat" to "Space Invaders." Home computers like Radio Shack's TRS-80 or Mattel's Intellivision start at $300, will play arcade games, and can be programmed for various functions from personal astrology to income tax preparation and word processing.

Video Cameras. They're not a threat to Kodak yet, but video cameras are coming on strong. A basic b&w unit like Sanyo's VC1400 lists for under $200 and can turn the livingroom into a soundstage. Color cameras have crashed from $5,000 just a few years ago to as low as $750 for RCA's CC005.

Satellite Receivers. As the sign says over Steppenwolf's mysterious door: "Magic Theater: Entrance Not for Everybody." Satellite receivers are expensive, cantankerous, and let you home in on as many as 70 TV channels! Prices start at $2,500 and go up, and up, and up.

Projection TVs. With videodisc players providing high-quality TV images, the multinational electronics companies are jumping on the big-screen bandwagon. The result? A new generation that are cheaper, brighter, and less obtrusive than their forebears. List prices begin at $2,495 for Advent's 50″ screen.

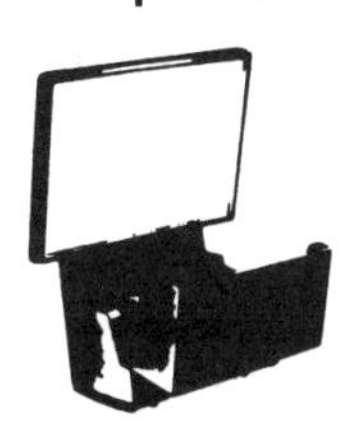

Video Furniture. Designed to accommodate the special needs of the VCR, videodisc-player, or home computer-owner, video furniture can be both attractive and affordable. Gusdorf offers one unit with side-by-side storage for home video and audio systems. List price is $350.

That's Entertainment—

PART 1

THE BEST IN VIDEO

In 1977, a small Midwest company called Magnetic Video had a crazy idea: Why not license popular films, record them on videotape, and sell them at $50 a pop to owners of videocassette recorders (VCRs). The entertainment industry scoffed—who in their right mind would pay that much money for a movie on cassette—and then nearly keeled over when the cassettes started selling faster than *Star Wars* T-shirts.

Today, prerecorded cassettes are a $500-million industry, and video discs are trying their best to hop on the runaway bandwagon. The result is a glut in video programming so immense (6,000 prerecorded cassettes and hundreds of discs) that it would take you 5 years of watching your video machine 8 hours a day to see everything once.

By the time you finished, of course, you'd have another 5 years of prerecorded movies and programs to catch up with. This section of THE SOURCEBOOK is designed to save you all that time and money, not to mention the wear and tear on your machine. What follows is a connoisseur's guide to the

©Warner Home Video

programming currently available: the best movies, concerts, vintage TV shows, and instructional programs on the market.

Here is our list of the *creme de la creme* of video programming. How do these shows differ from the thousands of others you'll find catalogued by major videotape and disc distributors like Nostalgia Merchant, Magnetic Video, or Fotomat? The answer is simple. We'd gladly invest the time and money to rent these programs on tape or disc. As for the first 25 films on the list—we've called them "The Best of the Best"—we would go even further. These are all films for which we'd be willing to spend $79.95, or more, to make a part of our permanent library.

In selecting "The Best In Video" we've no doubt omitted a few of your favorite films. No two persons' tastes are exactly alike—and not every film is available in video. What we have done is compose a list of video programs—feature films mostly, but also concerts, children's shows, vintage TV, and others—which we feel merit a long look.

Many of the selections are obvious—acknowledged masterpieces that will appeal to most movie buffs. But a few choices may surprise you. Some you may never have heard of before. In most cases we've tried to include shows that are representative of over 70 years of film-making, not merely the best of the past couple of decades. You'll find films like Eisenstein's *Potemkin*—a landmark in the evolution of film-making technique—as well as John Belushi's *Animal House*, a landmark film of a different sort. The common denominator? In our opinion, each ranks among the best within its genre.

The movies and programs have been chosen solely for excellence, not popularity. Blockbusters are not necessarily great films. When we think of *The Exorcist*, which made more than $100 million at the box office, we're reminded of the old men's room graffiti: "Eat manure—50 billion flies can't be wrong."

Should you want to amass a collection of the most popular films only, go to your nearest library and consult the year's first issue of *Variety* (the entertainment-industry weekly newspaper). There you will find the top grossing films of all-time. Incidentally, *The Exorcist* is ranked fifth, but it just may be the top grossing film of the bunch.

You may note a few of the all-time classics are missing from these reviews. The problem being that these films are not, as yet, legally available in video. At the introduction to each category we've mentioned some of these films.

(For a list of legal videotape and disc program distributors, along with a general description of the type of shows you'll find in their catalogs, see page 66.)

©Universal Pictures

About the Critic

Jim Wright is video columnist and movie critic of **The Record**, the largest evening newspaper in New Jersey. In January 1981, he launched "On Video," the first weekly column focusing solely on video in a major daily newspaper.

The newspaper's top movie critic for more than four years, Wright has also written two children's books for Putnam's and countless articles for national magazines—on subjects ranging from quadrophonic stereo to Wankel engines.

His profiles for the Associated Press on Andy Rooney of **Sixty Minutes**, Frank Oz (the man behind Miss Piggy and Yoda), and Walter Williams (the creator of "The Mr. Bill Show" on **Saturday Night Live**) have been featured in major newspapers across the country. He has also interviewed such noted film figures as James Stewart, Shirley Temple Black, Shirley MacLaine, John Milius, Brian DePalma, Martin Ritt, and John Carpenter. Additional video reviews were written by Lou Lumenick.

THE BEST OF **THE BEST**

©RCA

The royal pair in "The African Queen."

The easiest way to start a fight between movie buffs is to try to get them to agree on what films are the top 25 of all time. The following selections for a permanent video library will no doubt raise a few hackles, but keep in mind the criteria that went into the selections:

- The films must be legally available on videocassette or disc, which eliminates such great films as *Gone With the Wind, Star Wars, Dr. Strangelove, Kramer Vs. Kramer, Night Of the Hunter, A Star Is Born,* (the Judy Garland version), *Badlands, The Bank Dick, Five Easy Pieces,* and *A Streetcar Named Desire.*
- The films must be somewhat representative of seven decades of moviemaking.
- They must reflect the best in a range of genres, from cop films to comedies, from melodramas to musicals.
- They must stand up to repeated screening (this is a particularly tough criterion for comedies such as *American Graffiti,* which begin to pall after the fourth or fifth screening for many viewers). Most of the films on the list should get better each time you watch them, as the subtle qualities in direction, character interaction, thematic development, or script become more pronounced.
- They must translate well to the small screen. For example, a film such as *Days of Heaven,* which features breath-taking 70mm visuals and superior sound in theaters, gets dwarfed on the home screen; it's a great film, but many of its qualities are lost on video.

With those qualities in mind, here's one reviewer's list of the top 25 movies available on video:

The African Queen (1951) Directed by John Huston; written by James Agee; starring Humphrey Bogart and Katharine Hepburn; running time 105 minutes; color; United Artists—Magnetic Video, RCA Disc, Laserdisc.

A few years ago when the American Film Institute surveyed 35,000 of its members on the top 50 movies of all time, this John Huston film finished an incredible fourth—behind *Gone With the Wind, Citizen Kane,* and *Casablanca.* The melodrama was intended purely as a vehicle for its two stars, Bogart and Hepburn, but its stature continues to grow.

The casting—Bogart as a drunken steamboat skipper, Hepburn as a prim English missionary—was indeed kismet. The plot—the two unlikely allies join forces to escape from advancing German troops in Africa during World War I—has been imitated but never matched.

Noted film critic James Agee provided the screenplay, and cinematographer Jack Cardiff shot it on location in the heart of Africa. You can take the movie at face value—a rip-roaring yarn with ample doses of action and humor—or as an allegorical battle of the sexes and class structures.

What makes the film so delightful on video is that it improves with several screenings—Bogart and Hepburn play off one another so perfectly.

Academy Awards: Bogart, best actor.

Annie Hall (1977) Directed by Woody Allen; written by Allen and Marshall Brickman; starring Allen, Diane Keaton, and Tony Roberts; running time 94 minutes; color; United Artists—Magnetic Video, RCA Disc, Laserdisc.

Considered at first to be Woody Allen's breakthrough film—with Allen going from glib young comedy director to complete film-maker—*Annie Hall* looms more and more as Allen's crowning achievement, a masterful dramatic comedy about life and love in the Seventies. Set in Manhattan, Allen's home turf, the film is the tale of an on-again, off-again affair between two neurotics.

Watch it the first time purely for fun and the second time for the lines you missed from laughing too hard the first time. After that, watch more for art than antics. Allen really flexes his muscles as a director, using tricks ranging from subtitles to animation in a total tour de force.

Academy Awards: Best picture; Keaton, best actress; Allen, best direction; and Allen and Brickman, best screenplay.

Birth of a Nation (1915) Directed by David Wark Griffith; based on Thomas Dixon's *The Clansman;* starring Lillian Gish, Henry B. Walthall, Mae Marsh, and Miriam Cooper; running time 127 minutes; black and white with tints; Epoch—Blackhawk (this version has an added 7-minute prologue), Budget Video, Reel Images.

This film could also be called the birth of an art form—brilliant, controversial, and innovative. An epic about the American Civil War and the Reconstruction Era, the melodrama is regrettably marred by Griffith's blatant anti-black propagandizing (the Ku Klux Klan is portrayed heroically), but the film still succeeds as a technical marvel.

In *Birth of a Nation,* Griffith helped pioneer such then-obscure techniques as rapid editing and slow dissolves, and the Civil War action scenes—the battle of Bull Run,

"Birth of a Nation"—and birth of an art form.

©Blackhawk Films

the siege of Atlanta—shine.

On a movie screen, the film is a strong historical document (however askew its perceptions). On video—where you can replay all the key scenes—it's a primer on movie-making.

Bonnie and Clyde (1967) Directed by Arthur Penn; written by David Newman and Robert Benton; starring Warren Beatty, Faye Dunaway, Estelle Parsons, Gene Hackman, and Michael J. Pollard; running time 105 minutes; color; Warner Brothers—Warner Home Video, MCA Disc.

In 1967, with America knee-deep in Vietnam and the nation's youth approaching the winter of their discontent, along came a movie that turned the American myth-making machine upside down. Two murderous small-time hoodlums from the Depression—Bonnie Parker and Clyde Barrow—go on a killing spree in the Dust Bowl and come out looking like Maid Marian and Robin Hood, perfect anti-establishment heroes.

A bloody ballet, the film melds fine performances by Dunaway, Beatty, Parsons, and Hackman, crisp direction from Penn (play that final slow-motion death scene a few times to appreciate his skills), a sardonic screenplay by Benton *(Kramer Vs. Kramer)* and Newman *(Bad Company)* into a perversely riveting entertainment. WCI's cassette version and the MCA disc have trimmed the film's length by six minutes.

Academy Awards: Parsons, best supporting actress; Burnett Guffey, best cinematography.

Bridge on the River Kwai (1957) Directed by David Lean; written by Carl Foreman; starring Alec Guinness and William Holden; running time 161 minutes; color; Columbia—Time-Life Video.

David Lean's epic story of a Japanese prisoner-of-war camp in Burma during the World War II is a crackerjack action entertainment, with one of the most rousing finales in film. The melodrama can also be taken as an anti-war statement, about a man so consumed by orders—even the enemy's—that he has completely lost sight of the cause he was fighting for (an affliction most recently detected in the Watergate burglars).

The melodrama showcases the inestimable talents of Guinness, who plays a British POW brainwashed into supervising the construction of the titular bridge, a vital link in Japanese troop movements. Only Holden, as the leader of a commando squadron, has a chance to stop Guinness before the Japanese Army goes chugging through—but can he burn his bridges behind him? Since the film runs over 2½ hours, it's best seen over several sittings.

Academy Awards: Best picture; Guinness, best actor; Lean, best director; Carl Foreman, best screenplay; and Jack Hildyard, best cinematography.

Guinness is on top in "Bridge On The River Kwai."

©Columbia Pictures Home Entertainment

©Warner Home Video

"Bonnie and Clyde:" Small-time hoodlums in an all-time classic.

©Warner Home Video

Casablanca (1943) Directed by Michael Curtiz; written by Howard W. Koch, Julius J. Epstein, and Philip G. Epstein; starring Humphrey Bogart, Ingrid Bergman, Claude Rains, Paul Henreid, Peter Lorre, and Sydney Greenstreet; music by Max Steiner; running time 102 minutes; black and white; Warner Bros.—Magnetic Video, RCA Disc.

This romantic melodrama is the prime example of the sheer magic of movies: an albeit trite tale of wartime intrigue in Northern Africa that is transformed into a classic on the strength of first-rate acting by the entire cast, a script full of intrigue, romance, and humor, and Curtiz's able hand at the helm.

The script has been imitated, parodied, and canonized so much that it may seem a trifle overblown now ("Of all of the gin joints..."), but Bogart and crew manage to pull it off. Bogart plays an expatriate American saloonkeeper who harbors his former love and an escaped hero of the underground from the local Nazi-run constabulary.

The movie gets better as time goes by.

Academy Awards: Best picture; Koch, Epstein, and Epstein, best screenplay; and Curtiz, best director.

Chinatown (1974) Directed by Roman Polanski, written by Robert Towne; starring Jack Nicholson, Faye Dunaway, and John Huston; running time 131 minutes; color; Paramount—Paramount, Fotomat, Niles Cinema, RCA Disc.

One of the most under-rated films of the past decade, Polanski's *Chinatown* has such an abundance of riches that you need several viewings to realize what an achievement it is. The first time around, the show is Nicholson's, playing a Thirties L.A. detective who gets caught in an undertow of lies, corruption, and murder. On repeated viewings, the craftsmanship of Towne's script rises to the surface, particularly his development of the theme of appearance vs. reality. All the clues are there; none are what they seem.

It's also a first-rate Raymond Chandler-style detective story, a jigsaw puzzle in which Polanski fits each piece precisely into place. The tale gets timelier with each passing water shortage.

Academy Award: Towne, best screenplay.

©RKO

Welles raised "Kane" while Eastwood roared in "Dirty Harry."

Citizen Kane (1941) Directed by Orson Welles, written by Welles and Herman J. Mankiewicz; starring Welles, Joseph Cotten, and Agnes Moorehead; running time 120 minutes; black and white; RKO—Nostalgia Merchant.

Orson Welles's masterpiece—arguably the best film ever made—is the story of the rise and fall of powerful newspaper publisher Charles Foster Kane and, in effect, an exploration into the dark side of the American dream.

The narrative is told primarily in flashbacks, as Kane's life is pieced together through the reminiscences of the people who knew him. In the process, Kane's complex personality comes into full focus—a man torn by ambition, greed, and lost innocence.

The film is also superb technically—filled with innovations in editing, lighting, camerawork, and narrative structure. After 40 years, the film seems as crisp as the day it was made, and because of the advantages of video—particularly the ability to examine the film scene by scene, line by line—it is even more powerful now. A rosebud in full bloom.

Academy Awards: Welles and Mankiewicz, best screenplay.

The Deer Hunter (1978) Directed by Michael Cimino, written by Cimino et al., starring Robert DeNiro, Christopher Walken, John Savage, John Cazale, and Meryl Streep; running 183 minutes; color; Universal—MCA Videocassette.

After the debacle of Cimino's *Heaven's Gate,* critical opinion on this film has dropped a few notches, but it remains one of the most powerful dramas of the Seventies. The controversial Vietnam War epic brings home the insanity of war on the shoulders of three young GI's from a Pennsylvania steel town.

The film's central image—a devastating game of Russian roulette—was the perfect metaphor for the American involvement in Southeast Asia. Those who have condemned the film as fascist or racist (for its depiction of the Vietnamese) have missed the point—and the film's impact. DeNiro, Walken, Streep, and Savage led a first-rate ensemble cast. Once again, the film's length might make it advisable to watch in several sittings.

Academy Awards: Best picture; Walken, best supporting actor; and Cimino, best direction.

DeNiro gave it his best shot in "The Deer Hunter."

Dirty Harry (1971) Directed by Don Siegel, written by Harry Julian Fink, Rita M. Fink, and Dean Reisner; starring Clint Eastwood, Harry Guardino, and Andy Robinson; running time 103 minutes; color; Warner Bros.

I know what you're thinking, punk. What's a violent cop flick like this doing on anybody's 25 best list? The answer is simple. *Dirty Harry* is terrific: laconic Eastwood at his best, action director Don Siegel in top form, and plenty of themes to chew on.

The story is the running battle between Det. Harry Callahan and a crazed killer who threatens to execute a person a day until the City of San Francisco pays him a million-dollar ransom. A classic confrontation between a man and the darkest side of his own personality, the film has often been dismissed as fascist claptrap by liberals who don't bother paying close attention to what the film is saying. Alas, the film presaged several real California tragedies, including the Zodiac murders and the Chowchilla bus kidnapping.

The film was written by committee—reportedly going through nearly a dozen rewrites—but what a finished product. Each scene perfectly sets up the next with a minimum of frills, and an opening that sets up the best socko ending in the movies.

The General (1927) Directed by Buster Keaton and Clyde Bruckman; starring Keaton and Marion Mack; running time 80 minutes (approx.); b&w; US—Budget Video, Video Dimensions, Reel Images, Blackhawk Films.

Generally regarded as Buster Keaton's masterpiece, this silent comedy-adventure film is about a young Georgia engineer during the Civil War who tries to enlist in the Confederate Army but is rebuffed because he's considered vital as a train engineer. He finally gets his chance for glory when he goes behind enemy lines to rescue a train that has been captured by Union troops.

The glum-faced Keaton, who co-directed with Clyde Bruckman, shows how golden the silents were. Excellent sight gags devised by Keaton and some nifty action scenes carry the film over the top.

The Budget Video and Video Dimensions prints are recommended because they run the full 80 minutes; other versions are shorter, and *The General* is short to begin with.

The Godfather I & II (1971 and 1974) Directed by Francis Ford Coppola, written by Coppola and Mario Puzo; starring Marlon Brando, Al Pacino, James Caan, Robert Duvall, Robert DeNiro, Diane Keaton, John Cazale, and Talia Shire; running time 171 minutes (I), 200 minutes (II); color; Paramount—Paramount Home Video, Fotomat, RCA Disc, Laserdisc.

Arguably the best film saga of the Seventies, The Godfather story transforms Mario Puzo's pulp novel into a multi-layered study of power and an awesome 20th Century American tragedy.

The plot focuses on the changing shape of the Corleone family, the rulers of the New York mafia, as the aging Godfather (Don Vito Corleone, the underworld equivalent of Citizen Kane) is replaced by his reluctant young son—who soon warms to the job with a vengeance. The second film in effect sandwiches the

©Universal Pictures

©RCA

Brando stars in a classic about Mob rule.

first, as Vito Corleone's young manhood and the exploits of his increasingly power-mad son Michael unfold.

The ensemble acting is so uniformly brilliant that it requires several screenings to appreciate the interplay of characters and emotions. The scope of the narrative—encompassing 60 years and several generations—is awesome. And, to top it off, it's one heckuva gangster movie.

Academy Awards: Best picture (I & II); Brando, best actor (I); DeNiro, best supporting actor (II); Coppola and Puzo, best screenplay (I & II); Coppola, best director (II).

The Graduate (1967) Directed by Mike Nichols; screenplay by Calder Willingham and Buck Henry; starring Dustin Hoffman, Anne Bancroft, and Katharine Ross; running time 105 minutes; color; Avco-Embassy—Magnetic Video, RCA Disc.

This comedy is to the youth of the mid-Sixties what the *Rebel Without a Cause* was to the Fifties generation, but what a difference a decade has made. Both films tell of young men spiritually adrift from society, but *The Graduate* gets its point across with acerbic humor rather than shallow melodrama.

The graduate of the title is Ben, a disaffected middle-class kid who returns home to Southern California from college to find a real world filled with bankrupt values ("There's a great future in plastics," he's told) and dubious morality (in the form of the lecherous Mrs. Robinson).

The performances by Hoffman and Bancroft, Nichols's direction, Willingham and Henry's script, and Simon and Garfunkel's songs (Dave Grusin did the soundtrack) add up to one special movie.

Academy Award: Nichols, best director.

Hoffman and Bancroft indulged in extracurricular activites in "The Graduate."

©Magnetic Video

The Grand Illusion (1937) Directed by Jean Renoir; starring Erich Von Stroheim, Jean Gabin, and Pierre Fresnay; running time 111 minutes; black and white; Continental—Reel Images, Budget Video.

Jean Renoir's best-known film is an antiwar drama set in a German prison camp in 1917 as French prisoners attempt to escape. The grand illusion, for Renoir, is that any war will be the last. As history unfolded after the film was made, even Renoir's depiction of the nature of warfare and its own noble code of honor would be driven asunder by World War II, Korea, and Vietnam.

Erich Von Stroheim, as the quintessential German officer and prison commandant, is in top form. The film's subtle strengths and its ironic message improve with repeated viewings. French with English subtitles.

The Grapes of Wrath (1940) Directed by John Ford; written by Nunnally Johnson; starring Henry Fonda, Jane Darwell, Dorris Bowdon, and John Carradine; running time 129 minutes; black and white; 20th C-F—Magnetic Video.

Adapted from John Steinbeck's prize-winning novel. John Ford's epic about the migration of poor Okie farmers who head westward during the Depression in search of a better life boasts superlative performances by Jane Darwell and Henry Fonda (arguably his best film role), a then uncommon exploration of social injustices, first-rate cinematography by Gregg Toland and music by Alfred Newman, and—as usual—Ford's sure-handed direction. It's worth several video screenings just for Fonda's performance.

King Kong (1933) Directed by Merian C. Cooper; written by James Creelman and Ruth Rose, starring Robert Armstrong, Fay Wray, Bruce Cabot, and Frank Reicher; running time 105 minutes; black and white; RKO—Nostalgia Merchant, RCA Disc.

What can you say about a movie starring a giant gorilla who has Fay Wray eating out of the palm of his hand? This early version is infinitely superior to the recent remake. It's a truly memorable monster film, with dazzling special effects for its time, true pathos from a mechanical gorilla, ample suspense, and solid performances by the cast.

The story is about a movie producer who travels to Skull Island, captures Kong, and brings him back to New York—where he escapes in time to do his now-classic routine on the Empire State Building. This *Kong* is actually better on video than it has been on most telecasts because six minutes of original footage have been restored.

©RKO

©Warner Home Video

The Maltese Falcon (1941) Directed and written by John Huston; starring Humphrey Bogart, Mary Astor, Sydney Greenstreet, Elisha Cook Jr., and Peter Lorre; running time 101 minutes; black and white; Warner Bros.—RCA Disc.

This is perhaps the ultimate private eye film—actually the third version of the Dashiell Hammett novel. The story is simple: Detective Sam Spade tries to unravel the mystery of the missing falcon statuette and runs across enough double-crossing con artists to populate Times Square.

The film is close to perfect—in its casting, its performances, its plot nuances, and its direction.

Modern Times (1936) Directed and written by Charles Chaplin; starring Chaplin, Paulette Godard, Henry Bergman, and Chester Conklin; running time 87 minutes; black and white; United Artists—Video Connection, Magnetic Video, RCA Disc.

This social satire is vintage Chaplin, as he plays an assembly-line worker trying to find happiness in a world run by amok machines. The film's message about the tyranny of the machine age seems a mite stale, but the comedy—a string of marvelous sight gags—is superior.

The film is also notable for the debut of Chaplin's voice, albeit in a garbled song. It is still considered technically to be his last silent film and—to many film buffs—his best.

©United Artists

Chaplin put his muscle into "Modern Times."

On The Waterfront (1954) Directed by Elia Kazan; written by Budd Schulberg; starring Marlon Brando, Eva Marie Saint, Rod Steiger, Lee J. Cobb, and Karl Malden; running time 108 minutes; b&w; Columbia—Time-Life Video.

A social-commentary film about dockside corruption, *On the Waterfront* is a winner largely on the strength of Marlon Brando's "Method" performance as a palooka who lost his shot at the title when the Mob made him throw a big fight. Ultimately, however, he comes up the champ as he defies the waterfront union boss after the death of his brother.

The drama is as effective as the other Brando-Kazan collaboration, *A Streetcar Named Desire,* and—thanks in large part to some great on-location camerawork—not nearly so theatrical. The supporting cast is fine, too, even if Malden's priest is a tad over-earnest.

Ordinary People

Ordinary People (1980) Directed by Robert Redford; written by Alvin Sargent; starring Mary Tyler Moore, Donald Sutherland, Judd Hirsch, and Timothy Hutton; running time 124 minutes; color; Paramount—Paramount Home Video, RCA Disc, Laserdisc.

This melodrama, based on Judith Guest's best-seller about the emotional disintegration of a Midwestern family after the death of their son, marks a spectacular directing debut by Redford. The film represents the best qualities in American movie-making, a work of substance delicately crafted and effectively acted. It has few pretensions and many insights into common people torn by uncommon situations.

Hutton got an Oscar for best supporting actor, but he gets solid support from Sutherland, Moore, and Hirsch. The film won an Oscar for best picture, and Redford won one as best director. The nature of the film, a highly personal film, makes it a natural for the TV screen—but it is light years ahead of the standard made-for-TV fare.

Potemkin (1925) Directed by Sergei Eisenstein, starring Alexander Antonov, Grigory Alexandrov, Vladmir Barsky, and Mikhail Gomorov; running time 67 minutes; b&w; Thunderbird—Budget Video, Reel Images.

Eisenstein's silent Russian classic reconstructs an incident during the anti-Czarist revolt of 1905, in which sailors on the Battleship Potemkin rebel, only to be confronted by Cossack troops on the Odessa steps.

The film is a purist's heaven, full of innovative techniques that revolutionized the art of film-making—-particularly the intercutting and overlapping of varied camera shots to prolong and dramatize crucial scenes, from the first act of defiance (a sailor on KP duty symbolically breaks a plate) to the confrontation on the Odessa steps. The film is one of the first times that the dramatic powers of the medium itself were fully achieved.

Potemkin's worth several playbacks on your video machine just to appreciate its technical achievements.

The Searchers (1956) Directed by John Ford; written by Frank S. Nugent; starring John Wayne, Jeffrey Hunter, Vera Miles, Natalie Wood, and Ward Bond; running time 119 minutes; color; Warner Bros.—Warner Home Video.

No 25-best list would be complete without a John Ford western, and *The Searchers* is equal to anything else Ford has done. The film has grown enormously in stature in the past five years after such top younger directors as Martin Scorsese, Paul Schrader, and John Milius paid homage to it.

There's another reason for its new found popularity: The more times you see it, the richer it becomes in its techniques and metaphors. And with the advent of video playback machines, film buffs could watch it as many times as they liked. They've been spreading the word ever since.

At its heart is the classic tale of the quest: a Confederate veteran and his nephew go on a five-year search for a young girl who was abducted by Commanches when they massacred her family. What they search for and what they find, however, are two different matters.

If you pay close attention, you'll find more on Ford's mind than a straightforward western: some understated commentary on prejudices, blind idealism, and inflexibility.

Singin' In The Rain (1952) Directed by Gene Kelly and Stanley Donan: written by Adolph Green and Betty Comden; starring Kelly, Donald O'Connor, Debbie Reynolds, Jean Hagen, Rita Morena, and Cyd Charisse; running time 103 minutes; color; MGM—RCA Disc.

The best movie musical of all time, *Singin' In The Rain* simply sparkles from the first frame to the last. The story is a spoof of the movie industry in the era when the talkies first began replacing the silents; as one actress's star ascends, another's is sinking.

The movie is many splendored: Kelly's dazzling choreography and dancing, terrific tunes by Arthur Freed and Nacio Herb Brown, fine performances all around, and crisp direction by Donen. To see Gene Kelly do the title number is worth the price of the disc itself.

©MGM/CBS Home Video

Two winners from the nifty Fifties: Kelly in "Singin' In the Rain," Wayne in "The Searchers."

To Kill A Mockingbird (1962) Directed by Robert Mulligan; written by Horton Foote; starring Gregory Peck, Mary Badham, Philip Alford, and Brock Peters; running time 129 minutes; b&w; Universal—MCA Cassette,Disc.

Robert Mulligan has transformed Harper Lee's dark tale of Southern racism into a moving piece of cinema. The basic story is about a small-town attorney's defense of a black man accused of raping a white woman, but the atmospheric drama explores bigotry in its most fundamental form, a reaction to anyone who is different—be he a black man, a retarded man (Robert Duvall in a brief but affecting role), or just somebody from another neck of the woods.

Peck, as the defense lawyer, gives his finest performance. The two children who play his troubled son and daughter are remarkable. Horton Foote's script, mixing mystery and suspense with human insights, is also a winner.

The film, with its great acting and insights, works especially well on video.

Academy Awards: Peck, best actor.

©MGM/CBS Home Video

The Wizard of Oz (1939) Directed by Victor Fleming, written by Noel Langley, Florence Ryerson, and Edgar Allan Woolf; starring Judy Garland, Frank Morgan, Ray Bolger, Jack Haley, Bert Lahr, and Margaret Hamilton; running time 102 minutes; b&w; and color; MGM— CBS Cassette.

This terrific musical fantasy wasn't considered much when it was released in 1939—it had to compete with films the likes of *Gone With the Wind, Of Mice and Men,* and *Stagecoach*—but once it began to appear annually on television, it gradually became regarded as a classic.

An Alice-in-Wonderland type of story about a Kansas schoolgirl who dreams she has been whisked away by a tornado to the magical land of Oz, the movie has a beautiful if overly schmaltzy storyline, terrific music ("Over the Rainbow," "Off to See the Wizard"), and swell performances by Garland and crew.

George Lucas obviously saw this movie before concocting *Star Wars*—Dorothy becomes Princess Leia, The Cowardly Lion becomes Chewbacca, and The Tin Man becomes C3P0.

Academy Awards: Best score and best song ("Over the Rainbow").

THE BEST IN DRAMA

©Paramount Home Video

Linda Manz contributed to the daze of "Heaven."

The dramas celebrated in this section range from *Ten Days that Shook the World* (1925) to *The Elephant Man* (1980), from sweeping epics the likes of *Dr. Zhivago* to such small personal dramas as *An Unmarried Woman* and *Norma Rae.* They all have one common denominator, however. They are all decidedly human. dramas.

The characters that people them can be the born losers of *In Cold Blood,* a young rebel without a cause, or Charlton Heston's holy Moses in *The Ten Commandments,* but they all have something worth saying about the human condition.

Dozens of other terrific films weren't considered for this list because they are currently unavailable on video. They are worth a mention, however, so you might keep an eye out for them should they appear on cassette. *The Apartment, Bang the Drum Slowly, The Beguiled, Body and Soul, Brief Encounter* (the original 1946 version), *Casey's Shadow, Death of a Salesman, Easy Rider, The Manchurian Candidate, Mean Streets, Once In Paris, The Reivers, Taxi Driver,* and *A Thousand Clowns.*

As for some dandy dramas that are available, read on:

All About Eve (1950) Directed and written by Joseph L. Mankiewicz; starring Bette Davis, Anne Baxter, Gary Merrill, Celeste Holm, and George Sanders; running time 138 minutes; b&w; 20th C-F—Magnetic Video.

Mankiewicz won Oscars for writing and directing this Hollywood drama about an aggressive young actress (Baxter) whose rise to the top is littered with broken promises and betrayed friendships. The acting by Baxter, Davis, and Sanders is top-notch. Look for a very young Marilyn Monroe, whom Sanders refers to as "a graduate of the Copacabana School of Dramatic Arts." To appreciate Mankiewicz's flair for sharp dialogue, watch the film in small but attentive doses.

Anatomy of a Murder (1959) Directed by Otto Preminger; starring James Stewart, George C. Scott, Ben Gazzara, and Lee Remick; running time 161 minutes; b&w; Columbia—Time Life Video.

Several cuts above the average courtroom drama, *Anatomy* features two of America's finest actors, Stewart and Scott, as antagonists in the trial of an Army lieutenant accused of murdering a bartender who allegedly had raped his wife. While Stewart prevails, justice may not in this cynical melodrama. Stewart and Scott alone are worth a couple of playbacks.

Billy Budd (1962) Directed and written by Peter Ustinov; starring Ustinov, Robert Ryan, and Terence Stamp; running time 123 minutes; b&w; Allied Artists—Allied Artists Video.

The movies sometimes have trouble transforming literature to the screen, but Ustinov's intelligent adaptation of Herman Melville's novel about a young seaman's battle against tyrannical officers is superb. The b&w film translates especially well to the small screen since it emphasizes ideas and themes more than plot. The film may be tough to get a hold of on tape now that Allied Artists Video is kaput. MGM/CBS bought the rights, and they may release it one of these days.

Darling (1965) Directed by John Schlesinger; written by Frederic Raphael; starring Julie Christie, Dirk Bogarde, and Laurence Harvey; running time 122 minutes; b&w; Paramount—Magnetic Video.

Schlesinger artfully directed this bitter drama about the shallowness of upper-class London. The film features a young Julie Christie as an egocentric woman who manipulates the men in her life to get ahead. The strengths of the film are reflected in the Oscars it won: best actress (Christie), best script (Frederic Raphael). Fortunately, those are strengths that hold up well on video.

Days of Heaven (1978) Written and directed by Terrence Malick; starring Richard Gere, Brooke Adams, Linda Manz, and Sam Shepard; running time 95 minutes; color; Paramount—Paramount Home Video, Fotomat.

If ever there were a reason to own a big-screen TV, this is it. The film about a trio of young migrant farm workers in the Texas panhandle just before World War I is a visual masterpiece. Alas, much of the beauty is diminished on the TV screen. The story is about a love triangle, but it is also a portrait of America as it enters the world of the Twentieth Century—and that still comes through on video.

David Lean's "Dr. Zhivago:" A big-screen epic is right on track.

©MGM/CBS Home Video

"Hitler's Children": Propaganda par excellence. ©RKO

Doctor Zhivago (1965) Directed by David Lean; written by Robert Bolt; starring Omar Sharif, Julie Christie, Geraldine Chaplin, Rod Steiger, Alec Guinness, and Tom Courtenay; running time 197 minutes; color; MGM—MGM/CBS Video.

Lean's epic romance set against the Russian Revolution holds up surprisingly well, and the cast is solid—Sharif has never been better. Although some of its grandeur gets lost on the small screen, *Doctor Zhivago* is perfect for video because it can be best appreciated over several sittings.

The Elephant Man (1980) Directed by David Lynch; starring John Hurt and Anthony Hopkins; b&w; Paramount—Paramount Home Video, Laserdisc, RCA Disc.

Produced by Mel Brooks and filmed independently of the hit Broadway play, this drama is the story of John Merrick, a grossly disfigured young man who goes from London freak-show attraction to the darling of British society. Hurt, under several pounds of makeup, is exceptional in the title role, relying on voice and broad physical gestures to communicate his character's inner turmoil. Hurt's performance—and the stark black-and-white photography—stand up nicely on video.

The film is a perfect companion piece to Lynch's first film, *Eraserhead,* which—alas—is unavailable on video.

From Here to Eternity (1953) Directed by Fred Zinnemann; adapted from James Jones's novel; starring Burt Lancaster, Deborah Kerr, Montgomery Clift, Donna Reed, Frank Sinatra, and Ernest Borgnine; running time 118 minutes; b&w; Columbia—Time Life Video.

This steamy melodrama features a terrific cast in a first-class story about the turbulent period just before World War II in Pearl Harbor. Before the Japanese launch their sneak attack, the soldiers stationed in Hawaii drop a few emotional bombshells of their own. The film won Oscars for best picture, best director, and best supporting actor and actress (Sinatra and Reed, whose portrayal of a prostitute is somewhat obscured by censorship of the day). Play back that beach scene a few times.

Hitler's Children (1943) Directed by Edward Dmytryk; starring Tim Holt, Otto Kruger, Bonita Granville, Kent Taylor, and Hans Conreid; running time 83 minutes; b&w; RKO—Nostalgia Merchant.

A sleeper if there ever was one, this drama is about a German-American girl thrown into a concentration camp in the early days of World War II after the Nazis can't indoctrinate her. Originally considered pure anti-German propaganda, these days it stands on its own merits. The melodramatic ending is worth a few encores on tape.

Hud (1963) Directed by Martin Ritt; written by Irving Ravetch and Harriet Frank Jr.; starring Paul Newman, Melvyn Douglas, Patricia Neal, and Bandon DeWilde; running time 112 minutes; b&w; Paramount—Discount Video Tapes.

Newman, in the title role, gives his best film performance in this post-western drama about a modern-day cowpoke who rebels against the values of his father (Douglas). Although Hud's not much of a role model, his young nephew (DeWilde, who played the wall-eyed kid in *Shane*) idolizes him. Neal won an Oscar as best actress; Douglas as best supporting actor. The film works well on video, thanks to the moody black-and-white photography and the power of the performances.

In Cold Blood (1968) Directed and written by Richard Brooks; starring Robert Blake, Scott Wilson, and John Forsythe; running time 133 minutes; color; Columbia—Time Life Video.

Once upon a time, Robert Blake was an exceptional dramatic actor. In this dark drama about a pair of criminals who murder a family in Kansas and go on the lam, Blake is outstanding. Brooks' adaptation of Truman Capote's nonfiction novel is relentlessly bleak but arresting. The harrowing hangman's scene, with all its Freudian overtones, is worth several playbacks.

©Paramount Home Video

John Hurt played the thin-skinned "Elephant Man."

Newman clicked in "Hud." ©Paramount Home Video

The Informer (1935) Directed by John Ford; written by Dudley Nichols; starring Victor McLagen, Preston Foster, Heather Angel, and Una O'Connor; running time 91 minutes; b&w; RKO—Nostalgia Merchant.

Based on Liam O'Flaherty's novel about the Irish Rebellion, this early John Ford film is about a dim-witted man who betrays his friend to collect a reward. Ford won an Oscar as best director, McLagen won one for best actor, Nichols won one for best screenplay, and Max Steiner won one for best music—not a bad showing at all. Savor that moody cinematography the second time around.

Lawrence of Arabia (1962) Directed by David Lean; written by Robert Bolt; starring Peter O'Toole, Omar Sharif, Alec Guinness, Anthony Quinn, and Jack Hawkins; running time 185 minutes; color; Columbia—Video Communications; Time Life Video.

O'Toole plays the title role of an enigmatic British soldier who unites feuding Arabian tribes to battle the Turks in World War I. A thorough Lean production, it features solid performances, beautiful on-location cinematography, and a sprawling plot. Watch it over a few days; otherwise the sand will start to get to you.

©Paramount Home Video

Altman's "Nashville": American rhythms and blues.

James Dean spoke for a generation in "Rebel."

©Warner Home Video

Nashville (1976) Directed by Robert Altman; starring Henry Gibson, Lily Tomlin, Ned Beatty, Keith Carradine, and Ronee Blakely; running time 159 minutes; color; Paramount—Paramount Home Video, Fotomat.

Altman's scathing look at American values, set during a five-day country-music jamboree in Nashville, interweaves the stories of 24 characters into a compelling drama. The music—which features Keith Carradine's Oscar-winning song "I'm Easy"—is uneven, but Altman's aim is generally on target. The complex narrative structure is a dazzling piece of film-making, and it only gets better with close scrutiny on video.

Norma Rae (1979) Directed by Martin Ritt; written by Irving Ravetch and Harriet Frank Jr.; starring Sally Field, Beau Bridges, and Ron Liebman; running time 114 minutes; color; 20th C-F—Magnetic Video.

This romantic melodrama is set in a small Alabama town where a union organizer attempts to mobilize workers at the cotton mill where Norma Rae and her parents toil. It's a very old-fashioned film, heavy on emotional wallop and short on stylistic flourishes, but it is quite affecting. Field won an Oscar in the title role.

The Magnificent Ambersons (1942) Directed and written by Orson Welles; starring Dolores Costello, Joseph Cotten, Tim Holt, and Agnes Moorehead; running time 88 minutes; b&w; RKO—Nostalgia Merchant.

Citizen Kane was a tough act to follow, but Welles came up with a winner—the story of an early 1900s Midwest family that clings to tradition in the face of change. It's based on the Booth Tarkington novel, and it's quite affecting—even if it does seem to run out of gas near the end because of studio interference. It also features strong acting and striking b&w cinematography that stands up well on a TV screen.

A Man For All Seasons (1966) Directed by Fred Zinnemann; written by Robert Bolt; starring Paul Scofield, Robert Shaw, Orson Welles, Wendy Hiller, and Susannah York; running time 120 minutes; color; Columbia—Columbia Pictures Home Entertainment.

A well-made, no-frills drama about the life of Sir Thomas More, *A Man For All Seasons* has all the ingredients of a winner; tight direction, a highly literate screenplay, virtuoso acting (by Scofield and the supporting cast), and fine cinematography (by Oscar winner Ted Moore). Scofield, Zinnemann, and the film itself also won Academy Awards—not bad for a costume drama.

Romeo fell for a Hussey in Zeffirelli's tearjerker.

©Paramount Home Video

Romeo and Juliet (1968) Directed by Franco Zeffirelli; adapted from Shakespeare; starring Olivia Hussey, Leonard Whiting and Michael York; running time 138 minutes; color; Paramount—Paramount Home Video, Fotomat.

Get out your handkerchiefs. Zeffirelli's adaption of Shakespeare's classic play of teen-aged love's labours lost is heavy on the syrup, but it is also quite lilting. The acting, cinematography, and Zeffirelli's direction are all fine. Not recommended for cynics or diabetics, but a charming video movie nonetheless.

Rebel Without A Cause (1955) Directed by Nicholas Ray; starring James Dean, Natalie Wood, Sal Mineo, and Jim Backus; running time 105 minutes; color; Warner Bros.—Warner Home Video.

James Dean is outstanding in this story of a youth of the Fifties who tries to get his moorings after his family moves to a new town. The film seems a mite dated these days—the my-parents-don't-understand-me routine has worn quite thin—but Dean's performance will always be the film's saving grace. It makes *Rebel* a natural for video viewing.

©Paramount Home Video

Swanson and Holden in "Sunset Boulevard."

Sunset Boulevard (1950) Directed by Billy Wilder; written by Wilder, Charles Brackett, and D.M. Marshman; starring Gloria Swanson, William Holden, Erich von Stroheim, and Nancy Olson; running time 110 minutes; b&w; Paramount—Paramount Home Video, Fotomat.

The classic black comedy about Hollywood, Wilder's *Sunset Boulevard* is the tale of an aging silent-film star who, refusing to admit that she's over the hill, hires a young screenwriter to mastermind her comeback. Swanson and Holden shine, and the film still has staying power on the small screen. Swanson and von Stroheim's roles are semi-autobiographical.

The Ten Commandments (1956) Directed by Cecil B. DeMille; starring Charlton Heston, Yul Brynner, Anne Baxter, Yvonne DeCarlo, and a cast of thousands; running time 219 minutes; color; Paramount—Paramount Home Video, Fotomat, RCA Disc.

God's advice to Moses—take two stone tablets and don't call me in the morning—is the cornerstone of the most opulent Biblical epic ever made. It's essentially the life story of Moses from the bullrushes to the doorstep of the Promised Land, with stunning visuals and special effects and good performances all around. The ultimate Hollywood spectacle. Unfortunately, it's better in theaters. Face it: On TV, a cast of thousands looks like a cast of ants.

Ten Days That Shook The World (1927) Directed by Sergei Eisenstein and Grigori Alexandrov; running time 104 minutes; b&w; Thunderbird Films—Reel Images.

Eisentein's silent epic recreates the Russian Revolution of 1917, from Lenin's secret return from exile to the attack on the czar's winter palace. The drama uses actual locations and many of the real participants. An effective companion piece to *Potemkin, Ten Days* polishes the editing techniques and use of symbolism Eisenstein developed in the earlier film. The Reel Images cassette has English titles.

©Magnetic Video

Sarrazin and Fonda "They Shoot Horses"; last Fox Trot in Atlantic City.

They Shoot Horses, Don't They? (1969) Directed by Sydney Pollack; starring Jane Fonda, Michael Sarrazin, Susannah York, and Gig Young; running time 120 minutes; color; Cinerama—ABC, Magnetic Video, Laserdisc.

A bleak but compelling tale of Depression-era losers who try to find fleeting fame and fortune in a grueling dance marathon in Atlantic City, *They Shoot Horses* makes some mordant comments on the human race. It gets sharp performances by Fonda, Sarrazin, York, and Oscar-winning Gig Young as the sleazy owner of the dance palace.

The Turning Point (1978) Directed by Herbert Ross; written by Arthur Laurents; starring Anne Bancroft, Shirley MacLaine, Leslie Browne, Mikhail Baryshnikov and Tom Skerritt; running time 119 minutes; color; 20th C-F—Magnetic Video.

When Hollywood finally rediscovered the "woman's film" in the late Seventies, it did so in high style. This story about two aging ballerinas (played by MacLaine and Bancroft) deals with friendship and jealousy, the choices between career and family, and the business of ballet. If you're not a fan of the dance, the ballet sequences may be a bit *tutu* much (since they don't forward the plot all that much). That's what the fast-forward button is for.

An Unmarried Woman (1979) Directed and written by Paul Mazursky; starring Alan Bates, Jill Clayburgh, and Michael Murphy; running time 124 minutes; color; 20th C-F—Magnetic Video.

Paul Mazursky's best film deals with a woman who tries to piece her life back together after her marriage shatters. The film's greatest strength is Clayburgh, who takes a role of a lifetime and delivers it impeccably. The film's only weaknesses are the standard psychiatrist's couch scenes and an ending that seems a little too storybookish (Bates as the artistic knight in shining armor) in the context of the story. It's too pat a solution to Clayburgh's predicament. Still, it's a solid "little" film that comes across well on video.

Heston from on high in "The Ten Commandments."

©Thunderbird Films

THE BEST
COMEDIES

American Graffiti

Choosing the 25 best comedies available on video is tricky business. There are so many comedies in so many styles—from the crude antics of *Animal House* to the sophistication of *The Philadelphia Story*—that chances are we've neglected a classic here or there. Sometimes there's no accounting for tastes.

We have tried to make the list as representative as possible, including at least one work by the greatest comedy actors, writers, and directors: The Marx Brothers, Charlie Chaplin, Laurel and Hardy, Jack Benny, Peter Sellers, Neil Simon, Woody Allen, and Mel Brooks,

A handful of comedies weren't considered because they're aren't currently available on video, but they're worth watching if they pop up on TV: *Horsefeathers, It Happened One Night, The Awful Truth, It's a Gift, She Done Him Wrong, Monty Python's In Search of the Holy Grail,* and *Citizens Band* (a.k.a. *Handle With Care).*

Our apologies to the fans of Jerry Lewis and Bob Hope, who will no doubt find the following list utterly hopeless.

Adam's Rib (1949) Directed by George Cukor; written by Ruth Gordon and Garson Kanin; starring Katharine Hepburn, Spencer Tracy, Tom Ewell, Judy Holliday, Jean Hagen, and David Wayne; running time 101 minutes; b&w; MGM—MGM/CBS Home Video.

Cukor's battle of the sexes casts Hepburn and Tracy as a husband and wife legal team who find themselves on opposite sides in the trial of a women who shot her husband. It's grounds for delight. Crisp dialogue, crisp direction. Play those Hepburn-Tracy duels again.

"Adam's Rib": Tracy and Hepburn in top form.

©MGM/CBS Home Video

Airplane (1980) Directed and written by Jim Abrahams, David Zucker, and Jerry Zucker; starring Robert Stack, Julie Hagerty, Robert Hays, Kareem Adbul-Jabbar, and Lloyd Bridges; running time 88 minutes; color; Paramount—Paramount Home Video, Fotomat, Laserdisc.

A great spoof of all those hokey airport melodramas, *Airplane* is the strange saga of an airliner that comes down with a massive case of food poisonings. It's a good flick to rent, since the humor gets a little ragged after a couple of screenings.

American Graffiti (1974) Directed by George Lucas; written by Gloria Katz and Willard Huyck; starring Richard Dreyfuss, Candy Clark, Ron Howard, Paul LeMat, Harrison Ford, Cindy Williams, Bo Hopkins, and Suzanne Somers; running time 112 minutes; color; Universal—MCA, Fotomat, MCA Disc (stereo).

Before there was *Star Wars* came Lucas's *American Graffiti,* the story of the restlessness and fading innocence of American youth in the early Sixties. Set in a small California town on high-school graduation night, the comedy used radio rock 'n' roll as a framework for the action—cruising the main drag, carrying on, and coming of age. Solid performances all around—*Graffiti* launched many a career—and Lucas's direction is superb. Both the MCA tape and disc versions are the 1978 reissue, several minutes longer than the 1974 original and (in the case of the disc) in stereo sound.

©Paramount Home Video

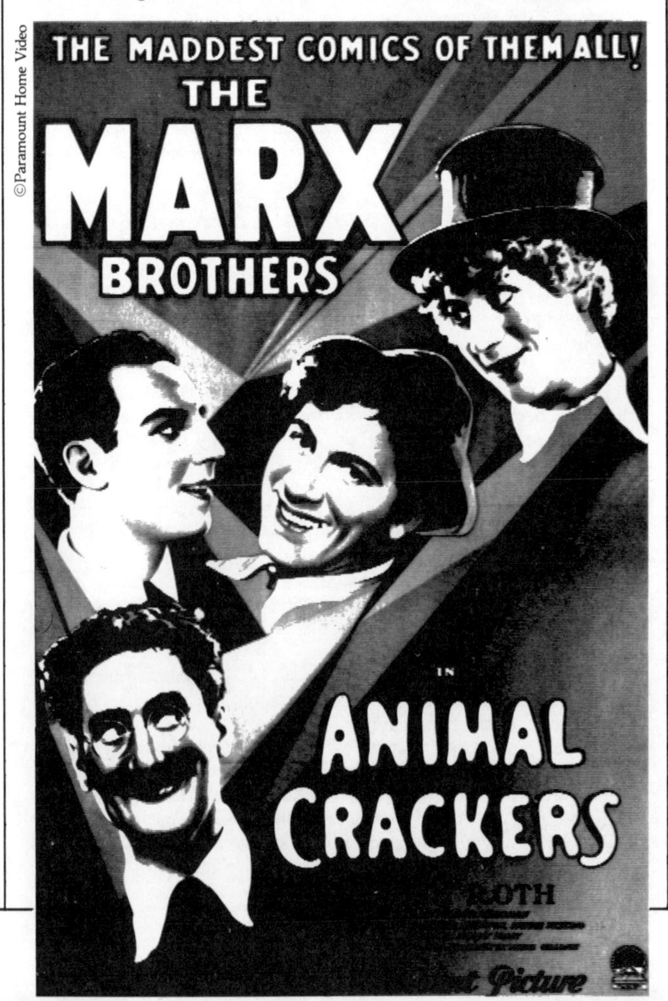

Animal Crackers (1930) Directed by Victor Heerman; written by Morrie Ryskind; starring the Marx Brothers, Lillian Roth, and Margaret Dumont; running time 97 minutes; b&w; Paramount—MCA Videocassette and MCA Disc.

The plot—crooks after a valuable painting—is a tad thin, and it's presented basically as a filmed stage play, but the Marx Brothers' rapid-fire exchanges are right on target. Hooray for Captain Spaulding and the whole crew.

Animal House (1977) Directed by John Landis; written by Harold Ramis, Douglas Kenney, and Chris Miller; starring John Belushi, Tim Matheson, John Vernon, Donald Sutherland, and Thomas Hulce; running time 109 minutes; color; Universal—MCA Videocassette, MCA Disc.

The quintessential gross-out movie of the Seventies, this fraternal farce is rude, crude, and lewd in the best sense of the words. It's the snobs against the slobs of Delta House, with Belushi singlehandedly breaking the Ten Commandments. The crass humor holds up surprisingly well on video—perhaps because you won't be embarrassed by laughing at the sick humor if you're in the privacy of your own home.

©Universal Pictures

Bedtime For Bonzo (1951) Directed by Frederick de Cordova; written by Val Burton and Lou Breslow; starring Ronald Reagan, Diana Lynn, Walter Slezak, Lucille Berkely, and Herbert Heyes; running time 83 minutes; b&w; Universal—MCA Cassette.

This movie may not prove that Ronald Reagan was presidential timber, but his presence in this skimpy social comedy is the main attraction. The plot? To prove that environment shapes character, a couple raises a champanzee as a baby. Fred de Cordova, now the producer-director of The Tonight Show, helmed this monkey business.

Being There (1979) Directed by Hal Ashby; written by Jerry Kozinski (from his novel); starring Peter Sellers, Shirley MacLaine, Jack Warden, and Melvyn Douglas; running time 130 minutes; color; United Artists—MGM-CBS Home Video.

Humor doesn't come any drier than in this portrait of a doltish gardener who becomes an advisor to presidents when people mistake his bizarre comments for insights. *Being There* is a fable about the Age of Television that has more charm than chuckles—a subtle commentary on an era when the press manufacturers celebrities and TV is the national nursemaid. You almost feel guilty watching it on video.

Breaking Away (1979) Directed by Peter Yates; written by Steve Tesich; starring Dennis Christopher, Dennis Quaid, Paul Dooley, Jackie Earle Haley, Daniel Stern, and Barbara Barrie; running time 100 minutes; color; 20th C-F—Magnetic Video.

This film has a refreshing style of humor that should appeal to anyone beyond the age of bubblegum cards—uplifting without being hokey, funny without being crass. It's about four dead-end kids out of high school, out of work, and out of sync with the snobby students on a nearby college campus. *Breaking Away* holds up just fine after repeated video viewings. It just might become a classic.

©Universal Pictures

City Lights (1931) Directed and written by Charles Chaplin; starring Chaplin, Virginia Cherrill, Harry Myers, Henry Bergman, and Jean Harlow; running time 81 minutes; United Artists—Magnetic Video.

An enchanting Chaplin once again plays his classic baggy-pants Tramp, becoming pals with a millionaire and befriends a poor blind girl. Call the movie sentimental, but Chaplin's pantomime is terrific. Play that spaghetti scene again.

Here Comes Mr. Jordan (1941) Directed by Alexander Hall; written by Seton I. Miller and Sidney Buchman; starring Robert Montgomery, Claude Rains, James Gleason, and Evelyn Keyes; running time 94 minutes; b&w; Columbia—Columbia Home Entertainment.

This Forties fantasy is about a prizefighter who dies in a plane crash because of a foul-up in Heaven's recordkeeping and gets sent back to Earth in the body of a just-murdered-millionaire. It's a little schmaltzy at times, but it works so well it's been imitated a dozen times. Warren Beatty's remake, *Heaven Can Wait*, comes darn close to matching it.

His Girl Friday (1940) Directed by Howard Hawks; written by Charles Lederer; starring Cary Grant, Rosalind Russell, Ralph Bellamy, Gene Lockhart, Ernest Truex, Clarence Kolb, Roscoe Karns, and Billy Gilbert; running time 92 minutes; b&w; Columbia—Budget Video, Cable Films, Video Dimensions, Reel Images, Thunderbird Films, and Video Warehouse.

This remake of Ben Hecht and Charles MacArthur's *The Front Page* is one of those rare birds—a movie that really did get better the second time it was made. The story focuses on a reporter (Russell) who wants to get married and retire, only to get snookered by her ex-husband and editor (Grant) into covering one last story. The cast is crackling, the pace is perfect, and the dialogue delightful.

I'm All Right, Jack (1959) Directed by John Boulting; written by Boulting and Frank Harvey; starring Peter Sellers, Ian Carmichael, Richard Attenborough, Margaret Rutherford, and Terry-Thomas; running time 104 minutes; b&w; EMI Videogram—Time Life Video, Fotomat.

The war between union and management gets a thorough thrashing in this British social farce about a rich nincompoop who lands a job in a warehouse. The candy factory sequence is utterly delicious, and the portrait of union labor—featherbedders *extraordinaire*—is so scathing you may never be able to punch a timeclock again. Catch it more than once.

The In-Laws (1979) Directed by Arthur Hiller; starring Alan Arkin and Peter Falk; running time 102 minutes; color; Warner Bros.—Warner Home Video.

This movie is so dumb that it's great. Falk and Arkin play an unwitting pair of international con men guilty of premeditated scene stealing. The action, which stretches from Manhattan to a Banana-Peel Republic, is silliness in the first degree.

It's A Wonderful Life (1947) Directed by Frank Capra; written by Capra, Frances Goodrich, and Albert Hackett; starring James Stewart, Henry Travers, Donna Reed, and Lionel Barrymore; running time 129 minutes; b&w; RKO—Nostalgia Merchant, Budget Video, Reel Images, Cable Films, Niles Cinema.

Stewart plays a hard-working banker in the American heartland who gets so depressed over life's failures that he decides to commit suicide. A guardian angel arrives to show him he's wrong. An unabashedly sentimental film about the American spirit, Capra's classic still holds up wonderfully well.

(Budget Video and Nostalgia Merchant's prints were taken from the original negative, so they're the best bets.)

The Ladykillers (1955) Directed by Alexander Mackendrick; written by William Rose; starring Alec Guinness, Cecil Parker, Katie Johnson, Herbert Lom, and Peter Sellers; running time 97 minutes; color; Continental—Time Life Video, Fotomat.

This British comedy features a stout Guinness and a band of crooks who plan the great train-station robbery, only to get tripped up by Guinness's little old landlady. The cast makes the most of sometimes skimpy material, and the grainy cinematography gives it a nice *Late Late Show* appeal on video.

A Letter To Three Wives (1948) Directed and written by Joseph L. Mankiewicz; starring Jeanne Crain, Kirk Douglas, Ann Sothern, Linda Darnell, Paul Douglas, and Jeffrey Lynn; running time 103 minutes; b&w; 20th C-F—Magnetic Video.

Mankiewicz is best known for his *All About Eve*, but the film he made just prior to it is top-drawer, too, and this one has more wry humor to it. The plot centers around three woman who discover that a neighbor (Celeste Holm, who narrates the story but is never on camera) has run off with one of their husbands. The problem is, they're not sure which one it is. The film translates nicely to the small screen, and the Linda Darnell-Paul Douglas courting scene is worth a few encores.

M*A*S*H (1970) Directed by Robert Altman; written by Ring Lardner Jr.; starring Donald Sutherland, Elliott Gould, Tom Skerritt, Sally Kellerman, JoAnn Pflug, Gary Burghoff, and Robert Duvall; running time 116 minutes; color; 20th C-F—Magnetic Video, RCA Disc.

This black comedy about a combat hospital during the Korean War is top-drawer Altman, and Gould and Sutherland are great as a pair of lecherous military MDs. If you have the movie on disc or cassette, you can skip the TV series. Sorry, but an Altman beats an Alda any day.

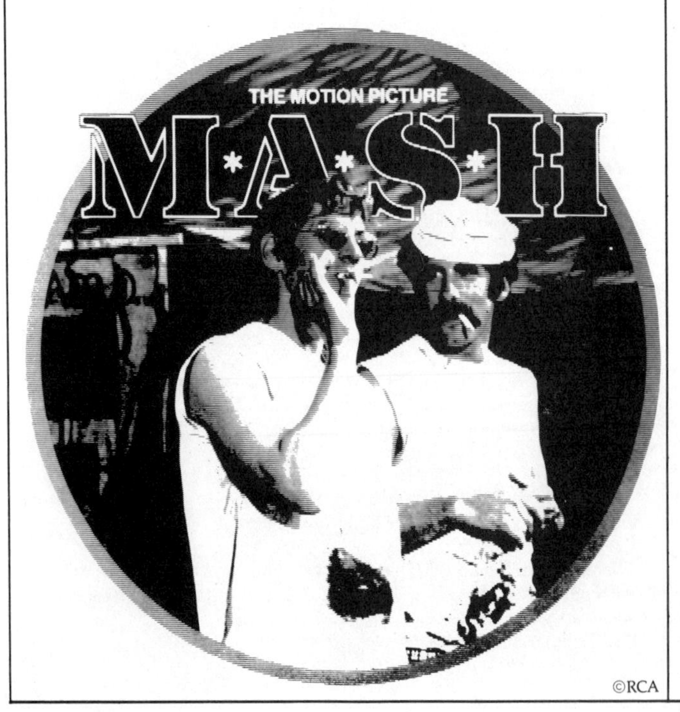

©RCA

©Columbia Pictures Home Entertainment

James Stewart, hayseed hero in "Mr. Smith."

Mr. Blandings Builds His Dream House (1948) Directed by H.C. Potter; written by Norman Panama and Melvin Frank; starring Cary Grant, Myrna Loy, Melvyn Douglas, and Reginald Denny; running time 93 minutes; b&w; RKO—Nostalgia Merchant.

Before you think about moving to the country, consult *Mr. Blandings.* A nifty social comedy about the ways of the woods, the film proves—once again—that Cary Grant is one of the best comedic actors ever.

Mr. Smith Goes To Washington (1939) Directed by Frank Capra; written by Sidney Buchman; starring James Stewart, Jean Arthur, Claude Rains, Thomas Mitchell, Edward Arnold, and Harry Carey; running time 129 minutes; b&w; Columbia—Columbia Pictures Home Entertainment.

Perhaps *the* vintage social comedy by Frank Capra, *Mr. Smith* presents James Stewart as a naive and idealistic young senator from Wisconsin who goes to Congress and gets taken in by the sleazy partisan politics of Claude Rains. Smith realizes the errors of his ways just in time to right wrongs and fall in love. The film takes an hour and a half to really get warmed up, but when it finally hits on all cylinders it really rolls. Good guys—and great movies— do come in first.

The Odd Couple (1968) Directed by Gene Saks; written by Neil Simon; starring Jack Lemmon and Walter Matthau; running time 106 minutes; color; Paramount—Paramount Home Video, Fotomat.

One of the many fine Lemmon-Matthau pairings, this comedy is Neil Simon at his slickest—a tale of two incompatible and incomparable pals who move in together. *Barefoot In The Park* (with Robert Redford and Jane Fonda) is even better Simon, and it has just become available on video. Keep an eye out for it.

The Philadelphia Story (1940) Directed by George Cukor; written by Donald Ogden; starring Katharine Hepburn, Cary Grant, James Stewart, Ruth Hussey, Roland Young, and John Halliday; running time 112 minutes; MGM—RCA Disc.

This sophisticated marriage-go-round features Hepburn, Grant, and Stewart as three sides of a Main Line triangle. Hepburn, is a snooty rich divorcee who can't decide whom to marry—her new fiance, a scandal-sheet reporter (Stewart) assigned to cover the wedding, or her playboy ex-husband. Tip-top on video.

The Producers (1968) Directed and written by Mel Brooks; starring Zero Mostel, Gene Wilder, Dick Shawn, Kenneth Mars, and Estelle Winwood; 88 minutes, color; Avco-Embassy—Magnetic Video, RCA Disc, Laserdisc.

Mel Brooks is like Bob Hope. You either love him or you hate him. The difference is that Hope is glib and predictable and Brooks is slightly bonkers. Watch *The Producers* at your own risk—a hallmark in bad taste—but play back that "Springtime for Hitler" song and dance more than once. Kenneth Mars, meanwhile, almost steals the show.

The Sting (1973) Directed by George Roy Hill; written by David S. Ward; starring Robert Redford, Paul Newman, Robert Shaw, Charles Durning, and Eileen Brennan; running time 129 minutes; color; Universal—MCA Videocassette, MCA Disc.

This flick could have qualified as a melodrama as well, a dandy Hollywood entertainment about a pair of conmen who get caught up in the ultimate shell game with a double-dealing gambler done in by his own greed. David Ward's script is a dandy, and Hill keeps the action rolling along. The only problem is the hammy acting by Redford and Newman, who wink at each other more than a pair of flirtatious fifth-graders. This is a perfect rental movie—it goes stale after a few viewings since so much of the film's zip comes from the surprise ending.

To Be Or Not To Be (1942) Directed by Ernst Lubitsch; written by Edwin Justus Mayer; starring Jack Benny, Carole Lombard, Robert Stack, and Stanley Ridges; running time 99 minutes; b&w; United Artists—RCA Disc.

One of the major reasons this black comedy has been included in this list is that it is one of the great unsung movies—with Jack Benny (in his best screen role) playing a Warsaw actor who impersonates a Nazi spy during Germany's invasion of Poland. Lubitsch is in top notch form. The film was Lombard's last—and near to her best—screen role. She died in a plane crash before the movie was released.

Way Out West (1937) Directed by James Horne; written by Jack Jevne, Charles Rogers, James Parrott, and Felix Adler; starring Stan Laurel, Oliver Hardy, Sharon Lynne, James Finlayson, and Rosina Lawrence; running time 66 minutes; MGM—Nostalgia Merchant.

The intrepid Laurel and Hardy hit the sagebrush to deliver a deed to a dead friend's gold mine to his daughter, who can't be found. Generally recognized as the duo's best film, *Way Out West* has terrific material and even better comedic timing.

(Nostalgia Merchant's cassette actually runs 86 minutes, because an added Zasu Pitts short, "Red Noses," runs 20 minutes.)

MGM presents
CARY GRANT
KATHARINE HEPBURN
JAMES STEWART
The things they do among the playful rich Oh boy!
The Philadelphia Story

©RCA

Shaw, Redford, and Newman: Three WASPs in search of a "Sting."

©Universal Pictures

THE BEST MOVIES FOR CHILDREN

bother most normal kids, and a film biography of Davy Crockett just wouldn't pass the muster without remembering the Alamo.

A few films for children weren't included because they aren't available on cassette or disc: *Star Wars, The Empire Strikes Back, Around the World in 80 Days, Cheaper By the Dozen, National Velvet, Sounder, Snow White, Cinderella,* and *Bambi.* Unfortunately, chances are slim to none that any of these will be marketed on cassette or tape soon.

©Walt Disney Home Video

MacMurray soars in "The Absent-Minded Professor."

If you've got kids, you probably took one look at this section, rolled your eyes, and thought: "Geez. The *last* thing my kids need is movies on video—they watch too much on the tube as it is."

You're probably right. If you can get your youngsters to read a book, build a model airplane, or clean the attic, more power to you. The reason for this section, however, is that if your kids are going to watch TV, at least they should be watching something better on Saturday mornings than poorly animated cartoons and commercials for fast foods, toys, and candy-coated cereals.

The criteria that went into the following selections is that they be good wholesome entertainment, that they appeal to small fry, and that they be interesting enough so that parents can watch without getting bored silly.

Only two of the movies—*The Bugs Bunny/Road Runner Movie* and *Davy Crockett*—have any real discernible violence. If you are very particular about TV violence, you may want to avoid these two. Otherwise, both *Bugs Bunny* and *Davy Crockett* are good entertainment: The cartoon violence in the Road Runner cartoons shouldn't

The Absent-Minded Professor (1960) Directed by Robert Stevenson; written by Bill Walsh; starring Fred MacMurray, Tommy Kirk, Keenan Wynn, and Nancy Olson; running time 97 minutes; b&w; Walt Disney—RCA Disc.

Back in the good old days, Disney Studios made some decent non-animated movies, and *The Absent-Minded Professor* is one of them. The story centers around that fabulous invention, flubber (sort of an anti-gravity goo) that enables MacMurray to take his Model-T airborne and outwit some dastardly spies. The special effects by Peter Ellenshaw and Robert A. Mattey were quite good for the time. It's silly but fun—and maybe a bit of nostalgia for Ma and Pa.

The Black Hole (1979) Directed by Gary Nelson; written by Jeb Rosebrook and Gerry Day; starring Maximilian Schell, Anthony Perkins, Ernest Borgnine, and Yvette Mimieux; running time 97 minutes; color; Walt Disney—Walt Disney Home Video.

This big-budget space fantasy did a bit of a bellyflop at the box office, but it's still fun for youngsters—lots of action, adorable robots, and weird spaceships. In other words, it's a third-rate *Star Wars.* But since *Star Wars* isn't around on video and *Star Trek: The Motion Picture* will probably bore most kids silly, *The Black Hole* is just about all she wrote. The plot deals with a mad scientist in outer space and the American astronauts who encounter him and dastardly robots on the edge of a black hole. If you're too old for this sort of stuff, play back some of those dandy special effects—especially the black hole finale—in slow motion.

The Black Stallion

The Black Stallion (1979) Directed by Carroll Ballard; written by Melissa Mathison, Jeanne Rosenberg, and William D. Wittliff; starring Mickey Rooney, Teri Garr, and Kelly Reno; running time 120 minutes; color; Omni-Zoetrope—Magnetic Video, Laserdisc.

This melodrama, based on the famous Walter Farley novel about a boy and his horse, is a bit plodding in spots, but it's one of the classiest family films around. Rooney and Reno are great, the story's just fine: The horse in question saves a little boy's (Reno's) life in a shipwreck and, with some coaching from Andy Rooney, the pair go on to run in the big horse race. Carmine Coppola's music is worth the price of a rental by itself. One minus: Caleb Deschanel's breathtaking cinematography gets wasted on the small screen.

Born Free (1966) Directed by James Hill; written by Gerald L.C. Copley; starring Virginia McKenna and Bill Travers; running time 95 minutes; color; Columbia—Columbia Home Entertainment.

The title song has been driven into the ground by now, but the movie itself is good clean-scrubbed entertainment about a Kenya game warden and his wife who raise three orphaned lion cubs and try to return them to the wild. The story is a little hokey, but the nature photography saves the day.

The Bugs Bunny/Road Runner Movie (1979) Directed by Chuck Jones and Phil Monroe; animated; starring Bugs Bunny, Daffy Duck, Elmer Fudd, the Road Runner, and Wile E. Coyote; running time 92 minutes; color; Warner Bros.—Warner Home Video.

This movie could also be called Chuck Jones' greatest hits, a compilation of the dandy old Warner Brothers cartoons including "Duck Dodgers In the 25th Century," "Duck Amuck," and "What's Opera,

Doc?" The animation and imagination are vastly superior to the junk on TV these days, and how can you top the cast? One note: These antics are a little much to take in one sitting. Catch the whole shebang over a weekend.

Charlotte's Web (1972) Directed by Charles Nichols and Iwao Takamoto; animated; voices by Debbie Reynolds, Agnes Moorehead, Paul Lynde, and Henry Gibson; running time 94 minutes; color; Paramount—Paramount Home Video, Laserdisc.
Hanna-Barbera's animation isn't the greatest, but it's tough to muck up a story as endearing as this E.B. White classic about a pig named Wilbur and his friendship with a spider called Charlotte.

©Magnetic Video

Kermit was fast on the drawl in "The Muppet Movie."

Davy Crockett, King of the Wild Frontier (1955) Directed by Norman Foster; written by Tom Blackburn; starring Fess Parker and Buddy Ebsen; running time 93 minutes; color; Walt Disney—Walt Disney Home Video, Fotomat.
Before Fess Parker played Daniel Boone and before Buddy Ebsen played Jed Clampett and Barnaby Jones, they were Davy Crockett and George Russell, a pair of Tennessee pioneers who battle renegades and Santa Ana's soldiers and fight for life, liberty, and the American way. Probably the first TV miniseries, the three-part Disney show was later incorporated into one movie. It's still good fun.

Grease (1978) Directed by Randall Kleiser; based on the long-running Broadway musical; starring John Travolta, Olivia Newton-John, Stockard Channing, and Frankie Avalon; color; Paramount—Paramount Home Video, Laserdisc.
One hesitates to recommend this sanitized musical because it's strictly for the under-14 set, with a paper-thin plot (young greaser falls in love with the sweet new girl in school) and tunes that sound more like disco than good old-fashioned rock 'n' roll, but what do kids care? The music has a good beat and you can dance to it. For grown-ups, it's jaunty music to do housework by.

Life With Father (1947) Directed by Michael Curtiz; written by Donald Ogden Stewart; starring William Powell, Irene Dunne, Edmund Gwenn, Zasu Pitts, and Elizabeth Taylor; running time 118 minutes; color; Warner Bros.—Budget Video, Cable Films, Reel Images.
This one is a great family fun about a New York City family in the 1880s—featuring an irascible father, his understanding wife, and their four precocious kids. Based on Clarence Day's boyhood reminiscences, the spirit is willing but the plot is a trifle thin (ol' Dad refuses to get baptized).

Miracle on 34th Street (1947) Directed and written by George Seaton; starring Maureen O'Hara, John Payne, Edmund Gwenn, and Natalie Wood; running time 96 minutes; b&w; 20th C-F—Magnetic Video.
Yes, Virginia, there is a Kris Kringle. And in this classic fantasy, he's a department store Santa who claims to be the genuine article, even if a little kid doesn't believe him. A good if sometimes overly sentimental story, it's best viewed during the holidays.

On Vacation With Mickey Mouse and Friends Directed by Jack Hannah; animated; starring Mickey Mouse, Minnie Mouse, Pluto, Jiminy Crickett, Goofy, and Donald Duck; 60 minutes; color; Walt Disney—Walt Disney Home Video, Fotomat, MCA Disc.
Since there's a dearth of Disney cartoons around, this short beauty will have to do. Jiminy Crickett is hosting a show but all the stars are vacationing. Jiminy goes after them, with amusing results. Mickey gets top billing, but the real star (as usual) is a duck named Donald.

The Muppet Movie (1979) Directed by James Frawley; written by Jerry Juhl and Jack Burns; starring Kermit, Miss Piggy, Edgar Bergen, Steve Martin, Milton Berle, Mel Brooks, and Madeline Kahn; running time 94 minutes; color; AFD—Magnetic Video, Laserdisc.
This film marked the movie debut of the Muppets, but it actually works better on the small screen (the cinematography wasn't too hot). The story takes Kermit from his lily pad to Hollywood, with various stops along the way. For romance, there's a not-too-kosher fling with Miss Piggy. The Muppets themselves are magical. Best scene: Play back Kermit's bicycle ride.

Kelly Reno in that dark-horse winner, "The Black Stallion."

©Omni-Zoetrope

Bugs starred in a 24-carrot cartoon feature.

©Walt Disney Home Video

Oh, God! (1977) Directed by Carl Reiner; written by Larry Gelbart; starring George Burns, John Denver, Teri Garr, and Donald Pleasance; running time 104 minutes; color; Warner Bros. —Warner Home Video, Fotomat.

George Burns stars in a Seventies version of an old-style Frank Capra human comedy with ecumenical overtones. Burns plays the title role, coming down from on high to ask grocery-store manager John Denver to spread the good word. Naturally, everybody thinks Denver is nuts—with lighthearted comedy to follow. Burns is the best thing about the flick and Denver is the worst, but it's solid family fare. Diehard true believers beware: The movie's just a tad irreverent.

Pete's Dragon (1977) Directed by Don Chaffey; written by Malcolm Marmorstein; starring Helen Reddy, Jim Dale, Mickey Rooney, Red Buttons, and Shelley Winters; running time 106 minutes; color; Walt Disney—Walt Disney Home Video, Fotomat.

Good old-fashioned Disney entertainment of a recent vintage, *Pete's Dragon* is a hokey story of a runaway orphan and a magical animated dragon named Elliott who gets him through more than few close scrapes in a Maine fishing village at the turn of the century. The supporting cast of adults is all-pro, with Jim Dale (of Broadway's *Barnum*) doing a nice turn as a snake-oil salesman. And Shelley Winters is always good for a chuckle. The original *Pete's Dragon* ran 135 minutes; the video version, nearly a half-hour less, cuts out several dead spots.

The Sound of Music (1965) Directed by Robert Wise; written by Ernest Lehman; music by Rodgers and Hammerstein; starring Julie Andrews, Christopher Plummer, Richard Haydn, and Eleanor Parker; running time 174 minutes; color; 20th C-F—Magnetic Video.

The hills are still alive with the von Trapps, an Austrian family trying to escape the Nazi terror just before World War II. A first-class film all-around, with tunes that'll have the family humming. Sure it's corny. So what? The movie is a bit long-winded, so several sittings are advised.

THE SOUND OF MUSIC

©Magnetic Video

Andrews was a mother superior in "Music."

Superman (1978) Directed by Richard Donner; written by Mario Puzo, Robert Benton, David Newman, and Leslie Newman; starring Christopher Reeve, Margot Kidder, Jackie Cooper, Gene Hackman, Ned Beatty, and Valerie Perrine; running time 128 minutes; color; Warner Bros.: Warner Home Video.

Christopher Reeve as the puckish Clark Kent carries this movie version of the Man of Steel, which is about as uneven as a film can get but great good fun for the kids nonetheless. The problem with the movie is that too much story to tell—from Superman's escape from Krypton and his Smallville schooldays to his first day at The Daily Planet and his duels with Lex Luther. Since the producers cut a few corners on some of the special effects, *Superman* works just as well on the TV screen as it did in 70mm.

Watership Down (1978) Directed and written by Martin Rosen; based on Richard Addams' novel; animated; voices by John Hurt, Richard Briers, Michael Graham-Cox, and Zero Mostel; running time 92 minutes; color; Avco Embassy—Video Communications, Fotomat.

This off-beat cartoon epic is about a group of rabbits—led by the foresighted Fiver—who set out to find a new home. The story is low-key by most animation standards, but it has some charming moments and a whole new perspective on the furry kingdom. The movie may be a little too sophisticated for viewers under the age of 8 or 9, but it works well on video. It was named Time Magazine's family movie of the year in 1978.

©Warner Home Video

MEET THE KIDISC

Kids and television: It's always been an unhappy marriage. At its best—which is to say very rarely—TV can be a child's window on the world, broadening horizons with shows such as "Big Blue Marble" or teaching ABC's on "Sesame Street." But TV has also become an electronic babysitter, an entertainment box filled at times with mindless cartoon violence and sugar coated advertisements.

No matter the quality of the programming, one broadside against television has been hard to rebut; By nature, TV fosters passivity in children. They sit and they watch, but they don't participate. In the words of author Marie Winn, TV is "the plug-in drug."

Thanks to the flexibility and ingenuity that video discs provide, that situation is slowly beginning to change. The breakthrough in area of disc programming for children is called "The First National Kidisc," and it represents a new approach to education video.

"The Kidisc," the second in a new series of interactive programs developed for the Pioneer and Magnavox LaserVision disc players, uses the unique features of the LaserVision system (particularly the ability to watch a program frame by frame at the push of a button) to allow younger viewers to play and learn as they watch.

The best way to describe "The Kidisc" is a video book of projects for children, with games, magic, riddles, and instructional material that teach kids all sorts of educational pastimes—many of which became passe when TV became so popular.

The disc is divided into 25 chapters, including segments on how to make paper airplanes, learn the sign language of the deaf, dance the Irish jig, tie knots, speak pig latin, devise secret codes, and do rope tricks. The disc also tries to balance the learning skills. A target game develops hand-eye coordination. A segment on a trip to the zoo teaches children to identify various animals. Another segment, "101 Jokes and Riddles," stimulates the imagination.

If that sounds like a lot of material for one 27-minute-long disc, keep in mind that the program is designed to be viewed frame by frame during many of the segments, and actual viewing time is considerably longer.

For example, a chapter on constructing and flying paper airplanes may run only 40 seconds if played at normal speed, but a child watching that segment will flip through it frame by frame with paper in hand, duplicating each fold as it is shown on the screen.

By the time the chapter is over (probably 45 minutes later), the child will have made some nifty airplanes, and chances are that he or she will spend the next hour flying them—instead of sitting idly by the TV screen.

And that's the whole theory behind "The Kidisc"—to get kids to learn by *doing* instead of merely watching. The segments, developed by Lin Oliver (who has her doctorate in child education), are merely a starting point for learning. Once a child learns how to speak pig latin or how to do assorted magic tricks, you can bet that he or she will demonstrate the new feats to family and friends.

The target age group for "The Kidisc" is 11 and 12 year olds, but younger children should be able to enjoy most of its concepts. Parents are encouraged to watch, too, supervising some of the more challenging projects (such as making a xylophone with glasses filled with water) and offering guidance when need be.

"The Kidisc," produced by Bruce Seth Green, retails for $19.95.

The other interactive disc now on the market is "How to Watch Pro Football," which uses the same frame-by-frame capabilities to teach viewers how to analyze football games with the expertise of a former quarterback. It's worth catching when football is in season again. Future interactive discs include programs on cooking and exercise, as well as sequels to "The First National Kidisc."

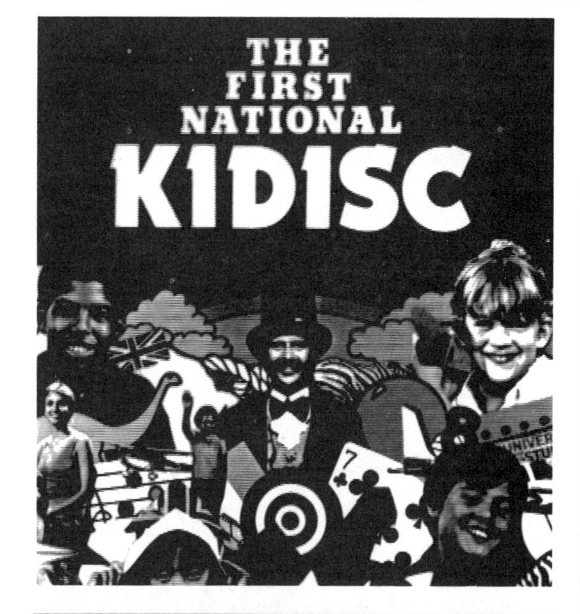

©MCA Videocassettes, Inc.

THE BEST OF THE WEST

If *The Searchers* is arguably the best western ever made, then a whole slew of horse operas are hot on its trail: *High Noon, Shane, The Wild Bunch,* the original *Stagecoach*. What puts them several notches above the typical Hollywood oater is that they transcend the sagebrush and come close to mirroring the spirit of the times in which they were made.

The old west has been America's prime source material for both heroes and mythology, and as the nation went through the turbulence of the Sixties and Seventies, the western embodied the fading of the American dream.

Hud, The Misfits, The Wild Bunch, Little Big Man, McCabe and Mrs. Miller, and the lesser westerns that followed weren't so much about the wild west as they were about the death of a spirit in America and the sense of national aimlessness created by the Vietnam War. For all intents and purposes the western as a genre died in 1974 when Mel Brooks lampooned it in *Blazing Saddles.*

Regrettably, a few westerns weren't considered for the following list because they haven't been available on video. The best examples are two Sam Peckinpah films, *Ride the High Country* and *The Ballad of Cable Hogue,* a Howard Hawks sleeper (starring the Duke) called *Rio Bravo,* Robert Altman's *McCabe and Mrs. Miller,* most of Clint Eastwood's spaghetti westerns, plus such popular sagebrush melodramas as *The Ox-Bow Incident, Little Big Man, The Magnificent Seven* (it should be released soon), and *A Man Called Horse.*

The westerns that have been selected for this section have one aspect in common: They all belie the famous 1929 quote by film critic Dwight Macdonald, who called westerns "the most vapid and infantile forms of art ever conceived by the brain of a Hollywood producer."

Butch Cassidy and The Sundance Kid (1969) Directed by George Roy Hill; written by William Goldman; starring Paul Newman, Robert Redford, and Katharine Ross; running time 110 minutes; color; 20th C-F—Magnetic Video, Laserdisc.

Purists may wince at this often cutesy story of the old Hole In the Wall Gang, but the direction is crisp and Newman and Redford play nicely off each other. Devised more as an entertainment than as a western, it features lots of humor and even a romantic musical interlude ("Raindrops Keeps Fallin' On My Head.") The plot follows the two legendary fun-loving train-robbers from the wild West to Bolivia, where they run afoul of the law for a final time. Play back that apocalyptic finale a few times.

Fort Apache (1948) Directed by John Ford; written by Frank S. Nugent; starring John Wayne, Henry Fonda, and Shirley Temple; running time 127 minute; b&w; Nostalgia Merchant.

Not quite a classic by Ford standards but a solid old-fashioned western nonetheless, *Fort Apache* pits cavalry officers Fonda and Wayne against each other—and, of course, against some feisty Indians. Wayne, as usual, stands tall, trying to make Fonda's foolishly ambitious Lt. Col. Thursday see the light about the hostile Apaches. For one of the first times in a major film, Indians were depicted as more than villainous heathens.

Gunfight at the OK Corral (1957) Directed by John Sturges; written by Leon Uris; starring Burt Lancaster, Kirk Douglas, Rhonda Fleming, Jo Van Fleet, and John Ireland; running time 122 minutes; color; Paramount—Paramount Home Video, Fotomat.

Wyatt Earp rides again in this John Sturges western, teaming with Doc Holliday to hook horns with the Clanton Gang in Tombstone, Arizona. The movie has a dull spot or two, but the shoot-em up finale is worth the wait. Lancaster and Douglas work just fine together.

Cooper and Kelly teamed in "High Noon."

©RCA

High Noon (1952) Directed by Fred Zinnemann; written by Carl Foreman; starring Gary Cooper, Grace Kelly, Thomas Mitchell, Lloyd Bridges, and Otto Kruger; running time 85 minutes; b&w; United Artists—Nostalgia Merchant.

Thanks to an intelligent script, Zinnemann's tight direction (the entire screen story takes place in an 85-minute span), and Cooper's Oscar-winning performance as a laconic lawman who—deserted by the townspeople—stands alone against four outlaws bent on revenge, *High Noon* ranks just a rung below *The Searchers* as the best of the West. Toss in a love interest with a sharp twist—newcomer Grace Kelly playing Cooper's Quaker, violence-loathing wife—and tick for tick *High Noon* is a standout. The grainy cinematography works quite well on video.

Hombre (1967) Directed by Martin Ritt; written by Irving Ravetch and Harriet Frank Jr.; starring Paul Newman, Frederic March, Richard Boone, Diane Cilento, and Cameron Mitchell; running time 111 minutes; color; 20th C-F—Magnetic Video.

Ritt, Ravetch, and Frank are one of the finest directing-screenwriting teams in Hollywood (they also teamed on such excellent films as *Hud* and *Norma Rae)*, and *Hombre* is as good an example as any of why the trio has been so successful. Ritt is an underrated, no-frills director who gets the most from his cast (Newman's best performances have been in Ritt films), and the husband-and-wife screenwriting tandem of Ravetch and Frank usually manages to transform standard stories into intelligent, socially relevant stories. In this case, Newman plays a half-breed who reluctantly helps a band of stagecoach travelers fight a vicious rustler.

Johnny Guitar (1953) Directed by Nicholas Ray; written by Philip Yordan; starring Joan Crawford, Mercedes McCambridge, Ernest Borgnine, and Sterling Hayden; running time 110 minutes; color; Republic—Nostalgia Merchant.

Don't let the title fool you. This offbeat horse opera, set in the days when men were men, is actually about two battling women—a feisty saloonkeeper and a ruthless land-grabbing banker—and good and evil in olden Arizona. Johnny Guitar, the object of their affections, plays second fiddle. Mommie Dearest and the rest of the crew manage to make some thin material memorable.

The Man Who Shot Liberty Valance (1962) Directed by John Ford; written by James Warner Bellah and Willis Goldbeck; starring James Stewart, John Wayne, Vera Miles, Lee Marvin, Edmond O'Brien, Andy Devine, and Woody Strode; running time 122 minutes; b&w; Paramount—Paramount Home Video.

What becomes a legend most? In this case, it's certainly not the truth. Stewart plays a tinhorn hero who gets the credit for killing the vicious Valance. Wayne actually did the deed but—as the movie says—skip the truth and print the legend. Stewart and Wayne, in one of their rare joint screen appearances, play perfectly off one another.

The Professionals (1966) Directed and written by Richard Brooks; starring Burt Lancaster, Lee Marvin, Claudia Cardinale, and Jack Palance; 117 minutes; color; Columbia—Time Life Video.

This hard-riding, hell-bent-for-leather western is the tale of a group of mercenaries who ride into Mexico to rescue a wealthy rancher's kidnapped wife and come back with more than they bargained for. The plot, filled with twists and double-crosses, will keep you hanging to the final reel. The cast, led by Marvin and Lancaster, is high caliber.

Red River (1948) Directed by Howard Hawks; written by Borden Chase and Charles Schnee; starring John Wayne, Montgomery Clift, Walter Brennan, and Joanne Dru; running time 125 minutes; b&w; United Artists—RCA Disc.

A great cattle drive along the Chisholm Trail is the backdrop for *Red River*, Howard Hawks' epic western that stars Wayne and Clift as a stubborn drover and his adopted son who feud over which way to take the cattle to Kansas City. One key to the film's success: the offbeat casting of Wayne and Clift—opposites in acting style and personality—to play the leads. Look for Shelley Winters in a bit role as a dance-hall girl.

Santa Fe Trail (1940) Directed by Michael Curtiz; written by Robert Bruckner; starring Errol Flynn, Olivia de Havilland, Ronald Reagan, Van Heflin, and Raymond Massey; running time 110 minutes; b&w; Warner Bros.—Thunderbird Films, Budget Video.

Errol Flynn made four westerns for Warners in the early Forties, and this one is the best of the bunch. Set on the Kansas range just before the Civil War, the story features such luminaries as George Armstrong Custer and Jeb Stuart and ends with the capture of John Brown—accuracy wasn't the film's strongest suit. It's still fun. Reagan, incidentally, plays Custer.

John Wayne was the consummate westerner in such John Ford classics as "She Wore a Yellow Ribbon" and "Stagecoach."

©Time-Life Video

Shane (1953) Directed by George Stevens; written by A.B. Guthrie Jr.; starring Alan Ladd, Jean Arthur, Van Heflin, Brandon de Wilde, and Jack Palance; running time 117 minutes; Paramount; Paramount Home Video, Fotomat, RCA Disc.

This George Stevens epic has all the trappings of a standard oater—troubled gunfighter trying to go straight, despicable cattle barons, and down-trodden farmers (who want to raise families, not cattle). And with its emphasis on old-fashioned virtues, *Shane* may seem corny at times, but watch it closely. Movies aren't crafted any better.

One fascinating aspect to watch for: Stevens' ability to set up all kinds of unspoken tensions between the gunfighter Shane and the sodbuster family he works for. Stevens also knows how to deliver a punch—witness the way he builds the drama in the escalating war between farmers and the cattlemen's hired hands. Scenes worth replaying: Jack Palance's gunfight with Elisha Cook Jr. and those incredible (and incredibly violent) barroom brawls.

She Wore a Yellow Ribbon (1949) Directed by John Ford; written by Frank S. Nugent and Laurence Stallings; starring John Wayne, Joanne Dru, John Agar, Ben Johnson and Harry Carey Jr.; running time 103 minutes; color; RKO—Nostalgia Merchant.

Probably John Ford's best cavalry picture, *Yellow Ribbon* features Ford's standard repertory company (led, as usual, by the Duke) and the glorious Monument Valley Landscape. Wayne plays a fort captain assigned to find out what the crafty Cheyennes are up to and, brother, does he find out. On video, the film gets dwarfed by the small screen, but it's a winner.

Stagecoach (1939) Directed by John Ford; written by Dudley Nichols; starring John Wayne, Claire Trevor, Thomas Mitchell, George Bancroft, and Andy Devine; running time 99 minutes; b&w;—Time Life Video.

At the time *Stagecoach* was released, one reviewer dubbed it a *Grand Hotel* on wheels–not a bad description. Wayne plays the outlaw The Ringo Kid, who joins six passengers on a stagecoach in the Monument Valley. En route to their destination are Indian attacks, fistfights, and the cavalry to the rescue. The movie made Wayne a star, but it's great in almost every department.

©Warner Home Video

3:10 To Yuma (1957) Directed by Delmer Daves; written by Halsted Welles; starring Van Heflin, Glenn Ford, and Felicia Farr; running time 92 minutes; b&w; Columbia—Time Life Video.

A no-frills plot—lawman Van Heflin holding outlaw Glenn Ford prisoner and holding off a mob until he and Ford can catch the train to Yuma—Daves' sharp direction, and good atmospheric cinematography by Charles Lawton Jr. keep *3:10* right on schedule. Also worth a look on similar themes: *Last Train From Gun Hill*, with Kirk Douglas.

Tumbleweeds (1925) Directed by King Baggott; written by C. Gardner Sullivan; silent; starring William S. Hart, Lucien Littlefield, and Barbara Bedford; running time 79 minutes; b&w; United Artists—Budget Video, Reel Images, Video Connection.

Ironically, the best silent western is about the end of the west, taking place in Oklahoma during the land rush of 1889, when the frontier was almost civilized. Hart plays Don Carver, a cowboy who tries to thwart a pair of evil land-grabbers. Lucien Littlefield, as Carver's sidekick (every cowboy needs a sidekick), adds the needed dashes of comic relief. Worth a few replays: that splendid land-rush scene at the end.

The Wild Bunch (1969) Directed by Sam Peckinpah; written by Peckinpah and Walon Green; starring William Holden, Ernest Borgnine, Robert Ryan, Edmond O'Brien, and Warren Oates; running time 127 minutes; color; Warner Bros.—Warner Home Video, MCA Disc.

Arguably Sam Peckinpah's best film, *The Wild Bunch* could be the most violent western ever made—a bloody saga set along the Mexican border just before the outbreak of World War I. An epic about the death of the west, the film has no heroes, only cutthroats who operate on either side of the law. Frame for frame, it's one of the bleakest portraits of human nature, opening with a group of Mexican children tormenting two scorpions and ending with a bloodbath. The film gets more resonant with each replay. Unfortunately, the video versions have cut six minutes from the picture's original-release form.

©Warner Home Video

"High Noon" may have been one of the western's finest hours, but Mel Brooks shot down the genre in "Blazing Saddles."

Alan Ladd put the wallop in the classic western "Shane."

©Paramount Home Video

THE BEST FOREIGN-LANGUAGE FILMS

This category, by its very nature, is a bit of a catch-all, encompassing everything from a 1941 Nazi propaganda film *(Triumph Of the Will)* to a slick 1978 French whodunit *(Cat and Mouse)*, so generalities on the body of selection aren't quite in order here.

An attempt was made to get a cross-section of foreign-language movies on video, but so many first-rate European films haven't been available that the choices are hardly representative. Notable in their absence are such great Ingmar Bergman films as *The Seventh Seal* and *Smiles Of a Summer Night* (which inspired the hit Broadway musical *A Little Night Music).* Francois Truffaut's *Day For Night* and *The 400 Blows,* and the magnificent recent German film, *The Tin Drum.* An excellent 1928 surrealistic film, *Un Chien Andalou,* directed by Luis Bunuel and Salvador Dali (and featuring the celebrated slit-eyeball scene), has not been included because it runs only 20 minutes. It's available from Budget Video.

Most of the films are in their native language with English subtitles, and the few that have been dubbed into English—such as *Swept Away* and *The Seduction of Mimi*—are generally worse off because of it. The advantages of a film in a foreign language with English subtitles are two-fold: If you speak the language, you get a truer film experience, and if anyone in the family has a hearing disability, the subtitles enable them to follow the stories much more easily.

"The Blue Angel": a professor obsessed.

©Thunderbird Films

Blue Angel (1930) Directed by Josef Van Sternberg; starring Emil Jannings and Marlene Dietrich; running time 93 minutes; b&w; German with English subtitles; Foreign—Thunderbird Films.

This melodrama, best known as the movie that made Marlene Dietrich a star, is the story of a staid professor (Jannings) who falls in love with a sultry nightclub singer (Dietrich) and who ultimately falls apart at the seams as a result—predating Wilbur Mills and Fanny Foxe by 45 years. Von Sternberg's direction, with some fluid camera movement, is exceptional for the period.

Cat and Mouse (1978) Directed by Claude Lelouch; starring Michele Morgan, Serge Reggiani, and Jean-Pierre Aumont; running time 107 minutes; color; French with English subtitles; Quartet Films—HomeVision.

Lelouch is better known for his romantic melodramas (notably *A Man and a Woman)* but he can be sharp with a straightforward mystery as well. *Cat and Mouse* is enormous fun, with an intricate plot, good suspense, and well-developed characters—a great film to rent for a Saturday night.

Diabolique (1955) Directed by Henri-Georges Clouzot; starring Simone Signoret, Vera Clouzot, Paul Meurisse, and Charles Vanet; running time 107 minutes; b&w; French with English subtitles; Clouzot—Budget Video.

Hitchcock didn't have a monopoly on bathtub scenes, as you'll see in this elaborate suspense story about a schoolmaster's mistress and wife who plot the most dastardly of deeds. It's sordid, but breathtaking. Lock the doors before you slide in the cassette.

8 1/2 (1963) Directed by Frederico Fellini; starring Marcello Mastroianni, Claudia Cardinale, and Anouk Aimee, running time 135 minutes; b&w; dubbed; Foreign—Time Life Video.

One of the great works of pure cinema, *8 1/2* is about a successful film director (a Fellini alter ego) battling self-doubts, sexual hangups, and religious guilt. The film is laced with bizarre fantasies and warped memories—often disjointed but affecting. Mastroianni, as the main character, is excellent. *8 1/2,* which won the 1963 Academy Award for best foreign-language film probably works best in small doses.

Hiroshima Mon Amour (1959) Directed by Alain Resnais; written by Marguerite Duras; starring Emmanuelle Riva, Eiji Okada, and Bernard Fresson; running time 88 minutes; b&w; French with English subtitles; Foreign—Budget Video.

One of the first films of the French New Wave, *Hiroshima Mon Amour* is an unorthodox bit of cinema that may require several viewings even to get the gist of. On the bare bones of the plot—a Frenchwoman who had a Nazi lover during World War II travels to Japan to exorcise her guilt by having an affair with a Japanese man—Resnais hangs his tale of Nuclear Age despair. If you're thinking of renting this film, you might try to borrow it from one of the video clubs with an open-ended rental plan. It's the type of film you watch, then come back to a few weeks later.

M (1930) Directed by Fritz Lang; starring Peter Lorre, Ellen Widmann, and Inge Langut; running time 95 minutes; b&w; German with English subtitles; Foreign—Budget Video.

A creepier film than M you're unlikely to find. Fritz Lang's classic is the story of a manhunt for a psychopathic child-killer in Dusseldorf. Lorre is at his most macabre, a fiend who seems childlike and vulnerable. The grainy b&w cinematography translates well to the small screen.

A Man and a Woman (1966) Directed by Claude Lelouch; starring Anouk Aimee and Jean-Louis Trintignant; running time 102 minutes; color; French with English subtitles; Allied Artists—Allied Artists Video.

One of the most commercial foreign films of the Sixties, *A Man and a Woman* is Lelouch's slickly packaged romance about a widow and widower who find a brief measure of happiness together. Lelouch's use of color—rich oranges for the love scenes, sepia for flashbacks—is most effective. But that title music sold a heckuva lot more tickets. Because of the demise of Allied Artists Video, this film might be tough to obtain.

Rashomon (1950) Directed by Akira Kurosawa; starring Machito Kyo, Toshiro Mifune, and Massayura Mori; running time 90 minutes; b&w; Japanese with English subtitles; Foreign—Thunderbird Films.

Kurosawa, one of Japan's most outstanding film-makers, has one of his best ideas to work with in *Rashomon:* Four people involved in a rape-murder recount their slightly varying versions of what really happened. Truth, for Kurosawa, is purely a matter of perspective. The film was remade, with less success, by Martin Ritt as a western called *The Outrage.* Catch the real thing.

Rififi (1954) Directed by Jules Dassin; written by Rene Wheeler and Dassin; starring Jean Servais, Carl Mohner, Robert Manuel, and Dassin; running time 115 minutes; b&w; French with English subtitles; UMPO—Budget Video, Thunderbird Films.

Before *Topkapi, Oceans 11,* or any of the other slick caper films came *Rififi,* the archetype of the genre and a dandy tale of a French jewel robbery—and the four crooks who prove to be their own undoing. The 20-minute hold-up scene, done in complete silence, is worth a few replays.

Rules Of the Game (1939) Directed by Jean Renoir; starring Marcel Dalio, Nora Gregor, and Renoir; running time 110 minutes; b&w; French with English subtitles; Foreign—Budget Video, Video Communications.

Next to *The Grand Illusion, Rules Of the Game* is Renoir's best film,a biting satire of the decadent French upper class before World War II. The plot—the weekend carryings-on of amoral guests and servants at a country chateau—provides the framework for Renoir's skewering of French morality, and a disquieting blend of comedy and melodrama.

The Seduction of Mimi (1974) Directed and written by Lina Wertmuller, starring Giancarlo Giannini and Mariangelo Melato; running time 92 minutes; color; dubbed; New Line Cinema—Magnetic Video.

Lina Wertmuller, in retrospect, seems to have been a flash in the pan. The Italian director made some brilliant political farces in the Seventies, then suddenly fell from grace. In this one, Giannini plays a working-class fool who tries to beat the system and stumbles his way through a series of tight jams—very funny but always bittersweet. The film was remade as *Which Way Is Up?,* a Richard Pryor vehicle that ran out of gas.

The Seven Samurai (1954) Directed by Akira Kurosawa; starring Takashi Shimura, Toshiro Mifune, and Yoshio Inaba; running time 141 minutes; b&w; Japanese with English subtitles; Kingsley International—Budget Video.

Another classic from Kurosawa, *The Seven Samurai* is the story of a small Japanese farming village that hires seven mercenaries to fight the ruthless bandits that have been terrorizing the town. For Kurosawa, however, the heroes are not the samurai, who kill solely for pay, but the farmers—an idea expressed in a well-crafted epilogue. The film was remade as *The Magnificent Seven,* a western that was heavier on violence and much lighter on substance. Both versions are worth a look.

A Special Day (1977) Directed by Ettore Scola; starring Sophia Loren and Marcello Mastroianni; running time 106 minutes; color; Foreign—Time Life Video.

This film reunites Sophia Loren and Marcello Mastroianni in roles decidedly different from their sexy Sixties roles—Loren as a haggard housewife, Mastroianni as a homosexual misfit. The story of their unconventional romance is telescoped into one day in Rome as the storm clouds of World War II thunder on the horizon. The plot is a little pat, but the movie works.

Sorrow and the Pity (1972) Directed by Marcel Ophuls; documentary; running time 260 minutes; b&w; French with English subtitles; Foreign—Time Life Video.

An epic documentary about the Nazi occupation of France during World War II, this four-hours-plus film incorporates film clips and interviews with the survivors into a moving account of a country in collapse. Told with sensitivity and filled with insights into the human predicament, it's remarkable chunk of cinema. Because of the inordinate length, several sittings are recommended.

Swept Away (1975) Directed and written by Lina Wertmuller; starring Giancarlo Giannini and Mariangelo Melato; running time 116 minutes; color; dubbed; foreign—Columbia Pictures Home Entertainment.

Wertmuller's finest film is a farce about the battles of the sexes and of the rich and poor, stripped down to the barest essentials. The story is about a pampered wealthy woman and a poor Communist sailor who get marooned on a desolate Mediterranean island—with plenty of ironic twists. The film works better in Italian with subtitles, but you can't win 'em all.

Triumph Of the Will (1941) Directed by Leni Riefenstahl; documentary; running time 110 minutes; b&w; German with no English subtitles; foreign—Reel Images.

Leni Riefenstahl's *Triumph Of the Will* is the exemplar of the propaganda film. Commissioned by Adolf Hitler and set during the Sixth Annual Nazi Party Congress at Nuremburg, it gives frightening insights into the hypnotic power a tyrant can wield over a country. One early scene of Hitler's plane descending from the clouds (with Wagner's *Ride of the Valkyries* played in the background) inspired John Milius to write the helicopter attack scene for *Apocalypse Now.* Originally designed to rally the German masses around Hitler, *Triumph* now chronicles a man—and a country—totally out of control. The film, shot by 3 camera crews, was originally three hours long, but this 110-minute version covers the subject effectively.

Wages of Fear (1955) Directed by Henri-Georges Clouzot; starring Yves Montand, Charles Vanel, and Peter Van Eyck; running time 138 minutes; b&w; dubbed; Hal Roach—Budget Video, Reel Images.

This is one of the great suspense films, about three losers on the lam who agree to carry truckloads of nitro-glycerin through jungles and across rickety bridges in the wilds of South America to put out an oil-well fire. It's a bumpy road for them, but a great ride for viewers. William Friedkin remade the film as *Sorcerer* a few years ago, but the results weren't nearly so explosive.

©Budget Video

Even in French you'll find Peter Lorre unnerving.

THE BEST OF
TELEVISION

©Paramount Home Video

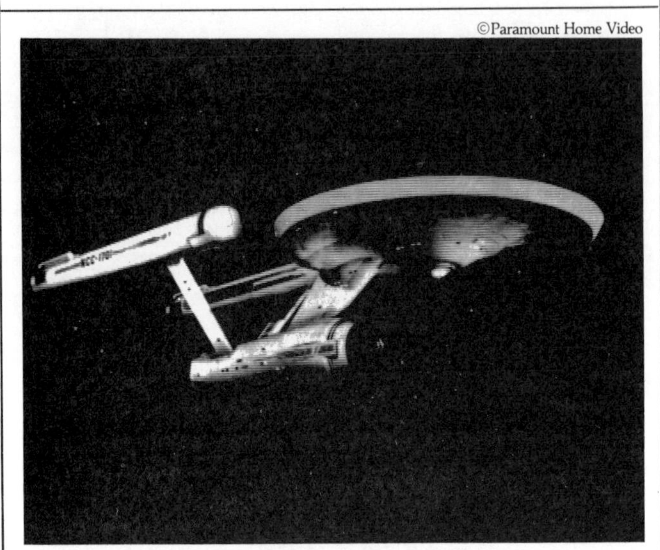

"Star Trek" was light years ahead of prime time TV.

There are more than 5,000 feature films available on video cassettes, but only a couple of hundred television programs are legitimately available on the home market. Because of complicated legal and financial tangles, you can't yet legally buy or rent "The Honeymooners," "The Twilight Zone," "Alfred Hitchcock Presents," or "Leave It To Beaver," let alone "Dallas" or "Happy Days."

The bulk of programs that are available are from the early 1950s, often kinescopes of live broadcasts filmed directly off TV screens. They provide a rare opportunity to look at television's formative years, but the quality of these tapes is often unavoidably poor. There are a few more recent programs available, including miniseries like "Shogun" and "Jesus of Nazareth."

A few made-for-TV movies available on video are worth a mention, too: *Duel* (the film that helped vault director Steven Spielberg to the bigtime), *Brian's Song* (the great tearjerker about a pro football player with cancer), and *The Killers* (Don Siegal's remake of the Ernest Hemingway story—see Page 56 for a capsule description).

The Best of Amos 'N' Andy (1951-53) Starring Spencer Williams, Tim Moore, Alvin Childress, Ernestine Wade, and Lillian Randolph; running time 120 minutes; b&w; CBS—Nostalgia Merchant.

While hardly a model of racial enlightenment, this long-suppressed series is fondly remembered by many fans and should find a new generation of aficionados, thanks to home video. The title notwithstanding, this adaptation of the phenomenally popular radio show revolves around the misadventures of George "The Kingfish" Stevens, the conniving hustler who heads the Mystic Knights of the Sea Lodge. He's played by Tim Moore, who shines in the four episodes offered here.

The Cisco Kid (1950s) Starring Duncan Renaldo and Leo Carillo; running time 53 minutes; color; ZIV—Home Video Tapes.

O. Henry's hero of the old West and his faithful sidekick, Pancho, were the subject of one of this popular syndicated series, one of the first filmed in color although it was originally shown in black and white. Likeable, if a bit moldy.

Early Elvis (1956) Starring Elvis Presley, the Tommy and Jimmy Dorsey Orchestras, Steve Allen, Milton Berle, Imogene Coca, Andy Griffith, Charles Laughton, and Ed Sullivan; running time 56 minutes; b&w; CBS and NBC—Reel Images.

If you've only see Elvis Presley in his mostly lackluster movie musicals, then this collection of TV appearances dating from his early stardom will be quite a revelation. They clearly show why his hip-wiggling sensuousness electrified bobby soxers and outraged parents. Reel Images also offers tape of Presley's 1968 comeback special (singing "Heartbreak Hotel" and "Hound Dog") and a 1973 concert broacast live from Hawaii.

Ernie Kovacs Starring Ernie Kovacs and Edie Adams; running time 30 minutes; b&w; Video Tape Network.

Nearly two decades after his death in an automobile accident, Ernie Kovacs is still revered as one of the most original talents in television comedy. Video Tape Network offers three programs that show why, including the famous sequences with a lady in a bathtub. Also included is a priceless spoof on TV westerns, including a Twilight Zone-style effort and cowboys in lederhosen and German accents.

Jesus Of Nazareth (1976) Directed and written by Franco Zeffirelli; starring Robert Powell, Anne Bancroft, James Mason, Laurence Olivier, Ernest Borgnine; running time 371 minutes; color; ITC/RAI—Magnetic Video, RCA Videodisc.

Zeffirelli stirred up a controversy when he said his TV program presented Jesus as an ordinary man, without emphasis on myths and miracles. There's plenty of spectacle in the film, but the director broke new ground with his intelligent approach. It's a moving experience for people of all faiths, and Robert Powell gives a stunning performance in the title role.

The Mary Tyler Moore Show (1970-77) Starring Mary Tyler Moore, Ed Asner, Valerie Harper, Cloris Leachman, Ted Knight, and Gavin McLeod; running time 120 minutes; color; MTM—RCA Disc.

A four-episode collection from one of the 1970's most influential situation comedies, the show that made staying home on a Saturday night a respectable pastime. Excellent writing, directing, and ensemble playing led to many Emmy Awards for this situation comedy about a single woman who works in the news department of a Minneapolis television station.

The Origin Of The Lone Ranger (1949) Starring Clayton Moore, Jay Silverheels, and Glenn Strange; running time 75 minutes; b&w; Wrather Enterprises—Nostalgia Merchant.

The first three episodes of the long-running western series recall how the masked man began his fight for law and order after becoming the sole survivor of an ambush. Rousing entertainment, *kemo sabe*. Nostalgia Merchant offers two other three-episode compilations from the series, both in color—"The Search" and "One Mask Too Many."

©Magnetic Video

Poster Courtesy Republic Film

The Republic serials spawned the masked man and his Indian sidekick.

Richard Nixon (1952-74) Starring Richard Nixon; running time 45 minutes; b&w; Various—Reel Images.

Would you buy a used car from this video star? The cassette features the former president's two most famous TV appearances, starting with his 1952 "Checkers" speech, in which he answers allegations about the misuse of campaign funds by discussing his daughters' cocker spaniel. In 1974, he becomes the first president to resign under threat of impeachment, saying "the country needs a full-time president and a full-time congress." Also included is a speech he made in New York City on Flag Day, 1957.

Shogun (1980) Directed by Jerry London; written by Eric Bercovici, based on the novel by James Clavell; starring Richard Chamberlain, Toshiro Mifune, Yoko Shimada, and John Rhys-Davies; running time 124 minutes; color; NBC/Paramount—Paramount Home Video.

A feature-film condensation of the extraordinarily popular 12-hour miniseries. An extraordinary dramatic and artistic achievement, with a fine performance by Chamberlain as a navigator shipwrecked off the Japanese coast in the early 17th century. The original TV version was more than a bit long-winded, but this version, shown in theaters overseas, does more violent cutting than the show's samurai warriors.

Star Trek (1966-69) Starring William Shatner, Leonard Nimoy, DeForest Kelly, George Takei, Nichelle Nicols, James Doohan; running time 60-120 minutes; color; Paramount—Paramount Home Video, Video Connection, All Star Video, Sheik Video, etc.

A large and vociferous cult has grown up around this unusually literate sci-fi program. Because the first season's episodes weren't copyrighted, they're available from a variety of sources, usually in two-episode combinations. Paramount Home Video is selling five "authorized" cassettes. According to Trekkies, the most collectable episodes include "City on the Edge of Forever," "The Trouble With Tribbles," and "The Corbomite Maneuver," Also highly recommended is "The Menagerie," a two-part episode that incorporates the series' pilot film, starring Jeffrey Hunter as the first commander of the Starship Enterprise.

"Beam me up, Scotty."

©Paramount Home Video

Ten From Your Show of Shows (1950s) Directed by Max Leibman; written by Mel Brooks, Neil Simon, Carl Reiner, Woody Allen, and others; starring Sid Ceaser, Imogene Coca, Carl Reiner, Howard Morris; running time 73 minutes; b&w; Walter Reade—Video Connection.

One of the best compilations of 1950s television comedy, including takeoffs on "From Here to Eternity" ("From Here to Obscurity"), silent films, and "This is Your Life." A highlight is the classic skit about a Bavarian clock factory, with mechanical figures that go haywire. Very funny stuff, originally performed live by four terrific performers.

You Bet Your Life (1950s) Starring Groucho Marx, George Fenneman, Harry Ruby, and William Peter Blatty; running time 120 minutes; b&w; NBC—Shokus Video.

Groucho's rapier wit spices the funniest game show of all time, complete with a duck demonstrating the "secret word" and his theme song, "Here Comes Captain Spaulding." The guests in these four episodes include songwriter Ruby, who collaborated on several of Groucho's films; and Blatty, who impersonated an Arab Sheik to make a few shekels years before he wrote "The Exorcist."

©NBC

"You Bet Your Life": The magic words of Groucho.

©Universal Pictures

Reagan played rough in "The Killers."

THE BEST IN SCIENCE FICTION—FANTASY

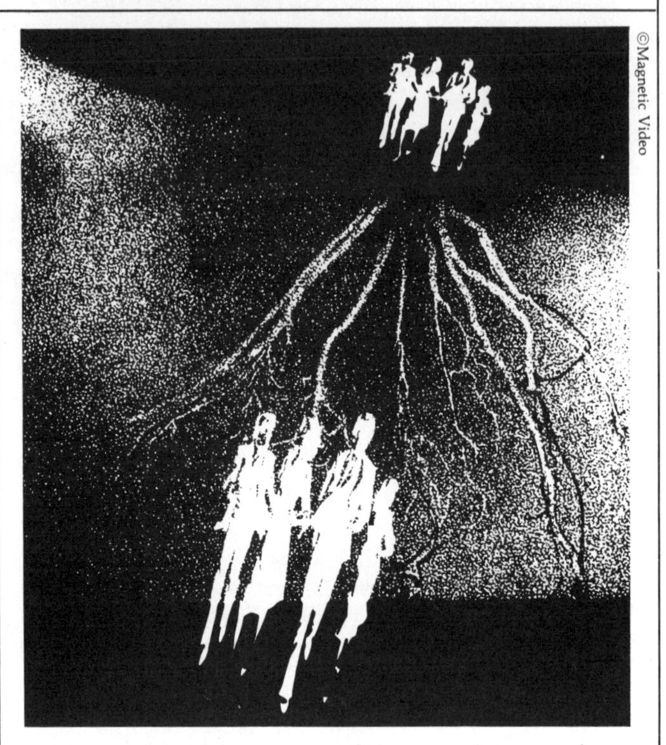
©Magnetic Video

Not until the late Seventies did science-fiction and space-fantasy films really escape the B-movie ghetto. There had been an occasional flourish from Stanley Kubrick, but few s-f films really succeeded in capturing America's imagination until the double-barreled attack of *Star Wars* and *Close Encounters* in 1977. Then Hollywood went crazy with big-budget—and, alas, low-mentality—space epics that haven't done much to elevate the s-f genre beyond the realm of comic-strip heroes.

Readers of serious science-fiction, of course, have always thumbed their noses at cinematic sci-fi, on the usually justifiable grounds that a literature founded on ideas gets trivialized, or even massacred, by an art form that thrives on action and visual effects.

In some cases, those coveted ideas managed to shine through on film, however, and one of the purposes of this section is to alert video viewers to the excellent older science-fiction films—such as *The Day the Earth Stood Still, Metropolis,* and the original *Invasion of the Body Snatchers*—which have been forgotten by the general public.

A few films are regrettably absent—*Star Wars,* of course, but also *Forbidden Planet* (a real gem, based on Shakespeare's *The Tempest), Alphaville,* and George Lucas' *THX-1138.* Hopefully, the latter three will be available on video very soon.

Alien (1979) Directed by Ridley Scott; written by Dan O'Bannon et al.; starring Tom Skerritt, Sigourney Weaver, Veronica Cartwright, and Harry Dean Stanton; running time 116 minutes; color: 20th C-F—Magnetic Video, Laserdisc.

A gut-wrenching experiment in primal terror, Ridley Scott's *Alien* was the third blockbuster extraterrestrial fantasy of the Seventies—along with *Star Wars* and *Close Encounters*—with a large helping of *The Thing* thrown in. Crewmen of an intergalactic freighter bring an alien being on board, and havoc reigns. The futuristic hardware and dazzling light shows are diminished on the small screen, but those horrifying special effects still work all too well. H.R. Giger's set designs are spectacular.

A Clockwork Orange (1971) Directed and written by Stanley Kubrick; starring Malcolm McDowell, Michael Bates, Adrienne Corri, and Patrick Magee; running time 137 minutes; color; Warner Bros.—Warner Home Video.

An exercise in future shock, Kubrick's *Clockwork* is set in Britain in a nightmarish time to come, when amoral hoodlums like McDowell's Alex terrorize the countryside. Interpret the brainwashing sequences as you wish. Kubrick is a tad too vague when it comes to meanings, and *Clockwork* ultimately is a triumph of style rather than substance.

Close Encounters of the Third Kind (1977) Directed and written by Steven Spielberg; starring Richard Dreyfuss, Teri Garr, and Francis Truffaut; running time 132 minutes; color; Columbia; Columbia Pictures Home Entertainment, Laserdisc.

Using the film technology of the Seventies, Spielberg has reshaped the flying-saucer flicks that helped pack so many Saturday matinees in the Fifties. Thriving on suspense, dazzling special effects, and a few swipes from Hitchcock, *Close Encounters* is s-f cinema of the first magnitude—with, of all things, adorable aliens. John Williams' music and the earthshaking sounds of the flying saucers sound better in a theater with Dolby stereo, but the film still works surprisingly well on video. The Columbia cassette is the reissued version of *Close Encounters,* featuring tighter editing and an on-board-the-mothership sequence that's been overly ballyhooed.

The Day The Earth Stood Still (1951) Directed by Robert Wise; written by Edmund H. North; starring Michael Rennie, Patricia Neal, Hugh Marlowe, and Billy Gray; running time 92 minutes; b&w; 20th C-F—Magnetic Video.

Just before Hollywood got smitten by bizarre aliens and terrifying flying sorcerers, Robert Wise made this benign little film—about an extraterrestrial being named Klaatu and his robot Gort, who fly to Washington, D.C., in their saucer to preach world peace. Klaatu gets shot for his trouble. This is a solid package all around, from Wise's lean direction and North's intelligent screenplay to the performance by Rennie and Neal. Bernard Herrmann wrote the soundtrack.

©RCA

©Warner Home Video

Invasion of the Body Snatchers (1956) Directed by Don Siegel; written by Daniel Mainwaring; starring Kevin McCarthy, Dana Wynter, Larry Gates, and Carolyn Jones; running time 80 minutes; b&w; Allied Artists—Nostalgia Merchant.

The question is, how could such an intelligent film get such a stupid title? Don Siegel has transformed a would-be schlock horror movie into a film of real substance, a parable about authoritarianism and the suppression of individuality. The setting is the quiet California town of Santa Mira, where strange things are happening—the residents are turning into soul-less creatures, and Kevin McCarthy thinks that some mysterious pods may be to blame. But can he save the rest of the world in time?

Siegel's finished film was altered by Allied Artists on the grounds it was too pessimistic, but Nostalgia Merchant says it now stocks that original version.

Metropolis Directed by Fritz Lang; starring Brigitte Helm, Alfred Abel, Rudolf Klein-Rogge, and Gustav Froehlich; running time 120 minutes; b&w; silent with musical score; Janus Films—Budget Video, Reel Images, and Thunderbird Films.

The future according to Fritz Lang, this German silent film depicts a world dominated by machines—driving the slave laborers to revolt. Loaded with bizarre characters (the villain Rotwag, the sensuous robot Maria) and some excellent special effects, the film is almost too cluttered for its own good. Take it slow—on video.

Planet of the Apes (1968) Directed by Franklin J. Schaffner; written by Michael Wilson and Rod Serling; starring Charlton Heston, Roddy McDowall, and Kim Hunter; 112 minutes; color; 20th C-F—Magnetic Video, RCA Disc.

This movie has generated so many humdrum sequels that you almost forget how clever the original was. The story of an astronaut (Heston) who gets suspended in a time warp and lands on a not-so-distant planet of the future where apes reign supreme and humanoids are reduced to grunting animals. The movie does wear thin after a viewing or two, so you might want to rent it. At any rate, the kids will love those chattering apes.

©Magnetic Video

Heston monkeyed around in "Planet Of the Apes."

The Thing (1951) Directed by Christian Nyby (although Howard Hawks reportedly did a lot of the direction); starring Margaret Sheridan, Kenneth Tobey, and James Arness; running time 87 minutes; b&w; RKO—Nostalgia Merchant.

The setting: the North Pole. The premise: A flying saucer has crashed and imbedded itself in the ice. The plot: A team of scientists rushes to the scene to investigate and discovers a frozen creature when they try to free up the saucer. But when they take the critter to a nearby army base, the thing gets thawed and proceeds to run amok. The Americans aren't sure what to do and get caught in a great debate. One of the leaders of the expedition (Kenneth Tobey) finally decides to skip the scientific dogma and try to nail the monster. The film has ample suspense and James Arness in the title role—and solid, albeit reactionary, science-fiction. The men of action win out over the men of ideas.

"Star Wars" spawned such intergalactic dreck as "Galaxina."

It was man against computer in Kubrick's "2001."

©MGM/CBS Home Video

2001: A Space Odyssey (1968) Directed by Stanley Kubrick; written by Kubrick and Arthur C. Clarke; starring Keir Dullea, William Sylvester, and Gary Lockwood; 141 minutes; color; MGM—MGM/CBS Video.

Stanley Kubrick's monolithic attempt at the ultimate science-fiction film is, in many spots, the most boring science-fiction film, but it has enough high points along the way to make it worthwhile—especially if your VCR has a rapid scan button. On the simplest level, the story is about a spaceship's journey to Jupiter and its upstart computer (HAL 9000), but Kubrick also throws in a capsule history of Mankind and the conflicts between Man and machine. Those special effects are—appropriately—out of this world. The music is special,too.

20,000 Leagues Under the Sea (1954) Directed by Richard Fleischer; written by Earl Felton; starring Kirk Douglas, James Mason, and Peter Lorre; 127 minutes; color; Walt Disney—Walt Disney Home Video

Walt Disney meets Jules Verne in this sci-fi melodrama, and it's a very happy marriage for both adults and children. Set near the turn of the century, the tale revolves around the nefarious Captain Nemo and his submarine, The Nautilus—plus some captured scientists who get in over their heads. Great good fun, and the special effects are anything but shabby.

The War of the Worlds (1953) Directed by Byron Haskin; written by Barre Lyndon; starring Gene Barry, Les Tremayne, and Ann Robinson; running time 85 minutes; color; Paramount—Paramount Home Video, Fotomat.

George Pal's low-budget version of Armageddon features Martians who invade California. The film has some nifty battle scenes (it won an Oscar for special effects) and some wonderfully hokey dialogue. When somebody mentions that the Martians could destroy the planet in six days, Ann Robinson deadpans: "The same time it took to make it." The film is obviously more warped fantasy than genuine science fiction, but a good Martian movie is hard to find these days. Check out those nifty Martian war machines in action again.

©Walt Disney Home Video

Things To Come (1936) Directed by William Cameron Menzies; written by H.G. Wells; starring Raymond Massey, Cedric Hardwicke, Ralph Richardson, and Ann Todd; running time 92 minutes; b&w; London Films—Budget Video, Cable Films, Media Home Entertainment, Reel Images, Video Warehouse.

What can you say about a 1936 movie that predicted a world war in 1940, a new glass-based society, and the first spaceship to the moon? This extravagant collaboration of producer Alexander Korda, director William Cameron Menzies, and writer H.G. Wells depicts society after a 76-year-long war (in the year 2036). Massey plays the visionary leader who tries to put the pieces back together. The film is uneven in its execution, but it works overall.

©Budget Video

"Things To Come"—modern insights in a 1936 film.

THE BEST IN HORROR

Karloff was born again in "Frankenstein."

©Universal Pictures

Considering the stampede of horror films that have trampled the big screen the past few years, you may be surprised to find a dearth of recent fright flicks in the following sampling of the best scary movies.

The reason is simple: The vast majority of these new schlock shockers have been cheap, crass commercial efforts that specialize solely in assaulting the senses with grisly special effects and stealing famous scenes from old thrillers—especially Hitchcock's shower scene in *Psycho.*

The best place to see a horror movie these days is at home on your VCR or disc player (the selection of vintage horror movies on the laser-disc system is first-rate), simply because most of the best fright flicks haven't been in a movie theater in years.

For the most part, the movies in this section have been selected because they put the craft of film-making before guts and gore. You'll probably find that the scariest scenes in these movies are the ones that cut away before the head gets severed or the body skewered. As director John Carpenter has remarked, the viewer will imagine a picture far more frightening than anything a director or makeup artist can put on the screen.

The operating principle of most successful horror films is the fear of the unknown, and the carnage of the recent fright movies seems to work against that notion. Shocking the audience with bloody corpses and preying on their basic fears—of death, of the dark, or of just being alone—are two entirely different matters. By and large, the following films specialize in the latter.

Some films weren't considered because they haven't been available on video: *Carrie* (which spawned a whole subgenre of telekinetic-weirdo pictures), *Freaks* (the most haunting horror film of all), the original *Dr. Jekyll and Mr. Hyde, The House Of Wax* (if it's ever released on video with good 3-D effects, don't miss it), *The Texas Chainsaw Massacre* (the best of its kind, regardless of its merits), and *What Ever Happened To Baby Jane?* (a horror film with an all-too-human montster).

Bride of Frankenstein (1935) Directed by James Whale; written by John L. Balderston and William Hurlbut; starring Elsa Lanchester, Boris Karloff, Colin Clive, and Valerie Hobson; running time 80 minutes; b&w; Universal—MCA Disc.

One of the best sequels ever made, *The Bride of Frankenstein* succeeds because—instead of merely capitalizing on the success of *Frankenstein*—it adds large dollops of black humor. In this tale, Baron Frankenstein is coerced into reviving his monster and building a girlfriend for him as well. It runs but 80 minutes, and not a second is wasted. Elsa is something else.

Creature From The Black Lagoon (1954) Directed by Jack Arnold; written by Harry Essex and Arthur Ross; starring Richard Carlson, Richard Denning, and Julie Adams; running time 79 minutes; b&w; Universal—MCA.

From out of the primordial ooze comes the creature of the title, a finny fiend who terrorizes an archeological expedition in the heart of the Amazon. Play back a few times that lyrical underwater sequence where the monster and Julie Adams swim in tandem. Now, if only MCA could perfect its 3-D process...(see Page 44).

Curse of the Cat People (1944) Directed by Robert Wise and Gunther Von Fritsch; written by De Witt Bodeen; starring Simone Simon, Jane Randolph, and Kent Smith; running time 70 minutes; b&w; RKO—Nostalgia Merchant.

Another supposed sequel that works as well as the original, this Val Lewton production deals with a young child haunted by the vision of her dead mother. Don't be fooled by the exploitation-film title. It's got class to spare, and it picks up steam once it gets rolling. Robert Wise directed with a sure and gentle hand.

©Universal Pictures

Lugosi put the bite in "Dracula."

©Universal Pictures

Dracula (1931) Directed by Tod Browning; written by Garrett Fort; starring Bela Lugosi, Helen Chandler, David Manners, and Dwight Frye; running time 75 minutes; b&w; Universal—MCA Cassette and Disc.

Never has a movie spawned so many imitators (more than two dozen at last count), but for many, Bela Lugosi's count remains the finest. He has a way of getting under your skin. It's the classic tale from the crypt, and directed by a master—Tod Browning. Scenes worth a few replays: the opening in the crypt and the hysterical craziness of the mental asylum. Most of the remakes and imitations have been toothless.

Frankenstein (1931) Directed by James Whale; written by Garrett Fort, Francis Edward Faragoh, and John L. Balderston; starring Boris Karloff, Mae Clark, Colin Clive, and John Boles; running time 71 minutes; b&w; Universal—MCA Cassette and Disc.

The most misunderstood of movie creatures, Frankenstein's monster gets a fairly decent shake in this early Karloff-Whale collaboration, even if it is a radically different retelling of Mary Shelley's classic ode to immortality. Karloff is the key: He makes the monster a scary *and* sympathetic character. The stormy scenes in the castle bear repeating, and the best of the lot is Frankenstein's hunchbacked assistant dropping a healthy brain on the floor and picking up a diseased one by mistake. You can't win 'em all.

©Universal Pictures

Halloween (1978) Directed by John Carpenter; written by Debra Hill; starring Jamie Lee Curtis, Nancy Loomis, P.J. Soles, and Donald Pleasance; 93 minutes; color; Debra Hill—Media Home Entertainment.

This low-budget shocker has become the bane of movie critics across America, primarily because it has spawned so many inferior imitations they've been forced to endure. The story—about a crazed killer who returns to the scene of his childhood crime—is a bit much to handle sometimes, but Carpenter squeezes every jolt he can out of it. (He also composed the music.) This *Halloween* has a lot of tricks—and treats. The one unanswered question: How does the killer, who has been locked in a hospital for the criminally insane since he was a kid, know how to drive a car?

House on Haunted Hill (1958) Directed by William Castle; written by Robb White; starring Vincent Price, Elisha Cook Jr., Carol Ohmart, and Richard Long; running time 75 minutes; b&w; Allied Artists—Allied Artists Video.

As the title implies, this is an old-fashioned haunted house movie—the kind that used to scare the Milk Duds out of kids at Saturday matinees. Vincent Price is his usual self, but Elisha Cook Jr. is guaranteed to give you the creeps. It's the best of the William Castle gimmick horror films, which include *The 13 Ghosts* and *The Tingler.* The opening sequence is still enough to scare youngsters out of their Keds. Alas, the film (about a demented millionaire who offers party guests $10,000 if they can survive a night in the possessed house) slowly goes downhill from there. For a Fifties fright flick, it's the next best thing to *The Unearthly,* which unfortunately hasn't been available on video. Both are recommended for children of all ages.

©Home Media Entertainment

The Incredible Shrinking Man (1957) Directed by Jack Arnold; written by Richard Matheson; starring Grant Williams, April Kent, and Randy Stuart; running time 81 minutes; b&w; Universal—MCA Disc.

Ace science-fiction writer Richard Matheson teamed with Jack Arnold for this now predictable story (Lily Tomlin lampooned it in —what else— *The Incredible Shrinking Woman)* of a man who gets caught in a radioactive mist and starts shrinking physically and disintegrating emotionally. It's still entertaining, however, and the special effects are first-rate. Best sequence: A thimble-sized Grant Williams uses a sewing needle to battle a spider. As Steve Martin used to say, let's get small.

I Walked With a Zombie (1943) Directed by Jacques Tourneur; written by Curt Siodmak and Ardel Wray; starring Frances Dee, Tom Conway, and James Ellison; running time 69 minutes; b&w; RKO—Nostalgia Merchant.

This is another dandy Val Lewton production and, typical of his style, light on shock and heavy on atmosphere. The story (about a nurse hired by a tropical-plantation owner to care for his voodoo-obsessed wife) is, believe it or not, a rather loose adaptation of *Jane Eyre.*

Mighty Joe Young (1946) Directed by Ernest B. Schoedsack; written by Ruth Rose; starring Terry Moore, Robert Armstrong, and Ben Johnson; running time 94 minutes; b&w; RKO—Nostalgia Merchant.

This *King Kong*-inspired tale pales in comparison to the original, but it's strangely and ingenuously affecting anyway—probably because Joe Young has some admirable human qualities. The story centers around a young girl who reluctantly agrees to bring her pet gorilla from Africa to America, only to see it get exploited by a sleazy nightclub owner. Willis O'Brien, who did the *King Kong* special effects, created the effects here also. Best scene: when Joe goes bananas in the Golden Safari club. Younger kids will go for the orphanage scene.

Chaney was unchained in "Phantom of the Opera."

©Blackhawk

Night of the Living Dead (1968) Directed by George Romero; starring Judith O'Dea and Russell Steiner; running time 95 minutes; b&w; Image Ten—Budget Video, Media Home Entertainment, Niles Video, Thunderbird Films, Video Dimensions, Video Warehouse.

A great and gruesome late night movie, George Romero's *Night of the Living Dead* is a *grand guignol* exercise in excess. Cannibalistic zombies rise from their graves and lay siege to a country house. It's scary all right, and not a bad black comedy either. That eerie black-and-white cinematography was tailor made for the home screen. One word to the wise: If Romero's sequal, *Dawn of the Dead,* becomes available on video, skip it.

Phantom of the Opera (1925) Directed by Rupert Julian; starring Lon Chaney Sr., Norman Kerry, and Mary Philbin; running time 85 minutes; b&w (one color sequence); silent; Universal—Blackhawk Films, Cable Films.

Along with F.W. Murnau's *Nosferatu,* Rupert Julian's *Phantom* ranks as the silent horror classic. It's all about a betrayed and disfigured composer who lurks in the bowels of the Paris Opera House awaiting his revenge. Lon Chaney is as scary as his ghoulish makeup—and catch that classic unmasking scene again. The -*Phantom* remakes just can't compare.

Val Lewton scored the "Cat People" and an even better sequel—"Curse of the Cat People."

Cannibalism was in vogue in "Living Dead."

©Media Home Entertainment

SON OF 3-D

Remember the halcyon days of the early Fifties, when rooftops across the country first began sprouting TV antennas? The only people who weren't thrilled by the first home-entertainment revolution since Marconi invented the radio were the movie moguls: Why should people go to movies when they can be entertained in the comforts of their rec rooms? To combat the TV invasion, the major Hollywood studios began groping for ways to lure audiences back to the big screen.

In their search, they re-discovered a gimmick that had been around since 1915. The gimmick was 3-D, and for a few years America was looking at the world of film through blue- and rose-colored glasses.

Simply stated, the 3-D process projected a double image on the screen, and viewers wore either the color-tinted lenses (for black-and-white films) or polarizing lenses (for color) to achieve a three-dimensional effect.

Alas, 3-D process wasn't the saviour the studios had hoped for. The fad faded, and 3-D movies soon went the way of Hula-Hoops and "I Like Ike" buttons.

Every so often, 3-D would appear again briefly as a novelty act—in Andy Warhol's *Frankenstein* for example, and in R-rated exploitation films such as *The Stewardesses,* which grossed $28-million to become the most successful 3-D movie ever. Now with the growing popularity of videocassette players, a few old 3-D films are being dusted off and adapted for home viewing.

At present, two black-and-white 3-D films from the Fifties, *It Came from Outer Space* and *Creature from the Black Lagoon* (both retail for $65 with four sets of glasses), are available on video cassette from MCA. (Both cassettes can be rented at some video stores for as little as $10 a week.) And Home Theater of Hollywood, Calif., is presenting *The Stewardesses* ($80 with two sets of glasses) in 3-D and in color on cassette.

Not surprisingly, the major drawing card of all three is their novelty. There's something patently ludicrous (and delightful) about sitting around your living room watching a program through silly-looking glasses—especially if a seven-foot finny monster keeps popping out at you from the TV set.

Alas, video 3-D is still in its infancy, as evidenced by the quality of *Creature from the Black Lagoon,* a classic 1954 schlock shocker about a team of archeologists who encounter a web-footed behemoth in the wilds of the Amazon.

The problem with "Creature" on cassette is that MCA hasn't quite perfected the 3-D technology yet. On many of the cassettes the three-dimensional effect is sporadic because the two images aren't always in perfect register. And when the effect fails, one gets the impression the movie should have been called *Eye Strain from the Murky Lagoon.*

According to a spokeswoman for MCA, the company is aware of the problems and has gradually stopped distribution of the cassettes until the gremlins are removed. She also advises that viewers who already have copies of the two 3-D movies should adjust their color TVs (the process doesn't work with black-and-white sets) as follows to achieve the best 3-D results:

"We recommend your contrast, brightness, and color contrasts be turned up higher than normal, the room be darkened during viewing, and you allow adequate time for your eyes to adjust to the picture."

MCA eventually plans to release an improved version of both cassettes. In the meantime, check the quality of the cassette before you rent or buy.

The biggest developments in 3-D have been coming from 3-D Video of North Hollywood, Calif., which developed the first broadcast of a 3-D movie for Select TV, a subscription-television network with outlets in Los Angeles, Milwaukee, and Ann Arbor. On Dec. 19, 1980, Select TV broadcast *Miss Sadie Thompson* with Rita Hayworth in 3-D, as well as the 3-D Three Stooges short, *Spooks.*

According to Jim Butterfield, chief scientist for 3-D Video, the company is negotiating with other subscription and cable networks to broadcast 3-D films.

Butterfield adds that his company is now working on plans for 3-D video discs. The company hopes to be involved with the reissue of *Creature from the Black Lagoon* as well.

©1982 THE HOME VIDEO SOURCEBOOK

CONCERTS IN VIDEO

Video rock, from Jimi to Jagger. ©Home Media Entertainment

Rock music on video has growing pains. A spate of performers have released original programming on cassettes and discs in the past couple of years, from Abba to Rod Stewart, from James Taylor to the Grateful Dead. Each has tried to come up with an approach that uses the medium to the fullest, and the results are decidedly mixed.

Video music is a tricky proposition. Although it provides the added dimension of visuals, it also suffers the limitations of a TV-sized presentation (unless the viewer is the lucky owner of a large-screen projection unit).

On video cassette machines and RCA's early SelectaVision disc system, the sound is less than superb: monaural sound, and usually from one mediocre TV-cabinet speaker. For Pioneer and Magnavox's laser-disc players, several musical programs are available in stereo, and the sound system can be hooked into a stereo unit. The music sounds great, but unless the speakers are next to each other above or below the TV screen (thereby reducing the effect of stereo), the sound doesn't mesh that well with the visuals—mouths move on the screen, but the vocals come out of left field.

Costs also vary. Music discs sell for between $20 and $30. Music-oriented cassettes sell for twice that—a steep price for an hour to 90 minutes of music—but rentals make the cassettes somewhat more practical.

Two popular recent cassettes, *James Taylor in Concert* and *Fleetwood Mac* exemplify the two most common approaches to video rock music. *James Taylor* is a straightforward, no-frills concert performance; *Fleetwood Mac* (also available on laser disc) mixes concert footage with behind-the-scenes segments of the band relaxing and recording their hit album "Tusk."

Taylor, the downhome singer-composer of "Fire and Rain" and other hits, is presented in a 90-minute concert in 1979, and video seems a perfect medium for his music. Unlike a hard-rock concert, where being in the audience is part of the experience, his songs are Taylor-made for home viewing, with gentle melodies and lyrics that are worth a listen—he's a thinking man's John Denver. The camera work is also low-keyed, concentrating on Taylor, a pair of back-up vocalists, and the band. In keeping with the tone of the program, shots of the audience are few.

If you're a fan of James Taylor, this cassette is worth a look and a listen. Few fans may be ardent enough to buy the cassette, but it's not a bad bargain as a rental. And, for better or worse, it's one of the few music cassettes that won't drive the rest of the family out of the room. In all, Taylor does 19 songs, from a lilting "Don't Let Me Be Lonely Tonight" to a fast-paced "Summertime Blues."

Part One. ©Media Home Entertainment

Fleetwood Mac is a different proposition altogether, an hour-long program that might delight diehard admirers—you can watch these stars at work and at play, at home and in concert—but will probably drive casual fans up the wall. Sure, the group (composed of Stevie Nicks, Lindsay Buckingham, John McVie, Christine McVie, and Mick Fleetwood) sings a dozen of their hits, including "The Chain," "Angel," "Sara," "Tusk," and "Go Your Own Way," but many of the songs are abbreviated ("Think About Me" runs a minute and a half, for instance) and the performers come across like a bunch of self-infatuated, pampered jerks:

Vocalist Stevie Nicks tells how her life doesn't revolve solely around rock 'n' roll; she takes ballet lessons, too.

Vocalist Christine McVie says "I paid my dues"—as she relaxes on her spiffy sailboat (complete with piano).

Drummer Mick Fleetwood tells of the band's earlier hard times—as he sits by the pool on his gigantic Southern California estate.

For all the behind-the scenes chatter, the viewer actually learns very little about the performers, their creativity, or their musicianship. Says Fleetwood, "I kick the [crap] out of my drums." The visuals then cut to Mick in concert and, by golly, he's right—if that's something to be proud of.

Some other noteworthy performers in concert are available on videotape, however, including the farewell concert of Cream (Eric Clapton fans take note), *The Grateful Dead Movie*, Jackson Browne, *The Last Waltz* (starring The Band), *Rust Never Sleeps* (for hard-core Neil Young fans), *Rude Boy* (about The Clash), *Gimme Shelter* (featuring The Rolling Stones) and—the grand-daddy of them all—*Woodstock*.

©Media Home Entertainment

THE BEST MUSICALS

Musicals are fairly well represented in video catalogues, and no wonder: who can resist the temptation to replay Fred Astaire and Ginger Rogers dancing the Carioca in *Flying Down to Rio* or examining the climactic ballet in *An American in Paris* backwards and forwards (assuming your video machine has that capability).

Our list ranges from *42nd Street* to *All That Jazz* and includes several Broadway classics, as well as some of the best efforts by Judy Garland. Elvis Presley, Marilyn Monroe, and Shirley Temple. Some titles we'd like to see in video include *Golddiggers of 1933, Footlight Parade,* almost all of Bing Crosby's pictures, the Judy Garland-Mickey Rooney series, *Seven Brides for Seven Brothers,* the two Beatles pictures, and the composer biographies like *Night and Day* (Cole Porter) and *Rhapsody in Blue* (George Gershwin).

All That Jazz (1979) Directed by Bob Fosse; written by Fosse and Robert Alan Aurthur; starring Roy Scheider, Jessica Lange, and Ann Reinking; running time 123 minutes; color; 20th C-F—Magnetic Video, Laserdisc.

A terrifically imaginative musical tragicomedy—and a jarring collision between reality and fantasy, dance and drama, sex and love. Roy Scheider has the most complex role of his career as a self-destructive movie and stage director, a chain-smoking, pill-popping workaholic and womanizer. You may not want to replay the open-heart surgery segments, but give that dazzling "On Broadway" routine a few reprises.

Cabaret (1972) Directed by Bob Fosse; written by Jay Presson Allen, from a novel by Christopher Isherwood; music and lyrics by John Kander and Fred Ebb; starring Liza Minnelli, Michael York, Joel Grey, Helmut Griem, and Marisa Berenson; running time 123 minutes; color; ABC/Allied Artists.

Liza shines in this imaginative recreation of the Broadway musical about decadent Berlin between the wars. It has more on its mind than just song and dance, and—parents take heed—the depiction of assorted sexual permutations is not for the kiddies.

Curly Top (1935) Directed by Irving Cummings; written by Patterson McNutt; starring Shirley Temple, John Boles, Rochelle Hudson, Jane Darwell, and Rafaela Ottiano; running time 75 minutes; b&w; 20th C-F—Magnetic Video.

This has been the only Temple vehicle available on tape, except for the *The Little Princess,* which really isn't a musical. Shirley sings *Animal Crackers in My Soup* and plays matchmaker for her sister and their wealthy guardian. Other Temples worth taping include *Rebecca of Sunnybrook Farm, Heidi,* and *The Blue Bird.*

42nd Street (1933) Directed by Lloyd Bacon and Busby Berkeley; written by James Seymour and Rian James, based on a novel by Bradford Ropes; music and lyrics by Al Dubin and Harry Warren; starring Warner Baxter, Bebe Daniels, George Brent, Dick Powell, Ruby Keeler, Ginger Rogers, and George E. Stone; running time 89 minutes; b&w; Warner Bros.—Magnetic Video, RCA Disc.

Choreographer Berkeley rescued movie musicals from virtual extinction with his kaleidoscopic dance routines. In this, his first major film, he's abetted by a bright cast and an irresistable chestnut of a plot about an ailing director who turns a chorus girl into a star on opening night. There are two spectacular production numbers: "Shuffle off to Buffalo," and the title song, staged on a set several blocks long.

Gentlemen Prefer Blondes (1953) Directed by Howard Hawks; written by Charles Lederer, based on the novel by Anita Loos; music and lyrics by Jule Styne and Leo Robin; starring Marilyn Monroe, Jane Russell, Charles Coburn, George (Foghorn) Winslow, Tommy Noonan, Norma Varden, and Elliot Reid; running time 91 minutes; 20th C-F—Magnetic Video.

Monroe and Russell are perfectly teamed as two husband-hunting, seagoing showgirls. Lots of laughs and a bright score that includes "Diamonds are a Girl's Best Friend." Best scene: Marilyn gets stuck in a porthole, and disguises the fact with Winslow's help.

Jailhouse Rock (1957) Directed by Richard Thorpe; written by Guy Trosper; songs by Jerry Leiber and Mike Stoller; starring Elvis Presley, Judy Tyler, Mickey Shaughnessy, Vaughn Taylor, and Dean Jones; running time 96 minutes; b&w; MGM—MGM/CBS Home Video.

Most of Presley's musicals are disappointingly bland, but this—his second movie, made before he entered the Army—comes close to capturing the essence of his animalistic appeal. After being sent to prison for accidentally killing a man, Presley takes up singing and guess what? He becomes a star. Rousing entertainment, with particularly good dancing to the title tune.

©Magnetic Video

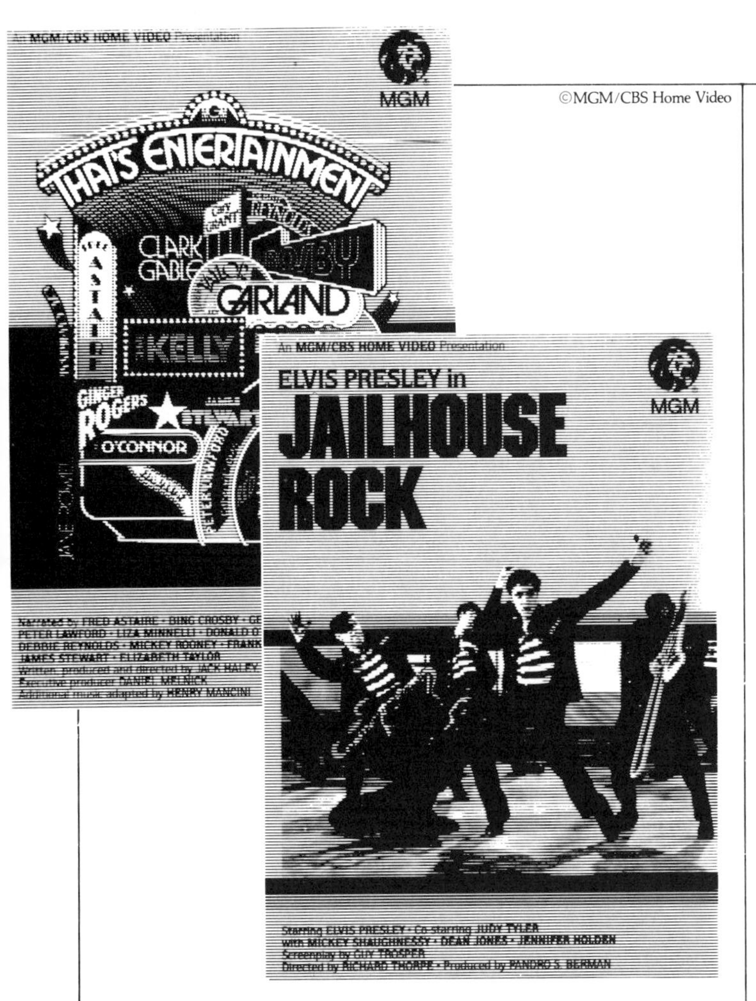

©MGM/CBS Home Video

Meet Me In St. Louis (1944) Directed by Vincente Minnelli; written by Irving Brecher and Fred Finklehoffe, based on stories by Sally Benson; starring Judy Garland, Margaret O'Brien, Tom Drake, Leon Ames, Mary Astor, Lucille Bremer, Harry Davenport, and Marjorie Main; running time 102 minutes (electronically compressed from 113 minutes); color: MGM—MGM/CBS Home Video.

An absolutely charming classic about a turn-of-the-century family that's rocked by father's decision to move to New York City. The score includes "The Trolley Song," "The Boy Next Door," and "Have Yourself a Merry Little Christmas."

My Fair Lady (1964) Directed by George Cukor; written by Alan Jay Lerner, based on a play by George Bernard Shaw; music and lyrics by Lerner and Frederick Loewe; starring Rex Harrison, Audrey Hepburn, Stanley Holloway, Wilfrid Hyde-White, and Gladys Cooper; running time 175 minutes; CBS/Warner Bros.—MGM/CBS Home Video.

So who cares if Marnie Nixon dubbed Hepburn's songs, and the film is basically a transcription of the Broadway classic? Harrison's letter-perfect performance as diction expert Henry Higgins is worth a spot in anyone's home library. The brilliant adaptation of Shaw's *Pygmalion* boasts a well-nigh perfect song, including "I Could Have Danced All Night," "Get Me to The Church on Time," and "The Rain in Spain."

That's Entertainment (1974) Directed and written by Jack Haley Jr.; starring Fred Astaire, Gene Kelly, Elizabeth Taylor, Mickey Rooney, Frank Sinatra, Bing Crosby, Liza Minnelli, James Stewart, Debbie Reynolds, Donald O'Connor, Judy Garland, Clark Gable, Esther Williams, and Eleanor Powell; running time 122 minutes (electronically compressed from 137 minutes); b&w; and color; MGM—MGM/CBS Home Video.

A thoroughly delightful catalogue of MGM musicals from 1929 ("The Broadway Melody)" to 1957 ("Gigi"). Seven stars introduce clips from scores of classics, including "The Wizard of Oz," "High Society," "The Harvey Girls," "Gone With the Wind," and "An American in Paris." Our favorite sequence features unlikely musical performers, including Stewart crooning "Easy to Love," Gable hoofing to "Puttin' on the Ritz," and Joan Crawford doing an unbelievably bad song-and-dance routine.

Top Hat (1935) Directed by Mark Sandrich; written by Dwight Taylor and Alan Scott; music and lyrics by Irving Berlin; starring Fred Astaire, Ginger Rogers, Edward Everett Horton, Helen Broderick, Eric Blore, and Erik Rhodes, running time 100 minutes; b&w; RKO—Nostalgia Merchant.

It's hard to single out any one Astaire-Rogers musical, but this one stands out because of its fine Berlin score and its unabashedly corny (but fun) plot about a woman who mistakes a dancer for a paid divorce co-respondent in Venice. Catch the dancing shadows that introduce the big production finale, "The Piccolino." Nostalgia Merchant offers six other Astaire-Rogers films; *Swingtime* (1936, a great Jerome Kern score) and *Shall We Dance* (1937, with songs by George and Ira Gershwin) are particularly worth investing in.

West Side Story (1961) Directed by Robert Wise and Jerome Robbins; written by Ernest Lehman, based on the play by Arthur Laurents; music by Leonard Bernstein, lyrics by Stephen Sondheim; starring Natalie Wood, Richard Beymer, Russ Tamblyn, Rita Moreno, and George Chakiris; running time 155 minutes; color; United Artists—Magnetic Video, Laserdisc.

This winner of 10 Academy Awards tells the story of a ghetto Romeo and Juliet, embroiled in a Manhattan gang war in the late 1950s. Some of the casting is questionable, but the plot, music, and dance numbers mesh perfectly. The unforgettable score includes "Maria," "Tonight," and "America."

Yankee Doodle Dandy (1943) Directed by Michael Curtiz; written by Robert Buckner and Edmund Joseph; songs by George M. Cohan; starring James Cagney, Joan Leslie, Walter Huston, Rosemary DeCamp, Richard Whorf, George Tobias, S.Z. Sakall, Jeanne Cagney, and Irene Manning; running time 126 minutes; b&w; Warner Bros.—Magnetic Video, RCA Disc.

Cagney deservedly won an Academy Award for his delightful impersonation of the feisty Irish-American composer-entertainer, George M. Cohan. Cagney's hoofing is delightful and the film boasts a fine supporting cast. Cohan's patriotic and sentimental songs are rousingly staged. Play that title number again.

©RKO

©United Artists

"West Side Story," and a girl named Maria.

THE BEST
WAR MOVIES

©Magnetic Video

Upon hearing that Francis Coppola was spending $36 million to make *Apocalypse Now,* Clint Eastwood remarked that he could have invaded a small country for that price—and no doubt made a movie about it later. Hollywood's fascination for modern combat is quite simple: If you juggle the themes to suit the mood of the times, you just might come up with a winner.

The trouble is, writers and directors have juggled the plots and ideas so often that it's gotten to the point where what combat movies really need these days—heaven forbid—is a new war to focus on.

The appeal of war movies varies greatly. In some, it just provides a smooth adventure vehicle *(The Dirty Dozen* could as easily have been a western). In others, it provides a sounding board for ideas, witness Stanley Kubrick's *Paths of Glory.* And in others, audiences can have it both ways. In *A Bridge Too Far,* you get an anti-war message, lots of heavy artillery, and the glorification of America's fighting men—all virtually in the same breath.

The war films on video mentioned here provide a decent cross-section of Hollywood at war—from the biographies of heroes in the trenches *(Sergeant York, Hell To Eternity)* to the men who gave the orders *(Patton, The Desert Fox).* From World War I *(Sergeant York, All Quiet On the Western Front, Paths of Glory)* to Vietnam *(Go Tell the Spartans).*

Other war films worth a look if and when they become available on cassette and disc include: *Apocalypse Now, Attack!, The Lost Patrol,* Samuel Fuller's *The Big Red One,* and Sam Peckinpah's *Cross of Iron.*

TAKKATAKKA!

ALL QUIET ON THE WESTERN FRONT

All Quiet on the Western Front (1930) Directed by Lewis Milestone; written by Milestone, Maxwell Anderson, Del Andrews, and George Abbott; starring Lew Ayres; running time 103 minutes; b&w; Universal/Realart—MCA Cassette.

Based on Erich Maria Remarque's classic World War I novel, this early anti-war film casts Lew Ayres as Paul Baumer, a gung-ho young German recruit who finds that war is indeed hell on the western front. The film won the Oscar for best picture, and Milestone won one for best director. The biggest shortcoming of the cassette version is that it has been severely truncated—the running time of the original has been trimmed by nearly half an hour.

The Desert Fox (1951) Directed by Henry Hathaway; written by Nunnally Johnson; starring James Mason, Jessica Tandy, Cedric Hardwicke, and Luther Adler; running time 87 minutes; b&w; 20th C-F—Magnetic Video.

One particularly successful war-movie format has been the military biography, and director Henry Hathaway honed it into fine form in *The Desert Fox,* the story of Nazi Germany's General Rommel. The drama covers Rommel's slow fall from grace, from the tank warfare in Northern Africa to his return to Germany and the unsuccessful plot to overthrow Hitler. Rugged combat scenes—which start before the credits even get a chance to roll—and a fine performance by Mason make this flick click.

The Dirty Dozen (1967) Directed by Robert Aldrich; written by Nunnally Johnson and Lukas Heller; starring Lee Marvin, Ernest Borgnine, Charles Bronson, Jim Brown, George Kennedy, and Trini Lopez; running time 149 minutes; MGM—MGM/CBS Home Video, RCA Disc.

Robert Aldrich is at his most effective directing action films, and this violent World War II saga was right up his alley. The movie itself resembles one of those *Mission: Impossible* plots: Twelve incorrigible convicts are recruited for a dangerous mission behind German lines, and the bullets fly for the next two hours. The cast is packed with rugged action stars—Marvin, Bronson, Brown, Borgnine—and they deliver a right to the jaw. The film is decidedly long on bloodshed and short on substance, but some people like their war movies that way. Best scene to replay: the assault on Nazi headquarters.

WHAM

Go Tell the Spartans (1978) Directed by Ted Post; starring Burt Lancaster, Joe Unger, Jonathan Goldsmith, and Craig Wasson; running time 114 minutes; color; Avco-Embassy—Time Life Video.

The Deer Hunter and *Apocalypse Now* have evolved into the major Vietnam war movies, but a third one—*Go Tell the Spartans*—deserves a lot more attention than it has gotten. Lancaster turns in another grand performance as an Army major assigned to defend a small hamlet with a squad of battle-weary GIs and Vietnamese mercenaries. As luck would have it, they end up in the middle of a major Cong offensive. In the process, the village becomes a microcosm for Vietnam—inexperienced Americans fighting a battle with inadequate support, uncertain morality, and widespread corruption among the South Vietnamese. Ted Post, who also directed the Dirty Harry opus *Magnum Force*, handles the combat scenes crisply.

The Great Escape (1963) Directed by John Sturges; written by James Clavell and W.R. Burnett; starring Steve McQueen, James Garner, Richard Attenborough; Charles Bronson, and James Coburn; running time 170 minutes; color; United Artists—Magnetic Video.

While the Dirty Dozen was trying to invade the Fatherland, Garner, McQueen, and the rest of the crew were trying to get out. The setting is a German POW camp, and the plot deals with the various (and usually futile) escape attempts by the Allies. This action drama runs way too long, but you'll want to replay those classic McQueen motorcycles sequences over and over.

Hell To Eternity (1960) Directed by Phil Karlson; written by Ted Sherdeman and Walter Roeber Schmidt; starring Jeffrey Hunter, David Janssen, and Vic Damone; running time 132 minutes; b&w; Altantic Pictures—Allied Artists Video.

This melodrama is based on the real-life exploits of Guy Gabaldon, who—after being raised by Japanese foster parents—finds himself in the Marines singlehandedly fighting the Japanese army on a South Pacific island. B-movie whiz Phil Karlson and Hunter provide the punch, even if the movie runs too long for its own good. This war film may be tough to get on cassette because Allied Artist Video is kaput. The same goes for an excellent war film about contemporary mercenaries, *The Wild Geese*, also released by Allied Artists Video.

The Longest Day (1962) Directed by Andrew Marton, Ken Annakin, and Bernard Wicki; written by Cornelius Ryan, Romain Gary, James Jones, David Pursall, and Jack Seddon; starring John Wayne, Robert Mitchum, Henry Fonda, Robert Ryan, Rod Steiger, Robert Wagner, Paul Anka, Fabian, Tommy Sands, Richard Beymer, Red Buttons, Tom Tryon, Richard Burton, Peter Lawford, Sean Connery, and countless others; running time 170 minutes: b&w; 20th C-F—Magnetic Video, RCA Disc.

Probably the best of the big-budget, all-star war movies, *The Longest Day* depicts the Allies' invasion at Normandy in 1944 from every angle imaginable. The problem with using so many stars is that they never really have a chance to develop their characters—John Wayne remains John Wayne, Mitchum remains Mitchum, etc. Since *The Longest Day* is a longest movie (the credits alone would qualify as a short subject), catch it over several days.

©RCA

©Universal Pictures

Paths of Glory (1957) Directed by Stanley Kubrick; written by Kubrick, Calder Willingham, and Jim Thompson; starring Kirk Douglas, Adolphe Menjou, Ralph Meeker, and George Macready; running time 86 minutes; b&w; United Artists—RCA Disc.

Before Stanley Kubrick started going off on his own tangents, he was an equally adept craftsman of solid conventional dramas—first with *The Killing* and then with *Paths of Glory*, an outstanding anti-war film set during World War I. Using the courtmartial of three French soldiers accused of cowardice in 1916 as the focus, Kubrick delivers a scathing attack on the self-serving officers who lead their men to death with equal amounts of incompetence and corruption. The trench scenes are spectacular. The film, shot in moody black and white, holds up particularly well on video.

Patton (1970) Directed by Franklin J. Schaffner; written by Francis Coppola and Edmund H. North; starring George C. Scott, Karl Malden, and Stephen Young; running time 171 minutes; color; 20th C-F—Magnetic Video, RCA Disc.

George C. Scott is in high gear in this many-sided portrait of World War II Army general George S. Patton, a brilliant leader and strategist who pushed himself and his men over the limit fighting Nazis in North Africa and Sicily and battling the U.S. War Department. Francis Coppola wrote the screenplay with Edmund North *(The Day the Earth Stood Still)*. Scott won an Oscar for his bravura performance, Schaffner won one as best director, the film won the best-picture Oscar. At 171 minutes, it's a little long for one sitting.

Sergeant York (1941) Directed by Howard Hawks; written by Abem Finkel, Harry Chandler, Howard Koch, and John Huston; starring Gary Cooper, Walter Brennan, Joan Leslie, and Ward Bond; running time 134 minutes; b&w; Warner Bros.—RCA Disc.

With America teetering on the brink of World War II, along came Howard Hawks, Gary Cooper, and *Sergeant York*, giving an American World War I hero the Hollywood treatment. Cooper as York is grand—a shy hillbilly from Tennessee who ends up a hero when he singlehandedly captures a hundred Germans. Cooper won the Oscar for best actor.

Twelve O'Clock High (1949) Directed by Henry King; written by Sy Bartlett and Beirne Lay Jr.; starring Gregory Peck, Dean Jagger, Hugh Marlowe, Gary Merrill, and Millard Mitchell; running time 132 minutes; 20th C-F—Magnetic Video.

A no-nonsence World War II drama about an 8th Air Force squadron, *Twelve O'Clock High* presents Gregory Peck as a bomber-group leader who nearly cracks under the strain of combat. The aerial battle scenes are rip-roaring, and Peck and Jagger (who got an Oscar for best supporting actor) are in top form.

Von Ryan's Express (1965) Directed by Mark Robson; written by Wendell Mayes and Joseph Landon; starring Frank Sinatra, Trevor Howard, Brad Dexter, and Edward Mulhare; running time 117 minutes; color; 20th C-F—Magnetic Video.

Von Ryan's Express is another crackerjack escape movie, with ol' Blue Eyes playing a rough-and-ready American colonel who leads English POWs in a train getaway in the heart of Italy during World War II. The film's a tad shallow, but the action scenes—and Sinatra—keep things rolling along the right track. Catch that finale again on a replay.

FOTOMAT

One of the most original corporate approaches to the home video market is Fotomat, which a few years ago mobilized its 3,800 photo-processing stores into video sales and rental outlets. The company has more than 400 titles for rental and purchase, sells video-game cassettes and offers another innovative service for amateur camera buffs who want to preserve their home movies or 35mm slides on videotape. Fotomat is also the largest blank-videocassette retailer in the country.

The best way to get acquainted with the Fotomat operation is call the toll-free number 1-800-325-1111 (1-800-392-1717 in Missouri) to request the current free catalog and price list.

The choice of programs ranges from such recent hits as *9 to 5* and *The Stuntman* to some great vintage British comedies like *I'm All Right, Jack* and *The Ladykillers,* plus instructional cassettes, concerts, cooking lessons, and golf tips. The titles can be purchased for between $40 and $75 or rented for five days for between $6 to $12. If you keep the cassette longer, you'll be charged another $3 per day.

When you're ready to buy or rent a film call the toll-free number and give the requisite information (name, address, phone number, cassette format, and the location of your nearest Fotomat store). Any movies you may order should be ready for pick-up at a Fotomat a day or two later. Rentals require a $75 deposit, but you can use your Visa or Master Charge card, and nothing will be charged to your account unless you don't return the rented cassette within three weeks (when the rental fee and late charges will equal the purchase price of the cassette).

Fotomat's system is an effective way to expand a cassette library, since hassles are kept to a minimum and you can inexpensively rent a movie that you've been wanting to see but don't want to add to your permanent collection.

Some—like Fotomat's exclusive video offering of *I Go Pogo*—provided delightful and inexpensive family entertainment. This animated cassette was originally intended as a 1980 feature film based on the late Walt Kelly's famous "Pogo" comic strip, about a possum and his disarming animal pals from the Okefenokee Swamp. When the movie distribution deal fell through, Fotomat picked up the video rights to the film—the story of Pogo Possum's reluctant campaign for the presidency. It's a winner, offering not only some solid family entertainment but some nicely jaundiced insights into American politics.

The production is topnotch, and little seems to have been lost in the transition from movie screen to TV set. The style of animation, using dandy clay figures in a process called Flexform, is actually better suited to a 90-minute program than the standard two-dimensional animation since it retains its novelty longer. The characters move and talk like sophisticated versions of Gumby or the Pillsbury Doughboy, and they are delightful to watch.

The voices, provided by the likes of Jonathan Winters (as Mole, Porky, and Wiley Catt), Vincent Price (as Deacon,), Skip Hinnant (as Pogo), Stan Freberg (as Albert Alligator), Ruth Buzzi, Arnold Stang, and Jimmy Breslin, give the characters that extra dimension needed to sustain them over a feature-length cassette. Some of Winter's bits as the shy Porky Porcupine are inspired.

Another Fotomat service of interest to video buffs is the company's Super-8 and 35mm-slide video transfer service. For $15, Fotomat will transfer 400 feet of Super 8 film (approximately 27 minutes' worth) onto a video cassette that you provide when you order. Fotomat will clean and lubricate your film and splice it together according to your specifications, then transfer it to your tape—frequently color-correcting the image and compensating for over-or under-exposures. For $10, Fotomat will do a similar transferral of 100 35mm slides in any order you choose.

And if you've got an audio dub on your VCR, you can even set your silent movies or slides to music—a heckuva lot less boring than the usual home-movie or slide show. You can get a video transfer kit by writing Fotomat (4829 N. Lindbergh, St. Louis, MO 63044) or by picking one up at the nearest Fotomat store.

©1982 THE HOME VIDEO SOURCEBOOK

THE BEST IN **SUSPENSE**

Ever since the early days of movies in America when Pearl White was going through *The Perils of Pauline* in weekly serials, big-screen suspense has been leaving viewers hanging. Perhaps the best definition of a thriller came from the undisputed master of the form, Alfred Hitchcock, who once explained: "A suspense movie is giving the audience information in advance. A surprise can take 10 seconds. Anticipation can take an hour."

Hitchcock made so many excellent thrillers that we've given him a section all his own. The choices on these two pages are mostly modern, simply because audiences have become so sophisticated that it's tough for many older thrillers to keep 'em guessing.

Plenty of traditional thrillers are still being made, but the more successful ones in the past two decades have had added fillips—the artful playing with illusion and reality in *Blow-Up,* the stylistic flourishes of *Don't Look Now,* or even the monstrous mechanical shark in *Jaws.*

Suspense films have also developed a savage streak in the past decade—witness *Straw Dogs, Don't Look Now, The Silent Partner,* or any DePalma shocker. Catch these on video because if they make it to TV, there won't be much left of them.

Thrillers that haven't been available on video but are certainly worth considering if and when they become available include: Francis Coppola's *The Conversation,* Brian DePalma's *Obsession,* John Huston's *The List of Adrian Messenger,* John Frankenheimer's *The Manchurian Candidate,* W.S. Van Dyke's *The Thin Man,* and Blake Edwards' *Experiment In Terror.*

But what's the very best way to keep a viewer in suspense? Maybe we'll tell you later.

Blow Up (1966) Directed and written by Michelangelo Antonioni; starring David Hemmings, Vanessa Redgrave, and Sarah Miles; running time 102 minutes; color; MGM—MGM/CBS Home Video.

Antonioni's best film is about a fashion photographer in London in the frenetic Sixties who thinks he may have stumbled onto a murder scene. His only proof—some photos he took and enlarged—soon disappear. A neat thriller that is sometimes slow in developing, *Blow Up* has more on its mind than suspense. Antonioni plays with the realms of reality and illusion throughout, even if he does get a little too artsy for the picture's own good. Nonetheless, the film is always diverting and just might improve the second time through your VCR. One complaint: MGM/CBS has shortened the original version of the film by 8 minutes.

Capricorn One (1978) Directed and written by Peter Hyams; starring Elliott Gould, James Brolin, Brenda Vaccaro, and Hal Holbrook; running time 123 minutes; color; Warner Bros.—Magnetic Video.

This thriller by Peter Hyams—about NASA's faking of the first manned space mission to Mars and Gould's attempt to uncover the plot—is pure cotton candy for the brain. It's emptyheaded and occasionally downright silly, but these deficiencies are rather like ants at a picnic. If you can ignore them, you'll probably have a swell time.

The China Syndrome (1979) Directed by James Bridges; starring Jack Lemmon, Michael Douglas, and Jane Fonda; running time 122 minutes; color; Columbia—Columbia Pictures Home Entertainment, Laserdisc.

After decades of nuclear-disaster horror movies, *The China Syndrome* gets right to the core of the matter—not mutant ants or incredible shrinking men but a near meltdown at a California atomic power plant and the resultant cover-up. The film touches upon a lot of still-current controversies—nuclear safety, news manipulation by the media, corruption in high places—but at its heart lies a thriller more sophisticated than its politics. It also shows that you can make a suspense film with a minimum of violence.

Coma (1978) Directed and written by Michael Crichton; starring Genevieve Bujold, Michael Douglas, Elizabeth Ashley, Rip Torn, and Richard Widmark; running time 113 minutes; color; MGM—MGM/CBS Home Video.

Director-screenwriter Crichton gives this thriller—based on Robin Cook's best-seller about a sinister plot to sell human organs for transplants on the black market—the full Hollywood treatment. It has a star-studded cast, a slick approach, a Jerry Goldsmith soundtrack, and suspense that relies heavily on shadowy hitmen and narrow escapes. Best scenes on the replay: Bujold's perilous trip to the private hospital where the bodies are stored.

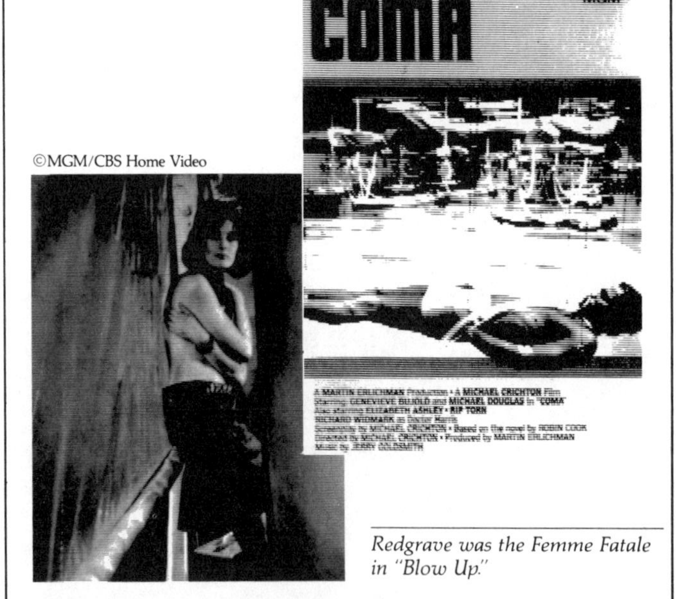

©MGM/CBS Home Video

Redgrave was the Femme Fatale in "Blow Up."

Don't Look Now (1973) Directed by Nicholas Roeg; written by Allan Scott and Chris Bryant; starring Donald Sutherland, Julie Christie, and Hilary Mason; running time 110 minutes; color; Paramount—Paramount Home Video, Fotomat.

Nicholas Roeg turns a Daphne Du Maurier short story (she also wrote *The Birds*) into a haunting mood piece—with a rip-roaring finale. Sutherland and Christie, recuperating from the tragic drowning of their young daughter, encounter two old women in Venice who claim they can communicate with the girl from beyond the grave. Soon after, Sutherland thinks he sees fleeting images of his daughter around the Venice canals, and he's foolish enough to follow. If you try to analyze this thriller too much, it won't make much sense, but it's still as scary as all-get-out. Put the kids to bed before you plug in the VCR. Voyeurs will probably want to replay those steamy love scenes.

©United Artists

From Russia With Love (1963) Directed by Terence Young; written by Richard Maibaum, Johanna Harwood, and Ian Fleming; starring Sean Connery, Robert Shaw, Pedro Armendariz, and Daniela Bianchi; running time 118 minutes; color; UA—RCA Disc.

In the good old days, Sean Connery played 007 and the James Bond movies actually hewed closely to Ian Fleming's pulp novels instead of relying on gimmicks and special effects. The early Bonds were rock-ribbed thrillers, and *From Russia With Love* was the best of the bunch. This time, secret agent Connery goes to Turkey to steal a Russian decoder and takes one incredible ride on the Orient Express—hounded by arch villain Robert Shaw. Play those terrific train scenes again.

Jaws (1975) Directed by Steven Spielberg; written by Peter Benchley and Carl Gottlieb; starring Roy Scheider, Robert Shaw, Richard Dreyfuss, Lorraine Gary, and Murray Hamilton; running time 125 minutes; color; Universal—MCA Cassette and Disc.

An *Enemy Of the People* by the sea, *Jaws* stars Roy Scheider, Richard Dreyfuss, and Robert Shaw as not-so-ancient mariners who set out to sea to tackle a killer shark after the finny fiend gobbles up several bathers in the resort town of Amity (the town fathers want to keep mum about the deaths—they wouldn't want to hurt tourism). Spielberg's direction is sharpest during the action scenes; the rest is somewhat waterlogged. One plus for the small screen: The mechanical shark doesn't look as fake.

The Sailor Who Fell From Grace With The Sea (1976) Directed and written by Lewis John Carlino; starring Kris Kristofferson, Sarah Miles, and Jonathan Kahn; running time 104 minutes; color; Avco-Embassy—Magnetic Video, Laserdisc.

If you're looking for an offbeat thriller that works more by mood than by might, you could try renting this film by Lewis John Carlino (he also directed *The Great Santini).* The story is about an American seaman who woos a young English widow—much to her sadistic son's dismay. It's definitely not for the squeamish (a cat gets dissected) or the prudish (Kristofferson and Miles have a torrid love scene), but there's nothing quite so terrifying as creepy kids.

The Silent Partner (1979) Directed by Daryl Duke; starring Elliott Gould, Susannah York, and Christopher Plummer; running time 103 minutes; color; EMC Film Corp.—Time Life Video.

This gritty suspense film pits Plummer against Gould in a deadly game of cat and mouse. Plummer plays a psychopathic bank robber who sets out for revenge when he discovers that a wimpy teller has short-changed him during the hold-up. *The Silent Partner* gradually turns into an exercise in terror and violence, topped off by a finale that's just shy of ingenious. The violence—occasionally excessive—turned off some critics, but this film is a real sleeper. Don't miss it.

Straw Dogs (1971) Directed by Sam Peckinpah; written by Peckinpah and David Zelag Goodman; starring Dustin Hoffman, Susan George, Peter Vaughn, and David Warner; running time 118 minutes; color; Talent Associates—Magnetic Video.

This Sam Peckinpah film is the tale of a meek American professor and his flirtatious wife who rent a country house in England, only to have a band of local thugs wage a terror campaign on them. It's a gruelling ride, building ever so slowly to a violence-riddled finale. The film is very discomforting and occasionally excessive, but it'll keep you in your seat till the fireworks finally erupt. Play back that last assault on Dustin's castle one more time: Rough stuff, and definitely not for the kiddies.

The Third Man (1950) Directed by Carol Reed; written by Graham Greene; starring Joseph Cotten, Trevor Howard, Alida Valli, and Orson Welles; running time 100 minutes; b&w; Selznick Releasing Org.—Budget Video, Thunderbird Films.

A slick, stylish thriller by Carol Reed, *The Third Man* is about an American writer who travels to Vienna after World War II to take a job with an old buddy—only to find that his friend has just died accidentally. Maybe. The film clicks in every department, from Reed's direction and the performances by Welles and Cotten to that haunting zither music in the background.

©Budget Video

Welles clicked in "The Third Man."

THE BEST OF ALFRED HITCHCOCK

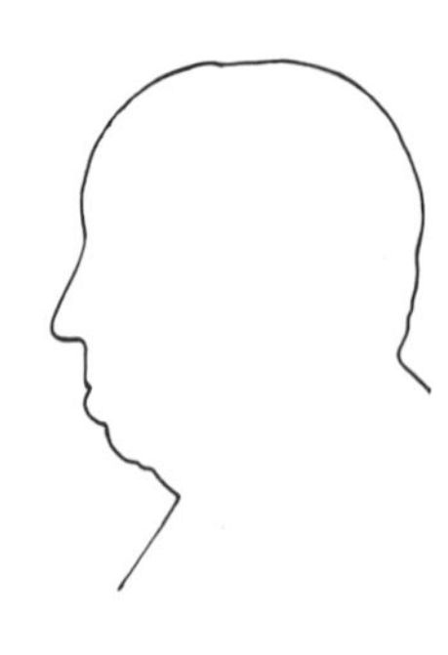

Alfred Hitchcock is undoubtedly the best-known movie director, so it logically follows that his career is better represented on video than any of his peers. More than a score of his titles are available, ranging from *The Lodger* (1926) to his final film, *Family Plot* (1976). Our selection is roughly divided between his spy classics *(North by Northwest, The 39 Steps,* and *Foreign Correspondent)* and chillers *(Rebecca, Psycho,* and *The Birds.)*

Other Hitchcock titles worth buying or renting include *Notorious* (1946), *The Man Who Knew Too Much* (1934), *Sabotage* (1936), *Young and Innocent* (1937) and *Frenzy* (1972). Several key titles aren't yet available on home video; these include *Shadow of a Doubt* (1943), *Lifeboat* (1944), *Stangers on a Train* (1951), *Spellbound* (1946), *Dial M for Murder* (1954), and *Marnie* (1964).

The Birds (1963) Directed by Alfred Hitchcock; written by Evan Hunter, based on a story by Daphne Du Maurier; starring Tippi Hedren, Rod Taylor, Jessica Tandy, and Suzanne Pleshette; running time 119 minutes; color; Universal—MCA Cassette and Disc.

As in many of Hitchcock's later films, the acting by some of the leads is poor, but that's really beside the point. The vaguely allegorical story about flocks of birds pecking at humans is terrifyingly realistic. And no wonder: in the attack scene, for example, Hedren actually allowed Hitchcock to unleashed winged attackers on her.

"The Birds": Terror from above.

©Universal Pictures

Peck had 'em "Spellbound," while Grant had great direction in "North By Northwest."

©RCA

Foreign Correspondent (1940) Directed by Alfred Hitchcock; written by Charles Bennett, Joan Harrison, James Hilton, and Robert Benchley, from a novel by Vincent Sheehan; starring Joel McCrea, Laraine Day, Herbert Marshall, Albert Basserman, Edmund Gwenn, George Sanders, Eduardo Ciannelli, and Robert Benchley; running time 120 minutes; b&w; United Artists—Time Life Video, RCA Disc.

A grab-bag of memorable sequences that can be savored at leisure: an assassination in the rain, a chase through a field of windmills, a murder attempt atop a cathedral, and a spectacular plane crash at sea. The story is thinly disguised propaganda about an American journalist who tangles with a German spy ring. The special effects and the sets are incredible.

The Lady Vanishes (1938) Directed by Alfred Hitchcock; written by Sidney Gilliat and Frank Launder, based on a novel by Ethel Lina White; starring Margaret Lockwood, Michael Redgrave, Dame May Whitty, Paul Lukas, Basil Radford, and Naunton Wayne; running time 97 minutes; b&w; Gaumont British—Budget Video, Cable Films, Media Home Entertainment, Video Warehouse, RCA Disc.

In the most famous of his "train" thrillers, Hitchcock has a great time with this story about a young lady's lonely struggle to convince her fellow passengers that an elderly schoolteacher has disappeared from the train. The film is vague about the nationality of the villains, but there's little doubt the Nazis are behind the exciting climactic shootout. And why is that Nun wearing high heels...?

North By Northwest (1959) Directed by Alfred Hitchcock; written by Ernest Lehman; starring Cary Grant, Eva Marie Saint, James Mason, Leo G. Carroll, and Martin Landau; running time 136 minutes; color; MGM—MGM/CBS, RCA Disc.

This classic about an advertising man who is marked for death after being mistaken for a CIA agent is a good starting point for a Hitchcock collection. It's even more fun when you can run the famous cornfield attack and chase across Mount Rushmore over and over again. The VistaVision sweep suffers a bit on a small screen, but you can still relish the wittiest script the master ever had to work with. Look for a sexual sight gag at the end.

©Warner Home Video

Psycho (1960) Directed by Alfred Hitchcock; written by Joseph Stefano, from a novel by Robert Bloch; starring Anthony Perkins, Vera Miles, John Gavin, Janet Leigh, and Martin Balsam; running time 100 minutes; b&w; Paramount/Shamley—MCA Cassette and Disc.

A natural for home video. For starters, you can study the shower scene—which TV censors usually make mincemeat of—and see that the knife never touches Janet Leigh's body. Hitchcock used the crew from his TV series to film this black comedy, which makes pointed observations about psychology and American mother worship.

Rebecca (1940) Directed by Alfred Hitchcock; written by Robert E. Sherwood and Joan Harrison, based on a novel by Daphne Du Maurier; starring Laurence Olivier, Joan Fontaine, George Sanders, Judith Anderson, Nigel Bruce, Gladys Cooper, Florence Bates, and Leo G. Carroll; running time, 130 minutes; b&w; United Artists/Selznick International—Magentic Video.

More typical of producer David O. Selznick *(Gone With the Wind)* than Hitchcock, this is still one of the master's most memorable films and the only one to win an Academy Award as best picture. Joan Fontaine copped an Oscar for her portrayal as the second wife of a wealthy aristocrat who is still obsessed with the memory of her deceased predecessor. The actress to watch is Judith Anderson, as the villainous housekeeper, Mrs. Danvers. If you can, compare the fiery climax with the closing sequence in *Citizen Kane.*

The 39 Steps (1935) Directed by Alfred Hitchcock; written by Charles Bennett and Alma Reville (Mrs. Hitchcock), based on a novel by John Buchan; starring Robert Donat, Madeleine Caroll, Godfrey Tearle, Lucie Manheim; running time 81 minutes; b&w; Gaumont British—Budget Video, Cable Films, Niles Video, Reel Images, Thunderbird Films, Video Dimensions, Video Warehouse.

Many film buffs argue this is Hitchcock's greatest film; it's certainly the best he made before coming to Hollywood in 1940. This quintessential spy thriller—remade twice by other directors—has an innocent hero accused of murdering a spy being chased across Scotland, handcuffed to a very skeptical young woman. There's also a memory expert and a missing-fingered villain.

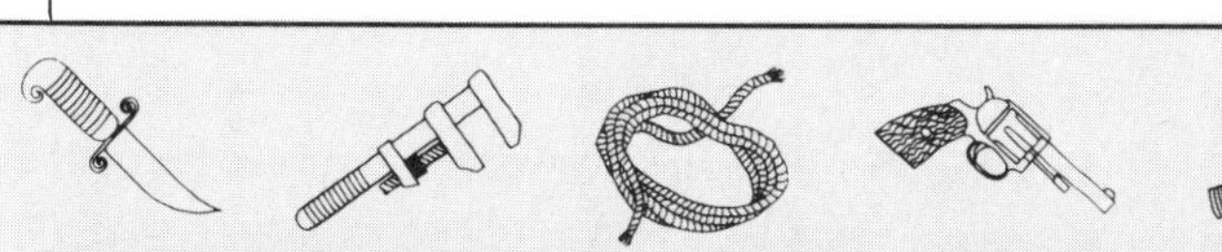

ALFRED HITCHCOCK

When Alfred Hitchcock died in the spring of 1980, a veil of hope accompanied the sorrow over his passing. Although the man who dazzled six decades of movie-goers was gone, many film devotees speculated that the five thrillers that he had withdrawn from circulation over the years might someday become available again.

None of the five—*Rear Window, Vertigo,* the 1956 remake of *The Man Who Knew Too Much, Rope,* and *The Trouble With Harry*—has been seen in 10 years. *Rope* hasn't been shown in the U.S. for three decades.

Herman Citron, Hitchcock's agent of 30 years and now representative for the late director's estate, has indicated that the films will be re-released "in the near future." Such a move would be a major cinematic event, given the intense public interest in Hitchcock's works. Books on Hitchcock's movies have been selling more briskly than ever, and film schools have developed curricula that focus solely on the British director's techniques.

Citron has steadfastly refused to give a specific timetable for the movies' return and has said that he isn't sure in what form the five will reappear—either on cassette, disc, or over network or cable TV.

According to Citron, the one place the films *won't* be seen again is in movie theaters. When asked if the films would lose some of the impact if they weren't shown on the large screen, Citron has been defensive. "It isn't any of your business," he told one reporter. "Other people have asked that, and we don't have to answer to anyone for what we do."

Whatever form the five films take, their return will be a significant event for the general public, as well as for film students and film-makers. Says Donald Spoto, professor of film at the New School and author of "The Art of Alfred Hitchcock": "These are five major works of film art by, I believe, the most important director in the history of the medium."

Spoto puts the films in this perspective: "I think that three of the five are among the half-dozen of Hitchcock's best. I think *Vertigo* is his greatest work and one of the most profound films ever made, and *Rear Window* and *The Man Who Knew Too Much* are also masterworks. *The Trouble With Harry* was a personal favorite of Hitchcock's, with a marvelous kind of quirky humor that many people will find engaging, *Rope,* of course, is an anomaly, but technically a work of genius."

Here's a brief synopsis of each film:

The Man Who Knew Too Much (the 1956 remake), starring James Stewart and Doris Day, is the adventures of an American doctor and his wife vacationing in Morocco who witness the murder of a French spy—who, in his dying breath, tells them of an upcoming assassination.

Rear Window (1954), starring James Stewart and Grace Kelly, focuses on a photographer with a broken leg who kills time by spying on his neighbors across the courtyard—and thinks he saw a murder.

Rope (1948), starring James Stewart, John Dall, and Farley Granger, is Hitchcock's biggest technical gamble—an 80-minute film shot as one continuous take; the plot deals with two young joy-killers who murder a companion, stuff his body in a trunk, and invite the dead man's relatives to a party, with the buffet served from the makeshift coffin.

The Trouble With Harry (1955), starring Shirley MacLaine and John Forsythe, is a macabre comedy about a bunch of Vermont villagers who discover a corpse in their backyard.

Vertigo (1958), starring James Stewart and Kim Novak, is the story of a detective who agrees to trail a friend's suicidal wife, falls in love with her himself, and becomes involved in one of the most haunting mysteries of all time.

THE BEST IN CRIME

Here's to you, Mr. Robinson.

Rat-a-tat-tat! With gats blazing, gangsters have mesmerized movie audiences from D.W. Griffith's *The Musketeers of Pig Alley* in 1917 to *The Godfather* (1972) and beyond. Al Capone, Bugsy Siegel, Lucky Luciano, and John Dillinger have all been portrayed (often under other names) in countless films.

The Godfather films and *Bonnie and Clyde* are discussed in the "Best of the Best List," but there are loads of other gangster classics on videotape, from 1932's *Scarface* to the 1964 version of *The Killers,* featuring a villain who's since come to occupy the Oval office. At this writing, none of the classic Warner Bros. gangster classics except *White Heat* are on the home video market.

Dillinger (1945) Directed by Max Nosseck; written by Philip Yordan; starring Lawrence Tierney, Edmund Lowe, and Anne Jeffreys; running time 90 minutes; b&w; Monogram—Allied Artists Video.

This was the first film that purported to be the biography of an actual gangster. Although they've played fast and loose with the facts, it's still a rousing, fast-paced portrait, with Tierney excellent in the title role. This low-budget production—the best to ever come out of Monogram—is better than the 1973 remake starring Warren Oates. This might be tough to get a hold of because Allied Artists Video went kaput.

The Killers (1964) Directed by Donald Siegal; written by Gene L. Coon, based on the story by Ernest Hemingway; starring John Cassavetes, Lee Marvin, Angie Dickinson, Ronald Reagan, Clu Gulager, and Claude Akins; running time 95 minutes; color; Universal—MCA Videocassette.

The 1946 film (starring Burt Lancaster) based on Hemingway's short story is justifiably more famous, but this loose remake is interesting on its own terms. This opening sequence in which the title characters terrorize a home for the blind is a classic by Siegel, who went on to direct *Dirty Harry.* This version is also notable for the final screen appearance of the president, in a very rare villainous role—at one point he belts Angie in the mouth. Originally shot as a TV movie, it was released theatrically because it was deemed too violent for home screens in the mid 1960s. How times change.

Pay Or Die! (1960) Directed by Richard Wilson; written by Wilson and Bertram Millauser; starring Ernest Borgnine, Alan Austin, Zohra Lampert, Robert F. Simon and Renata Vanni; running time 109 minutes; b&w; Allied Artists—Allied Artists Video.

A tough, convincing melodrama about a police detective who forms a special squad to combat the Black Hand in 1906 New York City. Good period flavor, nicely photographed. Good performances, especially Borgnine. Another Allied Artists beauty.

The Phoenix City Story (1955) Directed by Phil Karlson; written by Crane Wilbur and Daniel Mainwaring; starring Richard Kiley, Edward Andrews, John McIntyre, and Kathryn Grant; running time 100 minutes; b&w; Allied Artists—Allied Artists Video.

One of the best of the slew of exposes that followed the Kefauver Committee's investigations into organized crime. Kiley gives a fine performance as a young lawyer exposing corruption in his small home town. A good, gritty B-movie. Copies of this cassette are rare, so hold onto one if you can find it.

Scarface (1932) Directed by Howard Hawks; written by Ben Hecht, Seton I. Miller, John Lee Mahin, W.R. Burnett, and Fred Pasley, based on a novel by Armitage Traill; starring Paul Muni, Ann Dvorak, George Raft, Boris Karloff, Karen Morley, and Osgood Perkins; running time 90 minutes; b&w; United Artists—MCA Videocassette.

Possibly the best of the early gangster films, this holds up better than *The Public Enemy* or *Little Caesar.* Howard Hughes lavishly produced this vividly directed story about a thinly disguised Al Capone, who was reportedly pleased with Muni's bravura performance. The film also immortalized George Raft as Muni's coin-flipping confederate and contains suggestions of an incestuous relationship.

Take The Money And Run (1969) Directed by Woody Allen; written by Woody Allen and Mickey Rose; starring Allen, Janet Margolin, Marcel Hillaire, and Louise Lasser; running time 85 minutes; color; ABC/Palomar—Magnetic Video, Laserdisc.

Woody Allen's first film as a director affectionately sends up all the cliches of crime movies, including a hilariously muffed prison break. Uneven but often sidesplittingly funny. Best scene: When Allen gives a bank teller that illegible hold-up note that reads, "I have a gub."

White Heat (1949) Directed by Raoul Walsh; written by Ivan Goff and Ben Roberts, based on a story by Virginia Kellogg; starring James Cagney, Edmond O'Brien, Margaret Wycherly, Virginia Mayo, and Steve Cochran; running time 114 minutes; b&w; Warner Bros.—Vid America (rental only).

Cagney gives the performance of his career as a mother-obsessed gangster in this tautly directed thriller. *White Heat* revived the gangster genre (which had lain dormant during World War II) with new brutality and violence. The fiery climax is famous, as is the scene where the imprisoned Cagney learns of his mother's fate. Play 'em again.

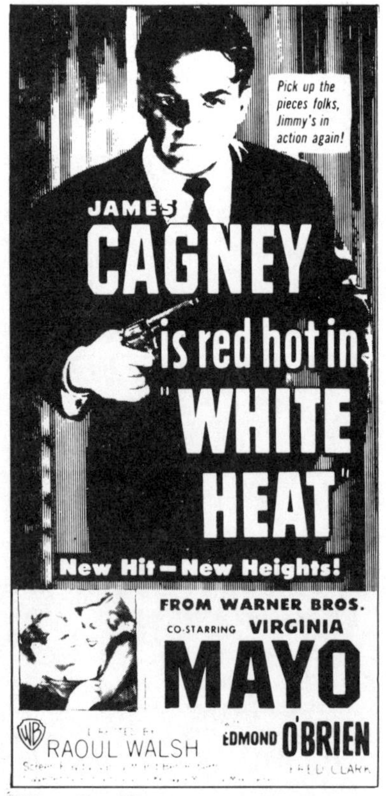

©Warner Home Video

THE BEST IN COP PICTURES

©Magnetic Video

Hackman connected.

If gangsters really ruled the movie roost in the Thirties, the police finally got their turn in the Sixties and early Seventies. Many of the cop flicks were simply updated westerns—witness *Assault on Precinct 13, Coogan's Bluff,* and *McQ*—but the ones that worked the best stalked their own spęcial turf. And, in a few cases, they had more on their minds than bullets and bad guys.

The choices here are three films based loosely on real stories *(Serpico, The Onion Field,* and *The French Connection)* plus the ultimate car-chase movie, *Bullitt,* the Oscar-winning *In the Heat Of the Night, Klute,* and—decidedly offbeat—*The Black Marble.*

Three others are worth looking for—Don Seigel's *Coogan's Bluff* (with Clint Eastwood) and *Madigan* (with Richard Widmark,) plus the great old Glenn Ford melodrama, *The Big Heat* (with thoroughly nasty Lee Marvin).

The Black Marble (1980) Directed by Harold Becker; written by Joseph Wambaugh; starring Robert Foxworth, Paula Prentiss, and Harry Dean Stanton; running time 107 minutes; color; Avco-Embassy—Magnetic Video.

The story of two mismatched cops on the trail of a demented dognapper, *The Black Marble* can best be described as a romantic comedy caper film—a definite change of pace from earlier Joseph Wambaugh cop sagas and, unfortunately, a sleeper that never quite woke up at the box office. Robert Foxworth and Paula Prentiss play a pair of cops who share a cop car and little else until they try to solve a case of the missing pedigreed schnauzer. Harry Dean Stanton is terrific as a loutish slob whose canine perfidy is matched only by his incompetence.

Bullitt (1968) Directed by Peter Yates; written by Harry Kleiner and Alan R. Trustman; starring Steve McQueen, Jacqueline Bisset, Robert Vaughn, and Robert Duvall; running time 113 minutes; color; Warner Bros.—Warner Home Video, MCA Disc.

Steve McQueen picks up where he left off on his motorcycle in *The Great Escape,* only this time he's a San Francisco cop driving a wild Mustang. The plot is nothing earth-shaking (McQueen is safeguarding a witness, and when the stool pigeon gets killed, McQueen goes after the killers on his own). But the brilliant car-chase sequences—achieved with rapid editing and telephoto lenses—are alone worth the cost of the cassette or disc. This *Bullitt* is top caliber.

©Warner Home Video

The French Connection (1971) Directed by William Friedkin; written by Ernest Tidyman; starring Gene Hackman, Fernando Rey, Roy Scheider, and Tony LoBianco; running time 104 minutes; color; 20th C-F—Magnetic Video, RCA Disc, Laserdisc.

In this cops-and-drug-dealers saga, director Billy Friedkin tries to top *Bullitt* with a high-powered chase involving a car and a runaway elevated train, and he just about pulls it off. The story is about two Manhattan undercover cops (Hackman and Scheider) who stumble onto a huge heroin-smuggling ring. Catching the culprits, however, is another matter. The film won Oscars for best picture, best actor (Hackman), and best director (Friedkin).

Klute (1971) Directed by Alan J. Pakula; written by Andy K. Lewis and Dave Lewis; starring Jane Fonda, Donald Sutherland, Charles Cioffi, and Roy Scheider; running time 114 minutes; color; Warner Bros.—Warner Home Video.

While Gene Hackman was chasing New York subways in *The French Connection,* Donald Sutherland was tracking a murderer who preyed on Manhattan callgirls in *Klute.* The next likely target? A foot-loose prostitute played by Jane Fonda. As the script would have it, the duo become a pair of unlikely lovers caught in a web of amorality and corruption. Aside from the jolt in the last reel, the film's strong points are the well-developed characterizations (Fonda won an Oscar) and the depiction of Manhattan's underbelly.

The Onion Field (1979) Directed by Harold Becker; written by Joseph Wambaugh; starring John Savage and James Woods; running time 126 minutes; color; Avco-Embassy—Magnetic Video.

In 1963, two L.A. cops stopped a pair of hoodlums in a beat-up jalopy for a routine search. The thugs got the drop on them and later executed one of the cops. The other somehow escaped. *The Onion Field* explores the shattered psyche of the cop who survived, and an American legal system that, in the name of justice, actually puts a criminal's victim on trial. You might want to catch the first half-hour or so on the replay—after that, the film gets tied up in grueling *juris imprudence.* Savage and Woods are tremendous.

Serpico (1973) Directed by Sidney Lumet; written by Waldo Salt and Norman Wexler; starring Al Pacino, John Randolph, and Tony Roberts; running time 128 minutes; color; Dino de Laurentiis—Fotomat.

In the anti-establishment atmosphere of the early Seventies, it was inevitable that Hollywood take a page from Peter Maas' nonfiction best-seller and turn its cameras on police corruption Manhattan-style. *Serpico* portrays the real-life tribulations of an honest cop who finds bribes and crooked patrolmen on every street corner. The trouble is, no one will listen. Pacino is charismatic as the cop who couldn't be bought, but the film does get tedious on the second replay.

THE BEST IN MOVIE MYSTERIES

The whodunit is tricky business on video. Usually it's the type of movie that you're better off renting—for the simple reason that the novelty is the mystery itself. Once you know whodunit, the movie loses its oomph.

The choices here are fairly standard ones—a recent Agatha Christie movie and an old one, a recent Sherlock Holmes flick and a vintage Basil Rathbone beauty, plus three other great ones—*Murder My Sweet,* the lilting *Laura,* and *Sleuth.*

A few to watch, should they pop up on the tube, are: *Witness For the Prosecution* (arguably the best Agatha Christie novel ever filmed), *Murder She Said* and *Murder At the Gallop* (two Agatha Christie romps starring Margaret Rutherford as the redoubtable Miss Marple), *Midnight Lace,* and Bogart in *The Big Sleep.*

And Then There Were None (1945) Directed by Rene Clair; written by Dudley Nichols; starring Walter Huston, Barry Fitzgerald, Louis Hayward, June Duprez, and Judith Anderson; running time 98 minutes; b&w, 20th C-F—Video Communications, Fotomat.

Agatha Christie's *And Then There Were None* has been filmed three times (twice as *Ten Little Indians),* but the first is still the best for a variety of reasons—sharper direction, a better cast, and that moody black-and-white photography. The plot is vintage Christie: Ten people are invited to a weekend idyll on a remote island, and each meets a sudden demise. But just try to guess the culprit.

Laura (1944) Directed by Otto Preminger; written by Jay Dratler, Samuel Hoffenstein, and Betty Reinhardt; starring Dana Andrews, Clifton Webb, Gene Tierney, Judith Anderson, and Vincent Price; running time 85 minutes; b&w; 20th C-F—Magnetic Video, RCA Disc.

Dana Andrews, Clifton Webb, and Gene Tierney star in this intelligent melodrama about a beautiful young woman whose death is shrouded in mystery. The film is Preminger's best directorial effort, and the cast is excellent. The movie made Webb a star. As with *And Then There Were None,* the black-and-white photography works particularly well on the small screen. And that music helps quite a bit. Dim the lights first.

Murder My Sweet (1944) Directed by Edward Dmytryk; written by John Paxton; starring Dick Powell, Claire Trevor, Anne Shirley, Otto Kruger, and Mike Mazurski; running time 95 minutes; b&w; RKO—Nostalgia Merchant.

Based on the Raymond Chandler private-eye classic *Farewell My Lovely,* Edward Dmytryk's atmospheric whodunit features song-and-dance man Dick Powell in a decidedly different role—Philip Marlowe. The film helped launch the trend toward *film noir* in the late Forties, and while Powell isn't any match for Bogart, he ain't half-bad. Check out that bizarre dream sequence again on the replay.

Murder On The Orient Express (1974) Directed by Sidney Lumet; written by Paul Dehn; starring Albert Finney, Ingrid Bergman, Lauren Bacall, Wendy Hiller, Sean Connery, Vanessa Redgrave, Michael York, Martin Balsam, Richard Widmark, and Jacqueline Bisset; running time 131 minutes; color; Paramount—Paramount Home Video, Fotomat.

This film started a new mini-industry in Agatha Christie movie mysteries—Hire a bunch of semi-over-the-hill stars (just like those hoary disaster movies did), put together a clever script, and leave the audience guessing. Some of the series—*Death On the Nile* and *The Mirror Crack'd,* for example—are rather lame, but *Orient Express* stays on the right track. But for heaven's sake, don't let anybody tell you who killed the man on the train before you slip the tape into your VCR. Without the big surprise at the end, *Orient Express* gets derailed.

The Scarlet Claw (1944) Directed by Roy William Neill; starring Basil Rathbone and Nigel Bruce; running time 74 minutes; b&w; Universal—Allied Artists Video.

No mystery list would be complete without a vintage Sherlock Holmes movie, and *The Scarlet Claw* is probably the best of the dozen that Rathbone and Bruce made for Universal in the Forties. In this one, Holmes and Watson travel to the wilds of Canada to solve the mystery of the marsh monster. Alas, getting a copy of *The Scarlet Claw* may not be elementary. It was released by Allied Artists Video, now as defunct as Professor Moriarty.

Sleuth (1972) Directed by Joseph Mankiewicz; written by Anthony Shaffer; starring Laurence Olivier and Michael Caine; running time 138 minutes; color; 20th C-F—Magnetic Video.

Scene for scene, twist for twist, this is probably the best mystery available on video. Anthony Shaffer's screenplay is a delight, and Olivier and Caine play off one another superbly. The plot involves a mystery writer whose real-life murder scheme goes awry—an idea that found a new home in Ira Levin's Broadway thriller *Deathtrap.* Watch *Sleuth* a second time to catch that boatload of red herrings.

©Universal Pictures

THE BEST IN SERIALS AND SHORTS

©Universal Pictures

Through the miracle of home video, you can recapture two types of moviegoing experiences that have become all but extinct since the rise of television. From the 1920s on, movie theater programs routinely included serials and shorts, which typically ran about 20 minutes apiece. About 220 serials were made during the talkie period—almost all of them cliffhangers that placed their heroes (and heroines) in nail-bitting situations at the end of every episode for 12 to 20 weeks. Since these serials infrequently turn up on television, home video offers about the only opportunity to enjoy them in their episodic glory. Most people will want to watch a couple at a time, although fanatics will revel in marathon showings that are longer than "Gone With the Wind."

Shorts are a relatively novelty on the home video scene, since their brief running time make an awkward package. There are some marvelous works by The Three Stooges, Charlie Chaplain, and Laurel and Hardy, but so far the Little Rascal shorts have yet to be released in home video.

The Adventures of Captain Marvel (1941) Directed by William Witney and John English; written by Ronald Davidson, Norman S. Hall, Arch B. Heath, Joseph Poland, and Sol Shor; starring Tom Tyler, Frank Coghlan Jr., William Benedict, and Louise Currie; running time 240 minutes; b&w; Republic—Nostalgia Merchant.

Regarded by buffs as the best of the serials, this 12-episode adventure has the comic book character—who becomes invincible by uttering "Shazam!"—battling the villainous Scorpion from Siam to America. He escapes death from a guillotine and falling out of a plane, among other things. Great fun, although a bit violent.

The Phantom Empire (1935) Directed by Otto Breuer and B. Reeves Eason; starring Gene Autry, Frankie Darro, Smiley Burnett, Dorothy Christy, and champion trick rider Betsy King Ross; running time 245 minutes (12 episodes); b&w; Mascot—Video Yesteryear.

One of the weirdest of the serials, possibly the only attempt to combine the Western and science-fiction genres. Singing cowboy Autry's efforts to keep his radio contract are distrupted by masked riders from the subterranean city of Murania, located 20,000 feet below his "radio ranch." He not only has to contend with the evil Queen Tika, but evil archeologists and primitive sci-fi hardware. A bit on the creaky side, but decidedly different.

Nyoka and the Tigermen (1942) Directed by William Witney; written by Ronald Davidson, Norman S. Hall, William Lively, Joseph O'Donnell, and Joseph Poland; starring Kay Aldridge, Clayton Moore, William Benedict, Lorna Gray, Charles Middleton, and Tris Coffin; running time 340 minutes (15 episodes); b&w; Republic—Nostalgia Merchant.

Outside of Pearl White and her silent "Perils of Pauline," Aldridge is one of the serials' most famous heroines. Lots of fun and thrills in darkest Africa as Nyoka searches for the fabled lost tablets of Hippocrates—the source of all medical knowledge, of course. She's helped by her pet monkey Jitters, as well as Moore, who would be playing "The Lone Ranger" on TV a few years hence. Also known as "Perils of Nyoka" and "Nyoka and the Lost Secrets of Hippocrates."

Flash Gordon Conquers the Universe (1940) Directed by Ford Beebee and Ray Taylor; written by George H. Plympton, Bail Dickey and Barry Shipman, based on the Alex Raymond comic strip; starring Larry (Buster) Crabbe, Carol Hughes, Charles Middleton, Frank Shannon, Anne Gwynne, Roland Drew; running time 240 minutes (12 episodes); Universal—Videotape Network, Thunderbird Films.

Forget the recent feature film, as if you haven't already. For millions of Americans, this is the definitive version of Alex Raymond's vision of the 21st century. In this, the third and final series (the first two aren't on tape), Flash again tangles with Ming the Merciless, who has unleashed The Purple Death on the earth. With Dale and Dr. Zarkov, he travels to Mongo, where he tangles with rock monsters and is on the verge of being disintegrated. And that's the end of only two chapters. Thanks to home video, you won't have to wait until next week to see how things turn out.

©Nostalgia Merchant

Moe, Larry, and Curly.

©Columbia Pictures Home Entertainment

The Three Stooges (1930s-1940s) Various writers and directors; starring Moe Howard, Larry Fine, Curly Howard, and Vernon Dent; running time 60 minutes (three shorts); b&w; Columbia—Columbia Pictures Home Entertainment.

With the possible exception of the Little Rascals, the Stooges starred in the longest and most successful series of shorts, from the mid 1930s through the late 1950s. Home video provides an unedited look at their fingers-in-the-eyes brand of slapstick, which has been pruned considerably by squeamish TV stations in recent years. Either you love or you hate the stooges; there's very little middle ground. Columbia so far has released three collections of their best work from the late 1930s and early 1940s.

Laurel and Hardy (1930s) Various writers and directors; starring Stan Laurel, Oliver Hardy, Jimmy Finlayson, Billy Gilbert; running time 90 minutes (four shorts); b&w; MGM/Hal Roach—Nostalgia Merchant.

Nostalgia Merchant, which sells seven full-length Laurel and Hardy classics ("Sons of the Desert" and "Way Out West" are the best), has also packaged six collections of their classic shorts. You can hardly go wrong with any of them. A good place to start is Volume One, which contains their only Academy-award winner, "The Music Box," in which they try to haul a piano up several flights of stairs. Hilarious.

The Tramp and A Woman (1915) Directed and written by Charles Chaplin; starring Chaplin and Edna Purviance; running time 45 minutes; b&w; silent with musical soundtrack; Essanay—Video Yesteryear.

Two shorts showing off Chaplin during his most famous silent period. In "The Tramp," the little hobo rescues the farmer's daughter from three evil hoboes, but goes off into the sunset when her boyfriend returns home. In "A Woman," Chaplin anticipates "Some Like it Hot" by 45 years. In order to get rid of his girl's suitors, he impersonates a woman and winds up wooing his girl's father.

THE BEST IN
ACTION FILMS

In "Deliverance," Ned Beatty got it in the end. ©Warner Home Video

This category is a catch-all for rousing, action-oriented movies that don't quite fit into any of the other niches. A few—*Deliverance, The Man Who Would Be King,* and *Rocky*—are outstanding action melodramas. *Enter the Dragon* is included because it's the best martial-arts drama on video, *Hercules* because it's the best spaghetti-toga epic, and *The Wild One* because it's probably the best motorcycle saga. Ridley Scott's *The Duellists* is a thunderous Napoleonic epic that's great for owners of large-screen televisions.

A few other action-adventure flicks are worth noting, even though they haven't been available on cassette or disc: *Grand Prix, Spartacus, The Vikings,* and a real sleeper called *Who'll Stop The Rain?*

Deliverance (1972) Directed by John Boorman; written by James Dickey; starring Burt Reynolds, Jon Voight, Ned Beatty, Ronny Cox, and James Dickey; running time 109 minutes; color; Warner Bros.—Warner Home Video, MCA Disc.

The story of four friends who take a weekend canoe trip in the wilds of Georgia and get a lot more than they bargained for. *Deliverance* is disturbing at times but it's unforgettable from the early "Duelling Banjos" interlude to the last rocky rapids. Voight and Reynolds are excellent—and look for James Dickey (the poet who wrote the screenplay and the novel it was based upon) as a redneck sheriff near the end.

The Duellists (1978) Directed by Ridley Scott; written by Gerald Vaughn-Hughes; starring Keith Carradine, Harvey Keitel, Albert Finney, and Christina Raines; running time 101 minutes; color; Paramount—Paramount Home Video.

The first feature film from the man who brought you *Alien, The Duellists* is based on a Joseph Conrad short story about a 14-year feud between two French officers during the Napoleonic Wars. The feud gets a little far-fetched after while—you start to pray that one of these guys finally knocks off the other—but the fighting scenes are top-drawer and Frank Tidy's cinematography is spectacular. A must if you have a projection-TV unit.

Enter The Dragon (1973) Directed by Robert Clouse; written by Michael Allin; starring Bruce Lee, John Saxon, and Jim Kelly; running time 90 minutes; color; Warner Bros.—Warner Home Video.

You can't have a listing of the top action-adventure films without including *Enter The Dragon,* the martial-arts movie that made Bruce Lee synonymous with kung fu and spawned a fistful of imitations. The plot (about a karate master hired by British intelligence to break up an opium-smuggling ring) wears rather thin, but it was never meant to be more than a vehicle for Lee's pyrotechnic fists and feet of fury. One gripe: Warner Home Video has lopped 9 minutes off the original theatrical version.

Hercules (1959) Directed by Pietro Francisci; starring Steve Reeves, Fabrizio Mioni, and Sylva Koscina; running time 107 minutes; color; Warner Bros.—Magnetic Video.

If *Enter the Dragon* touched off an unending supply of kamikaze karate flicks, then Steve Reeves and *Hercules* unchained a raft of Grecian-formula flicks about mythic muscle-flexers. Once again, the first was the best, with Reeves in great shape as the superhero of the title who trades in his immortality for the love of a good woman. There's plenty of hokum and bad acting, but that's part of the fun.

The Man Who Would Be King (1975) Directed by John Huston; written by Huston and Gladys Hill; starring Sean Connery, Michael Caine, and Christopher Plummer; running time 129 minutes; color: Allied Artists—Allied Artists Video.

One good rule of thumb for taping movies off television is: If the film stars Humphrey Bogart, Bette Davis, James Stewart, Cary Grant, or Sean Connery, hit the "record" button. Something in there is worthwhile. In this one, Connery teams with Michael Caine in Rudyard Kipling's tale of two adventurers in 1880s India who become gods in the eyes of a primitive tribe—only to get done in by their own greed. Ironically, Huston first planned to make the film 20 years earlier with Bogart and Clark Gable as the leads, but Bogart died before the project could get off the ground. This cassette is hard to find, as a result of Allied Artists' demise.

©Warner Home Video

"Hercules": The best of Grecian-formula flicks.

©Columbia Pictures Home Entertainment

Rocky (1976) Directed by John Avildsen; written by Sylvester Stallone; starring Stallone, Talia Shire, Burgess Meredith, and Burt Young; running time 119 minutes; color; United Artists—RCA Disc.

The sleeper hit of 1976 gets cornier with every replay, but it still goes the distance. Stallone was at his most effective here (before he became a celebrity), and he has written the perfect Cinderella story about boxing and the power of positive thinking. It's first-rate all-around, from the supporting cast and Bill Conti's brassy theme music to those brutal sequences in the ring. Still the most rousing scene: Rocky climbing his Everest, the steps of the Philadelphia Museum of Art.

The Wild One (1954) Directed by Laslo Benedek; written by John Paxton; starring Marlon Brando, Lee Marvin, Mary Murphy, and Jay C. Flippen; running time 79 minutes; b&w; Columbia—Video Communications (rental).

Early Brando still holds up well, whether he's a New Orleans Neanderthal, a punch-drunk Hoboken has-been, or—in this case—the leader of a pack of motorcyclists who terrorize a small town. Based on a true story but played for all the melodrama it can muster, *The Wild One* spawned an entire genre and made T-shirts and black leather jackets the attire of choice for a generation of young punks. It may be tough to get hold of a copy of this film (it has been rented for educational purposes only).

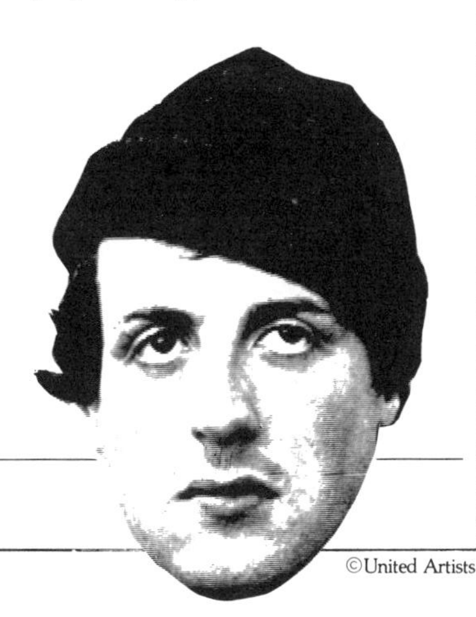

Stallone's "Rocky" went the distance.

©United Artists

THE BEST INSTRUCTIONAL PROGRAMS

In previous sections we've mentioned from time to time the concept of a home video *library*, and it's an apt description—slowly but surely, TV, home computers, and video are usurping the traditional roles of the printed page.

One section of the home video libraries that has just begun to develop is instructional or "how-to" programming, largely because it has been difficult to compete with reference books in terms of price and practicality. Although instructional video is already flourishing in the fields of medicine, industry, and education, for home use it has been limited mostly to cooking and sports "how-to" programs.

For Laser-Disc Machines:

MCA DiscoVision produces two excellent instructional discs—"The Kidisc" and "How To Watch Pro Football"—that use both the laser format and the medium of video to its fullest. "The Kidisc" is discussed on Page 29. "How To Watch Pro Football," the first interactive disc, is also a treat—especially for armchair quarterbacks. Because of the laser system's ability to scan frame by frame, you can read an entire NFL playbook cover to cover or time yourself to see how fast—and how well—you can diagnose an offensive play a pass or a run. A variety of NFL coaches, from Don Shula to Dick Vermeil, give you a thorough primer on the art of pro football. Highly recommended for gridiron fans.

For RCA Disc Systems:

"Caring For Your Newborn": Dr. Benjamin Spock gives his advice on solving the day-to-day problems of raising an infant. The program was filmed in homes using parents and their kids to demonstrate the various child-care techniques. (A cassette version is available from Vid America.)

"Julia Child—The French Chef, Volume I": The colorful master TV chef runs through her step-by-step recipes for roasted chicken, lasagna a la Francaise, strawberry souffle, and chocolate mousse.

"Scholastic Productions As We Grow": The publishers of *Scholastic Magazine* have designed a disc to foster learning techniques in kids ages three to six—using the features of the RCA disc system to full advantage. A teaching guide is available; parental participation is a good idea.

On Cassette:

"Belly Dancing": This series of five programs featuring Pam Van Dyke is designed to teach you to become an accomplished belly dancer. The first program, for example, explains such items of interest as basic belly-dance movement and hip circles. From Video Communications.

"Cooking a Chinese Meal": Joanne Hush, co-author of *The Chinese Menu Cook Book*, demonstrates how to use the exotic culinary equipment and techniques for Chinese cuisine—wok, cleaver, and stir-frying, for example—and shows how to prepare barbequed pork ribs, chicken egg-drop soup, and skewered fruit. From Fotomat.

"Judy Rankin's Golf Tips": This top pro golfer demonstrates a dozen golf tips, concentrating on chipping, putting, and getting the best shots out of sandtraps. From Video Sports Productions.

"On Tennis": Billy Jean King demonstrates the fundamental techniques of tennis, concentrating on such areas as the forehand and backhand strokes, the volley and lob, and the serve. Available through Fotomat.

THE BEST OF
THE X-RATED

©Home Media Entertainment

We're not too keen on X-rated movies, but they do make up a large chunk of the videocassette market whether the National Organization of Women or the Moral Majority like it or not.

To give you an idea of the extent of the popularity of X-rated videocassettes (no X-rated discs are available at present), consider these estimates by the Adult Film Association of America, a trade group that represents 60 film-production companies:

Between $40 million and $50 million worth of X-rated cassettes were sold in 1980—approximately 40 percent of all prerecorded cassette sales.

3,000 X-rated movies are available on cassette, selling for an average price of $70, or approximately $50 more than for the average feature-film cassette.

The all-time adult video blockbuster is *Deep Throat.* More than 300,000 copies of the celebrated film starring Linda Lovelace and Harry Reems have been sold. That's roughly $18 million worth of business on one title alone.

The adult-oriented cassettes are sold at most of the 10,000 video stores, nearly all of the 15,000 adult bookstores in America, and through a sizeable mail-order industry.

One reason for the overwhelming success of video porn is that adult-film producers were at the vanguard of the prerecorded cassette business, offering their films on video as early as 1978—thereby getting the jump on almost all of the major Hollywood studios.

In fact, in 1978 porno movies accounted for 70 percent of all prerecorded videocassette sales. That proportion has been gradually declining as more and more major studios are releasing their latest films on video. Although the porno share of the market has declined, overall sales figures for X-rated cassettes have continued to grow as more and more Americans have been purchasing VCRs.

And just as many mainstream programs have been made exclusively for the video market—"The Mr. Bill Show" and Blondie's "Eat To the Beat," for instance—so have adult features. Marilyn Chambers, a model who went from the cover of the Ivory Snow box to the porno hit "Behind the Green Door," is now starring in a series of adult features designed solely for home video, marketed under the title of "Marilyn's Erotic Fantasties."

Why the broad appeal of X-rated cassettes? As Chambers herself told Vernon Scott of United Press International: "People are willing to pay $100 for an X-rated title because they can see it in the privacy of their own home...It eliminates taking the chance of being seen at the box office of a porno theater while buying a ticket, for one thing. Some people are worried about their reputations."

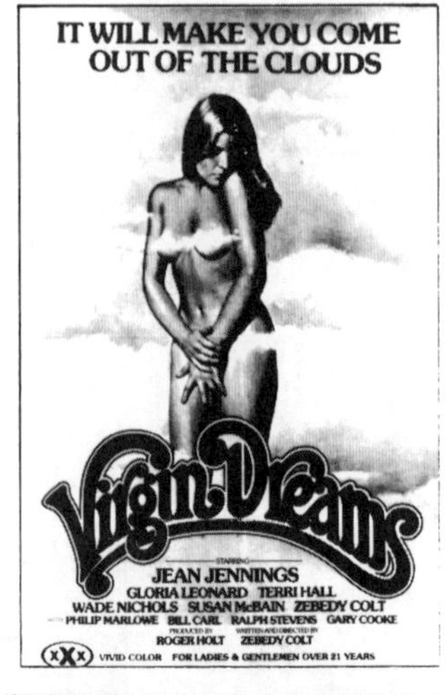

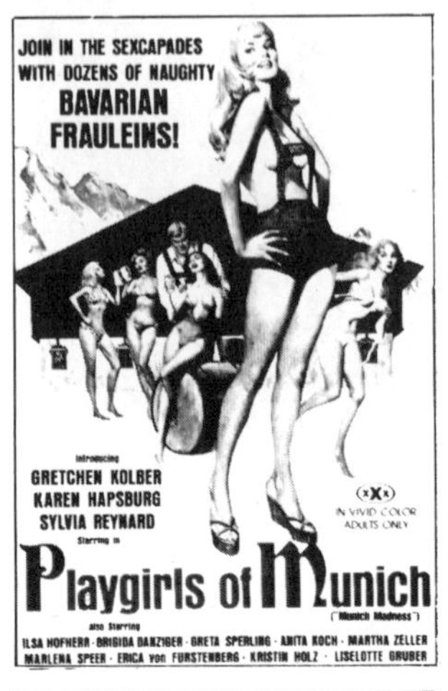

Chambers added that another major reason for the appeal of adult cassettes is "the excitement of the forbidden. If the controversy wasn't there, X-rated cassettes wouldn't sell."

A large segment of the porn-cassette business is rentals, since the novelty of an X-rated film tends to wear off quickly and $70 to $100 is a large investment in titillation. Many video stores, renting the adult cassettes for upwards of $10 a day, report that these cassettes account for 60 to 70 percent of their rental business. Says one video-store saleswoman, "The market for X-rated cassettes has been taking off, but audiences are getting more sophisticated. Customers demand the best in picture quality, editing, soundtrack, acting, and plot—just as they would with a regular movie. Even the appearance of the actresses has improved. Women with chipped nail polish or acne just don't cut it anymore."

And as with the major Hollywood pictures, some porn stars are bigger draws at the box office and video counter than others—recent favorites include Seka, Annette Haven, Marilyn Chambers, and Jessie St. James.

Which adult videocassettes are the best of the bunch? In deference to those who would like a brief rundown of the best of the "X," we've consulted a few experts in erotica. The following are their choices:

Barbara Broadcast
The Private Afternoons Of Angela Mann
Misty Beethoven

If you can have a good reputation in the porno business, Henry Paris has one. He directed these three high-quality porno films. *Barbara Broadcast* stars C.J. Laing, Annette Haven, Constance Money, and Susan McBaine in a story about a former hooker who gets deported from Puerto Rico and comes to the U.S. to promote her book. *Private Afternoons* stars Barbara Bourbon as an adulterous housewife. *Misty Beethoven* is a porno version of *Pygmalion*, starring Constance Money. All three are available from Quality X.

Debbie Does Dallas Jim Clark directed this 90-minute feature, starring Bambi Woods as a head cheerleader who gets her girlfriends to help raise money to go to Dallas for a shot at the bigtime. Their fund-raising activities are not the type sanctioned by the PTA. It's from VCX, available through Direct Video.

Deep Throat Gerard Damiano's contribution to pop culture stars the notorious Linda Lovelace, Harry Reems, and Carole Connors in a story about a woman with strange appetites. It's 65 minutes and available from International Home Video.

Emmanuelle One of the few X-rated movies released by a major studio (Columbia) this film by Just Jaeckin stars Sylvia Kristel as an ambassador's wife who finds the ultimate sexual experience in—of all places—Bangkok. It's 105 minutes long. Available from Columbia Home Video.

Fantasy Gerard Damiano, the man behind *Deep Throat*, wrote and directed this 90-minute exploration of the fantasies of patrons in a cafe on a resort island. Georgina Spelvin and Merle Michaels star. Available from VCX and Direct Video.

©Columbia Pictures Home Entertainment

©Direct Video

©Columbia Pictures Home Entertainment

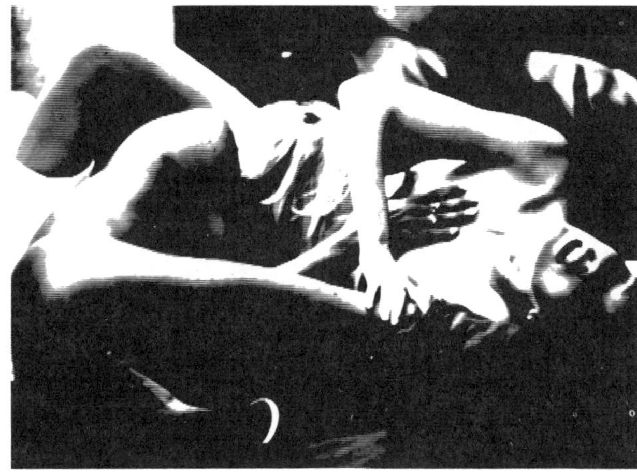

Sultry sex in "Emmanuelle," a rare flirtation between a major Hollywood studio and the sex-plotation genre.

MATURE FILMS

Fascination A recent hit from Quality X, this 105-minute production is a parody of all those "How To Pick Up Girls" books and the singles scene. Merle Michaels, Ron Jeremy, Candida Royale, Samantha Fox, Marlene Willoughby, Christi Ford, Arcadia Lake, Molly Malone, and Eric Edwards star in feature directed by Larry Revere.

Female Athletes Annette Haven stars in this 85-minute production from Video X Pix, playing a sports magazine publisher who demonstrates that women are men's equals in various physical endeavors. Johnny Wadd and Desiree West co-star.

Fiona On Fire From TVX comes this 85-minute skin flick about "the queen of classy trash," a drug addict who has been accused of murder. Amber Hunt, Gloria Leonard, Jamie Gillis, and Marlene Willoughby star.

Health Spa This 90-minute cassette, directed by Clair Dea and written by Marlene Burns, stars Abigail Clayton and Kay Parker in a tale of an X-rated gymnasium dedicated to some highly unusual exercises. It's available from Select Video.

Maraschino Cherry In the 83-minute feature from Quality X, Gloria Leonard teaches her younger sister the ins and outs of the prostitution business. Leslie Bovee, Constance Money, Annette Haven, and Jenny Baxter also star.

The Other Side of Julie Susannah French and John Leslie star in this 90-minute feature about sexual hi-jinx and revenge in the world of big business. Anthony Riverton directed from Colin Davies' screenplay. It's from VCX.

Secret Life Of A Willing Wife When a woman discovers that her husband is cheating on her, she decides to do some firsthand research to find ways to lure her hubby back. The Video X Pix release runs 78 minutes and stars Merle Michaels, Eric Edwards, and Vanessa Del Rio.

THE DISC LIST

A little competition never hurt anybody. The Magnavox/Pioneer LaserDisc system, which had a monopoly on video discography for more than a year before RCA's CED system arrived in March 1981, has had until recently a rather limited selection of discs available.

But when RCA joined the fray with a disc catalog loaded with recent movie hits and a few dozen classics, the LaserDisc people went out and got more ammunition—a heavy helping of winners from Magnetic Video and Paramount Home Video—to augment their basic Universal Pictures—oriented catalog.

With the arrival of such LaserDisc offerings as "Grease" and "The Godfather," the two rival systems have had overlapping titles for the first time. And should both systems flourish, insiders predict that the situation eventually will be akin to videocassette offerings, where the programming on VHS and Beta is nearly identical.

The wild card in the deck, of course, is yet a third incompatible disc system—VHD—which is bound to increase the programming scramble. The VHD camp hopes to have 600 titles in its program catalog by 1985, including many early Warner Bros. classics.

None of the three system's offerings can yet compete with the vast array of cassettes, but they're improving rapidly. And VCR owners have even reaped a dividend: After RCA obtained the disc rights to "Casablanca" and "The Maltese Falcon," that Bogart classics finally became available for purchase on cassette.

For a look at the offerings available on disc, read on—with one caveat. As with videocassettes, the titles available on disc are in a constant state of flux. Programs are added or dropped from month to month, so check with your video store for the latest offerings.

VHD Discs:

Bridge On the River Kwai
The China Syndrome
Close Encounter of the Third Kind
The Deep
Guns of Navarone
And Justice For All...
Tess
Star Trek, The Motion Picture
Ordinary People
The Godfather
Airplane
Saturday Night Fever
The Elephant Man

RCA Discs:

Adventures

Butch Cassidy and The Sundance Kid
The French Connection
The Longest Day
Patton
Tora! Tora! Tora!
The Dirty Dozen
High Noon
Escape From Alcatraz
Shane
The Boys From Brazil
Sands of Iwo Jima
The Black Stallion
Stalag 17

Comedy

M*A*S*H
The Seven Year Itch
Adam's Rib
The Philadelphia Story
Foul Play
Play It Again, Sam
The Bad News Bears
The Muppet Movie
Movie Movie
A Night At The Opera
Paper Moon
Ninotchka
Heaven Can Wait
Starting Over

Drama

Citizan Kane
The Hunchback of Notre Dame
Love Story
Looking For Mr. Goodbar
The Longest Yard
The Godfather
Romeo and Juliet
An Evening With The Royal Ballet
Giselle
Henry V
Hamlet
The Red Shoes
The Graduate
Rocky
Casablanca
The Ten Commandments
Hud
Sunset Boulevard
Clarence Darrow
The Mikado
A Doll's House

Inspiration/Information

Jesus of Nazareth
Victory At Sea
Caring For Your Newborn Dr. Benjamin Spock
Julia Child—The French Chef, Vol. I.
Tut: The Boy King/The Louvre
World of Wildlife
The Undersea World Of Jacques Cousteau Vol. I

Music

The Harder They Come
To Russia...With Elton
Gimme Shelter
Blondie—Eat To The Beat
The Grateful Dead In Concert
Don Kirshner's Rock Concert
Paul Simon
Jean-Pierre Rampal

Musical

Hello Dolly
Gigi
Meet Me in St. Louis
On The Town
Show Boat
Singin' In The Rain
Fiddler On The Roof
Saturday Night Fever
Lady Sings The Blues
Easter Parade
Grease
GI Blues

Mystery

North By Northwest
Laura

Science Fiction

The Thing
King Kong
Star Trek—The Motion Picture
Planet of the Apes

Sports

The Big Fights, Vol. I
Muhammand Ali's Greatest Fights
The New York Yankees' Miracle Year: 1978
Super Bowl XIV Souvenir Video Album
College Football Classics, Vol. I
Tennis Instruction Featuring Arthur Ashe and other Pros
The Miracle at Lake Placid, Winter
Wimbledon Highlights 79/80

Programs for Children

Charlotte's Web
Race For Your Life, Charlie Brown
A Charlie Brown Festival
Heidi
The Gold Bug/Rodeo Red & The Runaway
Terrytoons, Vol. I Featuring Mighty Mouse
As We Grow
Escape to Witch Mountain
Best of Big Blue Marble

Best of Television

Star Trek I
Star Trek II
The Count of Monte Cristo
Our Town
The Fugitive
Little House on the Prairie
Saturday Night Live
Autobiography of Miss Jane Pittman

Walt Disney

20,000 Leagues Under the Sea
Love Bug
The Bears And I
Candleshoe
Kidnapped
Disney Cartoon Parade, Vol. I
The Absent-Minded Professor
The Great Locomotive Chase
Old Yeller
Mary Poppins

Other Movies

Airplane
Ordinary People
The Pink Panther
Goldfinger
The Shootist
Fantastic Voyage
Farewell, My Lovely
Urban Cowboy
American Gigolo
Chinatown
Harold and Maude
War of the Worlds
The Great Escape
North Dallas Forty
Laurel and Hardy
City Lights
Barbarella
Death Wish
The Greatest Show on Earth
The Odd Couple
Swing Time
Stagecoach
The Magnificent Seven
Fun in Acapulco
The Elephant Man
From Russia With Love
Raging Bull
History of the World Part I
Friday the 13th
King Creole
The African Queen
Coming Home

LaserVision Discs:

Columbia Pictures Home Entertainment

California Suite
Chapter Two
The Chna Syndrome
Close Encounters of the Third Kind The Special Edition
Tess

Magnetic Video Corporation

African Queen
Alien
All That Jazz
Annie Hall
Apache
Autumn Sonata
The Black Stallion
Blue Hawaii
Butch Cassidy and the Sundance Kid
Carnal Knowledge
Carrie
The Day of the Dolphin
The Fog
The French Connection
Golden Decade of College Football
The Graduate
Hello Dolly
The King and I
Kotch
Let It Be
The Making of Star Wars
The Moon Is Blue
The Muppet Movie
Murder By Decree
Nine To Five
Notorious
The Omen
Phantasm
The Producers
Raise the Titanic
Return of the Pink Panther
The Rose
The Sailor Who Fell From Grace With The Sea
Saturn Three
Take the Money and Run
They Shoot Horses, Don't They?
Tom Jones
Tora, Tora, Tora!
West Side Story
Yankee Doodle Dandy

MCA Disco Vision

Abba
American Graffiti
Animal House
At Home With Donald Duck
Bernadette Peters in Concert
The Blues Brothers
Buck Rogers In The 25th Century
Cheech and Chong's Next Movie
Coal Miner's Daughter
The Electric Horseman
Flash Gordon
Fleetwood Mac
Heaven Can Wait
The Incredible Shrinking Woman
Jaws
The Jerk
Jesus Christ Superstar
Loretta Lynn in Concert
Mel Torme and Della Reese in Concert
Melvin and Howard
Mission Galactica
National Gallery: Art Awareness Collection
Neil Sedaka in Concert
Nighthawks
1941
Olivia
On Vacation With Mickey Mouse and Friends
Peter Allen in Concert
Prom Night
Psycho
Sergeant Pepper's Lonely Hearts Club Band
Shogun Assassin
Slapshot
Smokey and the Bandit
Smokey and the Bandit II
The Sting
The Touch of Love: Massage
The Wiz
Xanadu

Paramount Home Video

Airplane
American Gigolo
Barbarella
Bon Voyage, Charlie Brown
Charlotte's Web
Chinatown
Death Wish
The Elephant Man
Foul Play
The Godfather
The Godfather II
Grease
Heaven Can Wait
King Kong
The Longest Yard
North Dallas Forty
Ordinary People
Popeye
Saturday Night Fever
Star Trek—The Motion Picture
Starting Over
Up In Smoke
Urban Cowboy
War of the Worlds
The Warriors

Pioneer Artists

Joni Mitchell's "Shadows and Light"
The Kingston Trio/Limelighters/Glenn Yarborough 2-hour Concert
Leon Russell in Concert
Liza in Concert
Melissa Manchester in Concert
Paul Simon in Concert
Pippin
Royal Opera: Tales of Hoffman

VIDEO DISTRIBUTORS

Deciding what program you *want* to rent or buy is one thing. Finding it is something else.

Where do you get a videocassette version of *American Graffiti* or—let's take something a little more obscure—the 1940 Howard Hawks film *His Girl Friday?*

As for *American Graffiti,* a popular recent release, the chances are you'll discover a copy at your nearest video specialty shop. But *His Girl Friday* is a different story. That's why after the screen credits for every program in "The Best In Video" we've tried to list the studio that released the film *and* the principal distributors of the videocassette (and disc) version.

Take *American Graffiti.* As you can see, the picture was released by Universal Studios and is available in videocassette from MCA and Fotomat. It is one of the few hundred films that have also been released on videodisc—*stereo* disc at that. The video disc is also distributed by MCA.

American Graffiti (1974) Directed by George Lucas; written by Gloria Katz and Willard Huyck; starring Richard Dreyfuss, Candy Clark, Ron Howard, Paul LeMat, Harrison Ford, Cindy Williams, Bo Hopkins, and Suzanne Somers; running time 112 minutes; color; Universal—MCA, Fotomat, Disc (stereo).

Before there was *Star Wars* came Lucas' *American Graffiti,* the story of the restlessness and fading innocence of American youth in the early Sixties. Set in a small California town on high-school graduation night, the comedy used radio rock 'n' roll as a framework for the action—cruising the main drag, carrying on, and coming of age. Solid performances all around—*Graffiti* launched many a career—and Lucas' direction is superb. Both the MCA tape and disc versions are the 1978 reissue, several minutes longer than the 1974 original and (in the case of the disc) in stereo sound.

SCREEN CREDITS
VIDEO DISTRIBUTORS
RELEASING STUDIO
RUNNING TIME
REVIEW

What about *His Girl Friday,* a vintage comedy with a certain appeal to members of the Fourth Estate? We find it was released in 1940 by Columbia Pictures, but since it's in the public domain, the film is distributed in videocassette by 7 companies—Budget Video, Cable Films, Video Dimensions, Reel Images, Thunderbird Films, and Video Warehouse.

In the case of *American Graffiti,* you now know that if your dealer doesn't have a copy in your format, you can at least order one within a day or two from Fotomat. As for *His Girl Friday*, you've got 7 good leads.

To find out how to reach distributors like MCA or Budget Video just turn the page. Here you'll find capsule summaries of more than 30 of the leading video program distributors and mail-order retailers. Many of the distributors will also sell directly to consumers by mail order—in fact, you can usually order over the phone if you have a major credit card. A few, notably the distributors for the major film studios, won't sell directly tapes or discs to you. But most will help you locate a retailer or mail-order house that stocks the title you're after. As for videocassette rentals, please see "The Rental Alternative" beginning on page 169.

All Star Video

3483 Hempstead Turnpike
Levittown, NY 11756
516-752-7898

A small distribution house dealing mainly in public-domain material. All Star's inventory includes kinescoped performances by Judy Garland, Frank Sinatra, Elvis Presley, and others from the "golden" early days of television.

For jazz lovers All Star has compiled a series entitled "The Golden Classics of Jazz," featuring clips of jazz favorites like Billie Holliday and Louis Armstrong. Their catalog also contains a number of pop music and rock and roll videocassette compilations.

"All the King's Men"—a potential expose that helped move Columbia Pictures into the big-time during the late 1940s.

©Columbia Pictures Home Entertainment

Blackhawk Video

1235 West 5th Street
Davenport, IA 52808
319-323-9736

The Blackhawk catalog can be ordered for $1 and is a must for any serious videocassette collector.

For starters, Blackhawk has one of the best inventories of silent films and early film classics in video—as well they should. This Iowa-based company has been selling home-movie versions of Hollywood favorities for over half a century.

The current Blackhawk catalog lists over 1,200 films—from Hollywood's latest videocassette releases to D.W. Griffith's *Birth of a Nation.* There are hundreds of exclusive listings, some of which you *really* can't find anywhere else. And while it is true that perhaps the majority of Blackhawks' silent films are still available only in 8mm and 16mm movie film versions, more and more titles are being transferred to video.

Among the hundreds of rare films to which Blackhawk owns the exclusive distribution rights are 10 Gene Autry westerns from *Call of the Canyon* to *South of the Border, Alice Adams* starring Katharine Hepburn and Fred MacMurray, In Name Only with Cary Grant and Carole Lombard, *It* starring Clara Bow, *Little Annie Roonie* with Mary Pickford, *The Mark of Zorro* with Douglas Fairbanks, *Morning Glory* with Hepburn and Fairbanks, and dozens more featuring actors and actresses like Lillian Gish, Leslie Howard, Merle Oberon, John Wayne, and Sir Cedric Hardwicke.

Prices are reasonable, quality is excellent (at least considering the condition of many of the originals), service is prompt, and Blackhawk offers a full refund if you are not satisfied. For mail-order shopping, this company gets our highest recommendation.

Budget Video

490 Santa Monica Boulevard
Los Angeles, CA 90029
213-660-0187

You'll find Budget a good source for rare silent films, unusual foreign movies, and public-domain material. In the Budget catalog: *Algiers* with Charles Boyer and Hedy Lamarr, *Angel on My Shoulder* with Paul Muni and Claude Rains, *As You Like It* (1936), *Ballad of a Soldier* (Russian 1960), Eisenstein's critically acclaimed *Potemkin* (listed as *Battleship Potemkin),* Jean Cocteau's college cult favorite *Blood of the Poet, The Cabinet of Dr. Caligari* (1920), *Captain Kidd* with Charles Laughton, *Catherine the Great* with Douglas Fairbanks, *Cocaine Fields* (a 1937 antidrug exploitation film), *Daniel Boone* (1933), *Dementia 13* (Coppola's first, but rather unremarkable film), *Fanny* with Leslie Caron and Maurice Chevalier, *A Farewell to Arms,* and many, many more.

Quality has been generally good, and Budget is making serious committment to further improving their product. Prices are reasonable.

30,000 VIDEO PROGRAMS—THE VIDEO SOURCE BOOK

Although this section of THE SOURCEBOOK focuses on the best commercially-produced movies and programs, there's another realm of video programming that offers information, entertainment, and spiritual enlightment. The only trouble is, few video buffs know where to find out about it. Chances are that the information is as close as the nearest major library.

If you want to find out where the real offbeat action in video is, pick up a copy of *The Video Source Book*—it's also available by mail order from the National Video Clearinghouse. *The Video Source Book* ($64.94 hardcover; $59.95 paperback) lists 30,000 video programs, from major movies and documentaries to children's shows and off-the-wall instructional presentations.

•Ten pages of "how-to" programs, ranging from the mundane ("How to Buy Groceries," "How to Keep Awake while Reading," and "How to Buy a Vacuum Cleaner") to the esoteric ("How to Build an Igloo," "How to Get the Most out of Marginal Employees," and "How to Make Movies Your Friends Will Want to See Twice").

•Two non-kosher but delightful short subjects on pigs: "BLT" (a 14-minute program that follows the life cycle of a pig from babyhood to an ingredient in a bacon, lettuce, and tomato sandwich) and "Hush Hoggies Hush" (the four-minute story of Mississippi's Tom Johnson, who has spent 35 years teaching his pigs to pray before they eat).

•"Back Seat," a half-hour show in which New York City cabbies take you for a ride and tell you their favorite taxi adventures.

•"The Dizzy Patient," in which a group of doctors reviews the history of one such dizzy person and evaluates treatment.

•"How Do They Get Toothpaste in the Tube," a four-minute short narrated by Jonathan Winters and Jo Anne Worley.

•"Where Do Ballpoint Pens, Keys, and Combs Come From," a seven-minute program that answers those age-old questions.

•For the videophile who has everything, "Video Stockshots Volume I," which presents 27 stock scenes including an atomic blast, a forest in winter, heavy construction, and ducks on a pond.

•And perhaps the biggest bargain of all, "How to Find God," a half-hour religious program available on free loan from the Highland Church of Christ.

Every entry contains a description of the program, its application (home, school, cable TV), availability (rental, purchase, free loan), cast and credits, running time, release date, plus notes on whether it's captioned for the hearing-impaired and whether its available in any of 24 foreign languages.

The Video Source Book, 5½ pounds and 1260 pages, also has a subject index of 400 categories, cross-referencing, and a wholesaler/distributor index that lists where various programs can be obtained.

A rough breakdown of entries is 33% general interest and education, 23% health and science, 13% movies and entertainment, 7% fine arts, 6% juvenile, 5% business, 4% sports, and 3% instructional.

Other notes: 86% of the entries are available for schools and 40% for home use; 62% are available in the Beta format, 59% in VHS, and 1% on video disc.

For more information, write the National Video Clearinghouse, Dept. VB, 100 Lafayette Drive, Syosset, N.Y. 11791.

Classic X Video

Stevenson, CT 06109
800-243-9464

Mostly standard-fare pornography with an emphasis on the tacky—*Deep Throat, Cream Rinse, Voluptuous Vera, Birthday Babe, Passion Parlor, City of Sin, Eye Spy, Diary of a Nymph, Play Only With Me, Mother's Wishes,* and *Revolting Teens.* But for true porn historians, or perhaps just to bring back those days at the frat house, Classic X also offers a 7-volume series called "Naughty Nostalgia." The cassettes are organized by decade, beginning with a pair of stag flicks from the early 1920s titled "Mixed Relations" and "Bob's Hot Story" and culminating, on "Naughty Nostalgia #7," with "Damn, That May Be My Husband, Hide in the Bedroom Closet." Prices are low for this kind of material. Quality on some of the "Naughty Nostalgia" was probably pretty rank to begin with, and hasn't improved any with age.

Columbia Pictures Home Entertainment

711 Fifth Avenue
New York, NY 10022
212-741-4400

Columbia Pictures was founded in 1924—3 years before the first "talkie"—by a former salesman and movie shorts producer, Harry Cohn. Throughout the 1930s and most of the 1940s Columbia mostly produced entertaining, if uninspired, "second features" to accompany the blockbusters of the "big 5"—MGM, RKO, Fox, Warner, and Paramount—at the Saturday double feature.

Beginning in the late 1940s, however, Columbia Pictures came into its own with a string of box-office hits that included Robert Penn Warren's *All the King's Men*—a 1949 expose of the political corruption surrounding Louisiana's Senator Huey Long—and the lurid melodrama, *From Here to Eternity,* with Burt Lancaster, Frank Sinatra, Deborah Kerr, Montgomery Clift, and Donna Reed. With *On the Waterfront, The Bridge on the River Kwai, Lawrence of Arabia,* and *A Man for All Seasons,* Columbia Pictures emerged as one of the major Hollywood motion-picture studios.

Columbia Pictures Home Entertainment was founded in the late 1970s to exploit the growing demand for major motion pictures on prerecorded videocassettes. In addition to its own films—ranging from *Here Comes Mr. Jordan* (1941) and Frank Capra's *Mr. Smith Goes to Washington* to *Close Encounters of the Third Kind* and *The Deep*—Columbia Pictures Home Entertainment also distributes videocassette recordings of many Cinema 5 properties including *Gimme Shelter, The Garden of the Finzi-Continis, Seven Beauties, Putney Swope,* and *Z.*

In most instances the video quality of Columbia's recordings is excellent, with the video masters being transferred directly from the original film negatives. Final prices are established by the retailer, but Columbia's wholesale prices are generally in line with the other major studios. Columbia has recently sought to make rental agreements with video dealers selling its videocassettes.

Consumer Video Outlet

Box 143
Lawrence, NY 11559
800-645-2264

Another excellent catalog to have on file. Consumer Video Outlet is actually a club with a $25 annual membership fee, but their prices are among the lowest around. Over 500 major releases are available—both Hollywood and Consumer Video Outlet's discount prices start in the low $30 range for concert shorts like "Alice Cooper & Friends" ($33), "Farewell Concert by Cream" ($33), and "James Brown Live in Concert" ($33). You'll find *Night of the Living Dead* for $33.47, *Rio Lobo* for $38, *Attack of the Killer Tomatoes* for $36.97. Recent Hollywood releases range from a low of $40.97 for *China Syndrome* and *Chapter Two* to $44.14 for films like *Carnal Knowledge, Breaking Away, Change of Seasons, French Connection, The Graduate, Heartbreak Kid, Kotch, Paper Chase, Rebecca, Silver Streak,* and *Straw Dogs.* A few releases—*Cheech & Chong's Up In Smoke, A Clockwork Orange, The Dirty Dozen, Elephant Man, Hello Dolly, The King and I, Mary Poppins, Ordinary People, Rough Cut, Shogun,* and *Superman The Movie*— are priced in the $50 range, while a handful of films such as *The Godfather, Romeo and Juliet, Mutiny on the Bounty, Marathon Man, Looking for Mr. Goodbar, The Deer Hunter,* and *The Blues Brothers* cost over $60.

Direct Video

1800 North Highland
Suite 709
Los Angeles, CA 90028
800-423-2452

Direct Video's adult catalog is an ultra-slick, 4-color presentation of over 100 adult films. The titles—*Star Virgins, Debbie Does Dallas, Fantasy, 11, A Dirty Western, Defiance, Teen Age Cover Girl, Sex Boat, Charli On White Satin, Tangerine, Summer School,* and *Godiva High*—are certainly packaged and promoted a little more tastefully than your average pornography. Some even promise a modest, or immodest as the case may be, plot to supplement the sexual antics. There's also a 60-minute cassette of collected X-Rated cartoons. Tapes are $99.50. Buy 3, get one free.

Direct Video's "Film Classics" catalog is an equally classy listing of 40 b&w films—most of them, indeed, true classics. Titles include *The Third Man* with Orson Welles, Santa Fe Trail with Errol Flynn, *The Little Princess* starring Shirley Temple, *The Birth of a Nation* ("the classic of classics" starring Lillian Gish), *A Star Is Born* (1937), *Diabolique* (the 1955 French mystery classic), Catherine Deneuve in Roman Polanski's 1965 psychological horror flick *Repulsion, Night of the Living Dead,* and Marlene Dietrich's 1930 sizzler *The Blue Angel.* All films are $49.95 and claim to be "meticulously transfered to video tape from film masters."

Discotronics

713 North Military Trail
West Palm Beach, FL 33406
305-689-2202

Discotronics pioneered the video exchange system in 1977, the year prerecorded films first appeared on the market, and their video business has been growing ever since. The Discotronics catalog now lists over 600 titles—many available used at significant savings—with a heavy emphasis on adult titles.

Buy a new film from Discotronics and you'll end up paying full list price—which you can do at your video dealer and save the $3.00 charge for shipping and insurance. But trade in your films for a 50% credit, and buy Discotronics guaranteed defect-free used films, and you can get yourself some hefty bargains. Discotronics service is prompt and personal. New and used tapes are guaranteed. Defective tapes will be replaced within 14 days at no charge.

Disney Home Video

500 South Buena Vista Street
Burbank, CA 91521
800-423-2259

Walt Disney made his first Mickey Mouse cartoon in 1928, and the rest is history.

So far the Disney marketing people are holding back most of the really good features that can still be milked in the movie theaters—*Fantasia, Lady and the Tramp, Snow White and the Seven Dwarfs, Bambi, Pinocchio, Dumbo, Cinderella, Alice in Wonderland, The Jungle Book, Peter Pan,* and *Sleeping Beauty.*

What has Disney released on videocassette? Current titles include: *Pete's Dragon, The Black Hole, Mary Poppins, Love Bug, Escape from Witch Mountain, On Vacation with Mickey Mouse and Friends, Kids Is Kids, The Adventures of Chip 'N' Dale, Davy Crockett, 20,000 Leagues Under the Sea,* and *The Apple Dumpling Gang.* Quality is excellent, prices are somewhat higher than most, and you can rent Disney titles at Fotomat and many video specialty shops.

Fotomat

64 Danbury Road
Wilton, CT 06897
800-325-1111

Paramount, Disney, EMI, Columbia, RKO, Video Communications, and Warner Brothers films for sale or rent. More than 300 titles available. Good quality, reasonable prices, and 5,000-plus neighborhood outlets. (See Fotomat sidebar on page 51.)

Home Video Entertainment, Inc.

Box 32277
Cleveland, OH 44132
216-731-5228

A mail-order videocassette dealer with over 600 recent features, vintage movies and TV shows, double features, serials, rock concerts, cartoons, and sports events in their catalog. Discounts range from 6% to 13% off list price. Handles products distributed by MGM/CBS, MCA, Magnetic Video, Warner Brothers, Walt Disney, Paramount, Columbia, Reel Images, Nostalgia Merchant, MEDIA, and others. Charges $2 for the first tape, $1 per additional tape, for shipping.

"The Great Dictator"—Chaplin's comedy often carried a message that was sometimes deadly serious.

©Magnetic Video

Magnetic Video Corporation

(Video Club of America)
23689 North Industrial Park Drive
Farmington Hills, MI 48024
313-477-6066

The company that started it all in 1977. Magnetic Video is now a wholly-owned division of 20th Century-Fox, responsible for distributing videocassette recordings of films selected from the substantial Fox inventory. In addition, Magnetic Video has acquired properties from the Charlie Chaplin Library, United Artists, Brut Productions, and Avco Embassy Films—which makes Magnetic Video not only the oldest, but also the largest distributor of exclusive titles. You can buy Magnetic Video cassettes either directly through their club—Video Club of America—at a substantial savings, at many local video specialty stores, or through discount mail-order houses such as Marshall Discount Video Service. So far, Magnetic Video has refused to authorize its dealers to rent videocassttess, but most Magnetic Video titles can be rented anyway.

You'll find well over 300 films on the Magnetic Video list, covering most principal motion picture categories. *Alien, The African Queen, All About Eve, All that Jazz, Autumn Sonata, Beneath the Planet of the Apes, The Blue Max, Breaking Away, Brubaker, Butch Cassidy and the Sundance Kid, Carnal Knowledge, City Lights, Cleopatra, Darling, the Diary of Anne Frank, The Fog, The French Connection, Gentlemen Prefer Blondes, The Graduate, The Grapes of Wrath, The Great Dictator, Heidi, Kotch, Laura, Modern Times, The Muppet Movie, 9 to 5, Norma Rae, The Paper Chase, Patton, Rebecca, The Robe, The Rose, Silver Streak, Sleuth, The Sound of Music, Straw Dogs, Stunt Man, Sympathy for the Devil, They Shoot Horses Don't They?, The Turning Point, Twelve O'Clock High,* and *An Unmarried Woman* are among some of the most noteworthy titles.

Quality is excellent—or should be—since the videotape masters are made directly from original movie film negatives. Prices are reasonable, with substantial discounts available through the Video Club of America.

Marshall Discount Video Service

3130 Edsel Drive
Trenton, MI 48183
313-671-5483

Marshall's 14-page mimeographed "video catalog" isn't too impressive itself, but it does list 1,017 films released through Columbia Home Video, Magnetic Video, MCA, MEDIA, NFL Video Films, Niles Cinema, Nostalgia Merchant, MGM/CBS, Paramount, Video Gems, VCI, VCN, Disney, and Warner Home Video. A separate "catalog" lists 557 porno flicks you can order from Marshall. Prices for both the feature films and the adult movies are somewhat below average, but still no match for Consumer Video Outlet.

MCA Videocassettes, Inc.

445 Park Avenue
New York, NY 10022
212-759-7500

MCA is the videocassette distributor for Universal Pictures, as well as for Paramount films released between 1929 and 1949.

Hitchcock classics *The Birds, Frenzy,* and *Psycho* head the MCA list. Other titles include *American Graffiti, Animal Crackers, Animal House, Battlestar Galactica, Bedtime for Bonzo* with Ronald Reagan, *The Blues Brothers, Coal Miner's Daughter, The Deer Hunter, Dracula* (1931 and 1979), *Duck Soup, The Electric Horseman, Flash Gordon,* the original *Frankenstein, The Island, Jaws, The Jerk, Prom Night, The Seduction of Joe Tynan, Smokey and the Bandit, The Sting, Which Way is Up?,* and *Xanadu.*

Quality and pricing is on par with other major studios.

Media Home Entertainment

160 North Robertson Boulevard
Los Angeles, CA 90048
800-421-4500

MEDIA does not generally handle mail-order requests, but call their toll-free number for the name of the nearest store that sells MEDIA's tapes. You can also find most of MEDIA's exclusive films—including *The Groove Tube, Flash Gordon,* and *Halloween*—available through Home Video Entertainment, Inc., Niles Cinema, and others.

MEDIA's inventory contains over 100 titles with a sprinkling of classics like *The Third Man,* some horror films such as *Terror By Night, Horror Express,* and *Day of the Triffids,* some rock and roll retrospectives, and selection of X-Rated theme movies along the lines of *The Ribald Tales of Robin Hood, Flesh Gordon, Alice in Wonderland, Dracula Sucks,* and *Fairytales.*

MGM/CBS Home Video

CBS Video Enterprises
1700 Broadway
New York, NY 10019
212-975-5277

At one time Metro-Goldwyn-Mayer was the undisputed leader of the motion picture industry. Between its inception in 1920 as Metro Pictures, with such record-breaking hits as *The Four Horsemen of the Apocalypse* and *The Prisoner of Zenda,* and its sputtering decline in the 1960s with an occasional blockbuster like *Dr. Zhivago* and *Where Eagles Dare,* the MGM studios produced more films with more stars than any other studio in Hollywood. Distribution of those properties for the home video market is now the task of MGM/CBS Home Video, which was formed in 1980.

So far MGM/CBS have released a handful of outstanding titles on videocassette, but they have yet to really scratch the surface. Perhaps most appealing for musical fans is *That's Entertainment,* the 2-hour retrospective of show-stopping scenes from nearly 100 MGM musicals. Other MGM/CBS videocassettes releases include *The Wizard of Oz, 2001: Space Odessey, Dr. Zhivago, Ben Hur, Network, Coma, The Dirty Dozen, An American in Paris, The Sunshine Boys, Blow-Up, A Night at the Opera, Adam's Rib, Jailhouse Rock, Tom & Jerry, The Boys in the Band, Rude Boy, Rio Lobo, The Street Fighter, Electric Light Orchestra in Concert, James Taylor in Concert, The Nutcracker,* and *Giselle.* Merely an appetizer, we hope, for what is yet to come.

NFL Films Video

330 Fellowship Road
Mount Laurel, NJ 08054
609-778-1600

From Vince Lombardi's Green Bay Packers upsetting the Kansas City Chiefs 35 to 10 on January 15, 1969, in Superbowl I to Jim Plunkett guiding the wild card Oakland Raiders to their 27 to 10 victory over the Philadelphia Eagles, NFL Films Video has distilled the highlights of each Superbowl game onto a 24-minute videocassette. But beware, these are just the highlights—often taken from TV broadcasts and newsclips—not the entire game. There are also collections of "Football Follies," "The Son of Football Follies," and yearly "Team Highlights."

Niles Cinema

1141 Mishawaka Avenue
South Bend, IN 46624
800-257-7850

Niles Cinema's 64-page newsprint catalog with its more than 900 titles is another excellent resource for the videotape collector. Niles stocks videocassettes from all major studios and handles product from many of the smaller, more obscure video distributors as well. It's hard to name titles, since practically every major blockbuster released on video is in the Niles inventory, along with many of the more popular classics. One disadvantage: Niles charges list price for most of their titles. Return a defective videocassette within 10 days for a free exchange.

Nostalgia Merchant

6255 West Sunset Boulevard
Suite 1019
Hollywood, CA 90028
800-421-4495

Nostalgia Merchant's little 38-page catalog is one more must for old-time movie buffs. Part of the reason is that this firm was started by movie freaks, for movie freaks. Westerns, serials, and, of course, classics—including some of the film greats from the RKO collection—are the bread and butter of the Nostalgia Merchant inventory. Over 100 films, with hardly a loser among them. *Citizen Kane, Curse of the Cat People, Fort Apache, Gunga Din, High Noon, Hitler's Children, Hunchback of Notre*

Dame (1939), The Informer, Johnny Guitar, King Kong, The Magnificant Ambersons, Mr. Blandings Builds His Dream House, and *She Wore A Yellow Ribbon* are among the highlights.

Quality is tops, with prices just a little higher than elsewhere.

Paramount Home Video

5451 Marathon Street
Hollywood, CA 90038
213-468-5000

A nickelodeon player named Adolph Zukor founded Paramount Pictures in 1912 and quickly developed his studio into one of Hollywood's "big 5." Almost from Paramount's inception, film directors like Lubitsch, DeMille, and Wilder guided some of Hollywood's most popular actors and actresses including Valentino, Chevalier, Pickford, Hope, Colbert, Crosby, Dorothy Lamour, Alan Ladd, and the Marx Brothers. Paramount films were seldom noted for their intellectual stimulation, but as solid family entertainment they excelled. *The Ten Commandments* (1923 and 1956), *The Sheik, The Greatest Show on Earth, The Covered Wagon, Trouble in Paradise,* and *Going My Way* were typical of Paramount's best box-office fare under Zukor.

With video rights to the films produced between 1929 and 1949 assigned to MCA, Paramount Home Video has concentrated so far on marketing their box-office successes of the 1960s, 1970s, and 1980s, with an occasional DeMille epic thrown in for good measure. Among the tastiest morsels on the Paramount Home Video menu: *Airplane!, American Gigolo, Breakfast at Tiffany's, Catch 22, Charlotte's Webb, Chinatown, Days of Heaven, The Duellists, Elephant Man, Emmanuelle: The Joys of a Woman, Foul Play, The Godfather I and II, Goodbye Columbus, Gunfight at the OK Corral, Harold and Maude, Heaven Can Wait, Islands in the Stream, Shogun, King Kong, Little Darlings, The Longest Yard, Looking for Mr. Goodbar, Love Story, The Man Who Shot Liberty Valance, Marathon Man, Murder on the Orient Express, Nashville, North Dallas Forty, The Odd Couple, Ordinary People, Paper Moon, Play It Again, Sam, Popeye, Pretty Baby, Romeo and Juliet, Rosemary's Baby, Rough Cut, Saturday Night Fever, Serial, The Shootist, Stalag 17, Samson & Delilah, Star Trek: The Motion Picture, Starting Over, Sunset Boulevard, The Ten Commandments, Three Days of the Condor, True Grit, Urban Cowboy, War of the Worlds,* and *When Worlds Collide.* In addition, Star Trek fans will be delighted to know that 5 double-episode cassettes are available uncut and uninterrupted.

©Paramount Home Video

©Paramount Home Video

Quality X Video Cassette

356 West 44th Street
New York, NY 10036
800-223-7981

The quality is debatable, but the X isn't. *Bang, Bang, You Got It; Campus Girls, Danish Pastries, Easy Alice, French Shampoo, High Rise, Honey Pie, Lady on the Couch, The Love Bus, Maraschino Cherry, Misty Beethoven, Sexteen,* and *Sweet Cakes* ought to be enough to give you the general idea.

Red Fox Enterprises

Route 209 East
Elizabethville, PA 17023
717-362-3391

A mail-order house with a modest list of tapes, most of which can be found elsewhere. It's main attribute is that it carries Columbia Home Video's releases, but then so do Niles Cinema, Marshall Discount Video Service, and Fotomat.

Reel Images

Box 137
Monroe, CT 06468
800-243-9289

With over 400 public-domain early classics and foreign films in its 72-page newsprint catalog, Reel Images is yet one more mail-order distributor to add to your list if you've got a soft spot for rare golden oldies.

You'll find no recent commercial releases here, but lots and lots of "video yesteryear." Some of the titles, particularly the kinescoped TV shows like Dean Martin, Jerry Lewis, and Margaret DuMont on "The Colgate Comedy Hour" (1955) or some segments of "The Frank Sinatra Story" are of questionable video quality, and Reel Images tells you so. Even if you don't buy the films, the catalog makes excellent reading.

Quality is good to excellent, unless otherwise noted in the catalog. Most films are dubbed to order, so if you order several short titles they may arrive on the same cassette (at SP or Beta II, of course). Prices are about average.

RM Films International

Box 3748
Hollywood, CA 90028
213-466-7791

Distributes only Russ Meyer's films—*Beyond the Valley of the Dolls, UP, Supervixens, Lorna,* etc. Prices are high, artistic quality debatable.

Starlog Video

475 Park Avenue South
New York, NY 10016
212-689-2830

Vintage cult flicks, horror, sci-fi, and public-domain television.

Thunderbird Films

Box 65157
Los Angeles, CA 90065

Public-domain material, silent films, foreign films. *The Scarlet Pimpernel* (1934), *Saturday Night and Sunday Morning* (British 1960), *Reefer Madness* (1938), *Seance on a Wet Afternoon* (1960), *Reaching for the Moon* (1931) with Douglas Fairbanks, *Rashomon* (the 1951 Japanese classic), *Rain* (1931) with Joan Crawford, *The Private Life of Henry VIII* (1933), and *Metropolis* (1926) are examples of the Thunderbird collection. Quality of the older films is variable.

Time-Life Video

Harrisburg, PA 17105
800-662-5180

Lifetime membership is $15. Time-Life has a few real dogs in its inventory, but these are more than balanced out by dozens of excellent Hollywood and foreign films. *Bell, Book, and Candle, Bye, Bye Birdie, The Deep, Here Comes Mr. Jordan, King Kong* (1933), *A Man for All Seasons, Scenes from a Marriage, Swept Away, Z, King of Hearts* with Alan Bates and a nubile young Genevieve Bujold, *Visions of Eight* (1973 Olympics), *To Be Or Not To Be* (1942), *The Third Man, The Tall Blonde Man with One Black Shoe, Stolen Kisses* (Truffaut), *State of Seige, Stagecoach, A Special Day with Sophia Loren, The Sorrow and the Pity, The Silent Partner, Rust Never Sleeps* (Neil Young), and *Pumping Iron,* are among the titles available to Time-Life Video Club members.

TVX

1643 Cherokee Avenue
Hollywood, CA 92028
800-421-4133

TVX, as in X-Rated for TV. *Babylon Pink, China Girl, Chopstix, Chorus Call, Double Exposure of Holly, The First Time, The Health Spa, Inside Jennifer Wells, The Journey of O, Little Girls Blue, Sex World, Soft Places,* and *Star Babe* are among the titles that will surely never get the approval of Jerry Falwell. But we suspect you may find a few members of the Moral Majority sneaking a peek anyway.

VCX

7313 Varna Avenue
North Hollywood, CA 91605
800-423-2587

VCX, meaning X-Rated videocassettes, or so we assume. Slightly different titles—*Candy Lips, Cherry Truckers, China Lust, Eruptions, Finishing School, Hard Bargain, Lady Luck, Meatball, Masked Ball, Pastries, Pink Lips,* and *Pleasure Island*—but we don't doubt for a second that the plots are almost identical.

VidAmerica, Inc.

231 East 56th Street
New York, NY 10022
212-355-1505

VidAmerica maintains a small catalog of sports films, adult films, and a sprinkling of movie classics: *Ali—Skill Brains & Guts, The Bermuda Triangle, The Amazing World of Psychic Phenomena* hosted by Raymond Burr, *Sugar Cookies, Centerfold, NFL Football Follies, Boxing's Greatest Champions, Grudge Fights, The Greatest Comeback Ever* (the 1978 New York Yankees), *The Two Best World Series Ever, Baseball Fun and Games* with Joe Garagiola, *Oh! Calcutta!, Vanessa, A Spectacular Evening in Paris* with Lauren Hutton, *Emmanuelle in Bangkok, Flying Leathenecks* with John Wayne, *Greatest Heroes of the Bible, Room Service* with the Marx Brothers, and *Caring for Your Newborn* with Dr. Spock are among the films available for sale and one-week rentals.

Video Communications, Inc.

6555 East Skelley Drive
Tulsa, OK 74145
800-331-4077

Some exclusives—mostly RKO releases after 1950 and several works of epic producer Samuel Bronston including *King of Kings, El Cid, 55 Days at Peking, The Fall of the Roman Empire,* and *Circus World* with John Wayne. Other Video Communications films include *The Christmas Tree* with William Holden, *A Christmas Carol, Clash by Night* with Marilyn Monroe, and *Devil's Rain.*

Video Dimensions

43 East 10th Street
New York, NY 10003
212-533-5999

Public-domain material—especially bloopers, golden-era TV shows, cartoons, and some foreign films.

Video Gems

731 North La Brea Boulevard
Los Angeles, CA 90038
213-938-2385

A few exclusive titles, including the Andy Warhol films—*Andy Warhol's Dracula* and *Andy Warhol's Frankenstein*—and some of the Bruce Lee Kung Fu flicks, including *The Invincible, The Real Bruce Lee,* and *Bruce Lee's Greatest Revenge.*

VMC Video Company

7920 Alabama Avenue
Suite 201
Canoga Park, CA 91304
800-423-5106

More X—*Butterflies, The Ecstasy Girls, MASH'd, More than Sisters, Robin's Nest, Rockin' with Seka,* and *Sensations* are among them.

Warner's triumph—"The Jazz Singer"—sounded a new era in film making.

©Warner Home Video

Warner Home Video

75 Rockefeller Plaza
New York, NY 10019
212-484-6108

Warner Brothers gained admission to the "Big 5" through their spectacular gamble on talking pictures—specifically, Al Jolson's *The Jazz Singer.* Warner Brothers kept up the momentum through the 1930s, 1940s, and into the early 1950s with the box-office appeal of such big-name stars as Humphrey Bogart, James Cagney, Edward G. Robinson, Bette Davis, Errol Flynn, and Dick Powell, combined with a formula that mixed gangster movies and musicals.

But, as with the other major studios, the advent of television brought an era of gradual decline to Warner Brothers. Today the massive sound stages in Burbank, California, where Elia Kazan once directed James Dean, Julie Harris, and Raymond Massey in *East of Eden,* have been sold. And when they are used at all, it is most often by TV production companies like Lorimar, Ltd. and Columbia Television Pictures for weekly episodics such as "Little House on the Prairie," "The Dukes of Hazzard," and "Fantasy Island."

Meanwhile in Manhattan, a division of Warner Communications, Inc., called Warner Home Video has been charged with bringing the last 30 years of Warner films to market on videocassettes (Warner's films released between 1927 and 1948 are owned by United Artists, which distributes through Magnetic Video). Among the titles that have made it to the dealer's shelves, in addition to *East of Eden* are: *Blazing Saddles, Bonnie and Clyde, Gilda Live, Hooper, The In-Laws, Klute, The Life of Brian, A Little Roman, The Main Event, Mister Roberts,* and *The Searchers.*

CLOSED CAPTIONING EQUIPMENT AND WHERE TO FIND IT

A silent revolution has been occurring on television, and it is slowly making its way into video, too. It is imperceptible to the average viewer, but to the 16 million Americans with hearing problems, it could be the next best things to ears.

The advance is called closed captioning, a process developed by the Public Broadcasting Service that translates the audio portion of a TV program into captions that appear at various places on the TV screen—akin to subtitles on foreign films—on television equipped with a special decoding unit. This enables people with hearing impairments to follow the dialogue. And—if they have a VCR as well—they can tape the subtitles programs for future viewing.

Columbia has closed captioned its cassettes of three films, "Close Encounters," "The China Syndrome" and "Chapter Two"—with the captions visible only on TV sets with decoders. And while plans are in the works for other video-programming companies to follow suit, the cable movie network Showtime has been presenting the three Columbia films with closed captions.

Leading the way for closed captioning is the National Captioning Institute, a nonprofit, Virginia-based organization developed to caption TV programs for hearing-impaired viewers. TV producers provide the institute with videotapes of their programs prior to broadcast, and caption editors arrange the dialogue into subtitles, which are encoded on a magnetic disc.

The disc is then sent to TV broadcasters, who insert the disc's caption information on Line 21 of the TV picture—a line that does not carry picture information. The coded caption is then broadcast with the regular visual and audio portions of the show. The captions remain invisible on normal TV sets but are decoded into subtitles by sets with a captioning adapter.

Two types of decoders are available from Sears, Roebuck, and Co.: a TeleCaption captioning adapter ($260) that can be hooked to an ordinary TV, and 19-inch TeleCaption color portable set ($530) with a built-in decoder. Sears is selling the adapters at a no-profit price, giving $8 to the National Captioning Institute for every decoder sold. Sears has already sold more than 40,000 of the units. The cost of the units is tax-deductible.

Closed captioning on a regular basis has now expanded to well over 30 hours of regular programming a week on ABC, NBC, and PBS—from "Love Boat" to "Little House on the Prairie." More than 100 advertisers close-caption their commercials—from Ace Hardware to the Xerox Corporation.

According to David Kleeman of the National Captioning Institute, more and more owners of closed-captioning equipment are also hooking up VCRs to their TV sets—so that they can watch one closed-caption program and tape another at the same time. Although the dual installation is simple, Kleeman suggests that the adapter be hooked into the VCR and then the VCR into the TV set so that tapings will carry the captioning whether they're played through the adapter or not.

(CBS is developing its own alternative to closed captioning, a process called teletext, which should be available in the United States on limited basis sometime soon if the FCC ever acts on CBS's petition to establish national standards for teletext. In the interim, the network does not closed-caption programs and goes so far as to remove closed captioning from commercials it broadcasts.)

For more information on closed captioning, contact the National Captioning Institute, 5203 Leesburg Pike, Suite 1500, Falls Church, Va. 22041. The institute's phone number is (703) 988-2440 (voice the TTY). The institute also has a toll-free number that provides TV scheduling information and updates: (800) 336-4703 (voice and TTY).

The Big Time Players

SELECTING YOUR VIDEOCASSETTE RECORDER

"It's not true that life is just one damn thing after another," said Edna St. Vincent Millay. "It's one damn thing over and over." Which may be exactly what former FCC Commissioner Newton Minnow had in mind a couple of decades ago when he called broadcast television "a vast wasteland."

Much has happened since those early years of television. The red-hot competition between ABC, CBS, and NBC generates dozens of new TV specials, movies, and quality sports programs each week. And where the networks have given short shrift to cultural and educational programming, PBS and its affiliates (such as WGBH Boston, WNET New York, and KCET Los Angeles) have stepped into the breech with everything from "Monty Python" to "Masterpiece Theater." Include cable and pay-TV services with round-the-clock news, sports, nostalgia, classics, uncut feature movies, and concerts—and the problem isn't *what* to watch. The problem is *when* can you find the time to see all the programs you'd like.

Enter the videocassette recorder (VCR).

The VCR is a tape recorder for television. Attach a VCR to the antenna terminals on the back of your TV set and you can:

- Record any program while you are watching (and edit out the commercials while you are at it).
- Record one program while you watch another.
- Automatically record programs while you are doing something else, or even while you're out of town on a trip or vacation.
- Play back any of the thousands of videocassette tapes of movies, TV shows, concerts, and instructional programs that are available for sale or rental.

In addition, videocassette recorders can be bought with special-effects features like "freeze frame," "frame-by-frame advance," and "slow motion" that will let you be your own director. Buy a b&w video camera for under $200, or a color video camera for as low as $750 (see "Video Cameras" on page 144), and you can start producing your own videotapes as well.

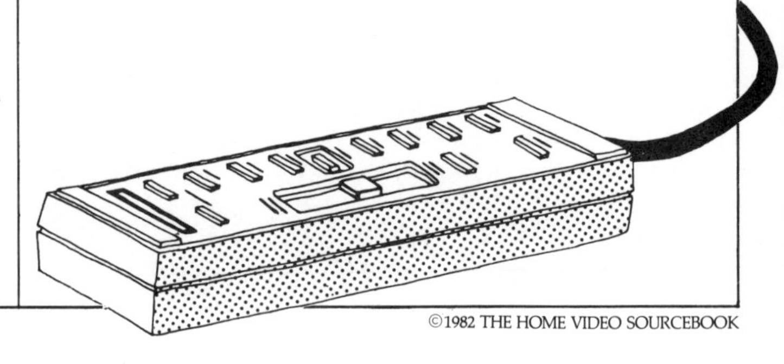

©1982 THE HOME VIDEO SOURCEBOOK

Beta format videocassette recorder.

In The Beginning

The first commercial video tape recorder was introduced by Ampex in 1956. Despite a price tag of $50,000—or about $200,000 in today's inflated dollars—Ampex sold over 80 of the cumbersome reel-to-reel units within a week to TV-station owners.

Nineteen years later the first practical consumer version of the video tape recorder made its debut. Called the Betamax and designed and manufactured by Sony, this device had a built-in 19″ TV set, used an easy-to-load videotape cassette, and sold for $2,300. Unfortunately, it never occurred to Sony that anyone would want to record more than one hour's worth of programming on a single videocassette.

The following year, when a rival Japanese electronics firm—known in America as JVC—introduced an alternative design that could record up to *two* hours on a single videocassette, the battle of the VCR formats had begun. Various technical factors probably gave the early Sony unit a slight edge in picture quality, but it was not enough to overcome the convenience—and economy—of being able to record two hours without changing the videocassette.

JVC's system, known as VHS for Video Home System, took off like a Saturn V rocket. Alas, the Beta and VHS systems use different size videocassettes, as well as different loading designs. A Beta cassette simply CANNOT be played in a VHS machine, and *vice versa.* Like plaintiffs in a Hollywood divorce, the formats are INCOMPATIBLE.

Note: *If you have children in the house, if you find changing a light bulb taxes your mechanical aptitude, or if you just don't care about recording off television, but would still like to see prerecorded programs at home, you may be a candidate for a VIDEODISC PLAYER. For an analysis of the pros and cons of the two basic home video mediums* *see page 128, "Videocassette Recorder vs. Videodisc Player."*

©1982 THE HOME VIDEO SOURCEBOOK

Beta or VHS?

Your first decision in selecting a videocassette recorder is whether to buy a Beta or VHS machine. Here we offer a word of advice. *FIND OUT WHAT FORMAT YOUR FRIENDS OWN.*

Beta and VHS are here to stay. Unlike the new videodisc formats—where the future is far from certain—both Beta and VHS are firmly entrenched, with thousands of programs available and millions of machines already in use. There are minor differences between the formats, but neither offers any significant advantage over the other.

As mentioned before, the first VHS videocassette recorders offered two hours of recording time to Beta's one. The result was a race to see who could squeeze the most programming hours out of their machines. Slower recording speeds were used (which lengthened playing time at the expense of image and sound quality), and techniques were developed for putting longer and longer tapes into the Beta and VHS cassettes. The race continues, but its significance, as far as the consumer is concerned, is minimal.

The current generation of Beta format machines (with the exception of two no-frills models you may still encounter: Sanyo VTC-9100A and Sears 5305) is equipped with two recording speeds, called Beta II and Beta III, that will allow you to record up to five hours on a single cassette. Most VHS videocassette recorders on the market have three speeds—known as Standard Play (SP), Long Play (LP), and Super Long Play (SLP)—which allow up to six hours of recording per cassette (and up to seven hours using new VHS tapes like the TDK T-150).

What about image quality? Most experts agree—for machines of comparable specifications there is little, if any, perceptible difference between formats at the Beta II (or X2) and the VHS Standard Play (SP) speeds. Both speeds will give you an image which is only slightly inferior to the TV original, and both speeds are the *industry standard* for all pre-recorded programs.

WHICH FEATURES ARE FOR YOU?

There are 3 features you absolutely must have on any VCR. These are the basic tape transport functions:

- **Play**
- **Stop**
- **Record**

Generally each of these buttons does what it says, regardless of the brand, model, or format you may choose.

Play. Loads the cassette onto the tape heads (while you wait for three to five seconds, depending on the model) and, if all goes well, translates the coded signals from the magnetic tape into a crisp and brilliant image on your TV screen, complete with monaural sound. It works whether the tape is commercially prerecorded, or has been home recorded. The videotapes you play can be in color or black and white. They will play on your machine just as long as they are of the same *format* as your unit, and were recorded at a *tape speed* with which your machine is equipped.

Stop. *Unloads* the tape from the audio and video heads (again, this takes a few seconds, depending on the unit). You can now eject the videocassette, or select another function such as rewind or fast forward. On unit equipped with piano key controls it is always necessary to press stop before going from one transport function to another. In units with electronic "soft touch" controls there is usually a built-in memory that allows you to go from one function or special effect to another *without* having to press the stop button.

Record. Usually pushed simultaneously with play. As a result your videotape will become loaded (once again, after a brief interval) around the *recording* heads. Whatever image is on your tape will be automatically erased as a new magnetic signal is painted onto the tape by the video recording head. The same goes for your soundtrack. You can record with a video camera, from your TV set, from another VCR (regardless of format), from a videodisc player, or even from a suitably equipped home computer.

In addition to these "basic" transport functions, your VCR will also have:

Fast Forward. This allows you to advance the tape at a rapid speed so you don't have to sit through an hour of "Road Runner" cartoons to get to your favorite performance of "Saturday Night Live."

Rewind. Does the same thing as fast forward, but in reverse.

Pause. This lets you stop a program temporarily *without* unloading the videotape from the playback/recording heads. Pause is extremely useful for brief interruptions, such as "pausing" to answer a quick phone call. It is also a great help in "editing" out commercials when recording TV broadcasts. You should be aware, however, that pausing puts enormous strain on your videotape. Most models feature an automatic release that returns the VCR to its previous function—play or record—after a few minutes have elapsed. Some stripped-down models *don't* have this safety feature. So be warned—"over-pausing" can ruin your tapes.

Audio dub. Lets you re-record the soundtrack without erasing the video image.

Tuner/timer. Permits you to select a TV channel and automatically record a designated time block at any point within the next 24 hours. The only reason you would not need a tuner/timer is if you are buying a portable VCR and have absolutely no interest in recording off television.

There are a number of features that you *can* live without on a VCR, although most increase the convenience and usefulness of your videocassette recorder.

Dew indicator. Warns of moisture on the video heads (which can ruin both your tape and the heads). If you live in a sultry climate this one could save you some big repair bills.

Slow motion. Lets you play back recorded material at reduced speed. Not really necessary unless you have a specific reason for it.

Counter. If you set the counter at zero every time you put in a new tape you can then keep track of where programs, or program segments, begin and end—as well as how much tape is left on your cassette—by "logging in" the numbers on the counter. You can also use the counter to estimate when to stop fast forwarding through commercial breaks in shows you have recorded. In reality, very few people have the discipline and the patience to use a counter effectively. Fortunately, recent special effects such as speed scan and electronic indexing have rendered the counter unnecessary on the higher-priced VCRs.

Electronic programmability. This feature extends the number of "events," or programs, you can automatically record, as well as the length of time for which you can leave the VCR unattended. An 8-event/14-day capability is now common on deluxe models. Programmability also increases how much you pay for your VCR, by anywhere from $100 to $200.

Extended playback. In the Beta format you will find most models offer an extra speed known as Beta III. In VHS the majority of units are equipped with 2 extra speeds—Long Play (LP) and Super Long Play (SLP). These extended playback speeds save money on videotape—since in both Beta and VHS you can only get 2 hours of programming on most cassettes at the "standard" speed—and give you the freedom to borrow any videocassette that has been recorded in your format. We recommend investing the extra bucks necessary to buy a machine with extended playback capability.

Speed scan. Allows you to search through a videotape at an accelerated speed *with* a visual image. A good feature for zipping through commercials or finding that "Saturday Night Live" segment without a lot of stopping and starting. In our opinion, speed scan is extremely convenient, but if price is a big consideration, you can easily live without it.

Electronic indexing. Automatically stops the tape at the beginning of each program as you fast forward through a videocassette. Another great convenience, but not essential.

Freeze frame and frame-by-frame advance. Freeze frame allows you to keep an image on the screen while you pause, frame-by-frame advance lets you analyze a sequence of frames on at a time. Great for second guessing the referee on those close calls.

Remote control. A real advantage for portable users, since with a remote controller tucked in your belt you can wear the VCR deck in a backpack. It's convenient on a console VCR, too, but maybe not worth an extra $50 to $100.

Electronic controls. Yet another terrific convenience item. Electronic controls are especially useful on portable models where pressing the old-fashioned piano keys can be very awkward. Not necessary, but nice.

VCR Loading Systems

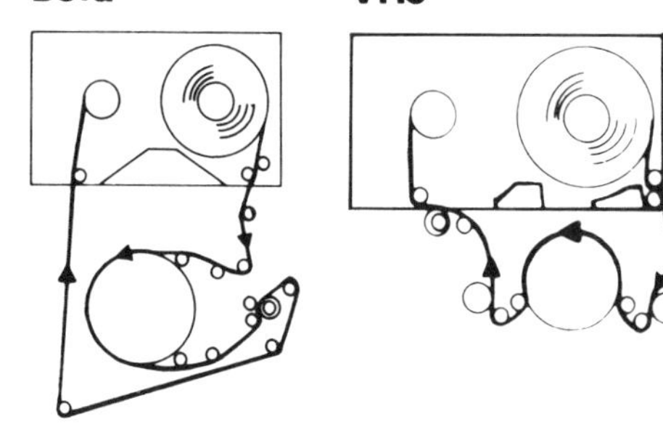

The VHS format's traditional advantage? Slightly longer playing times than Beta.

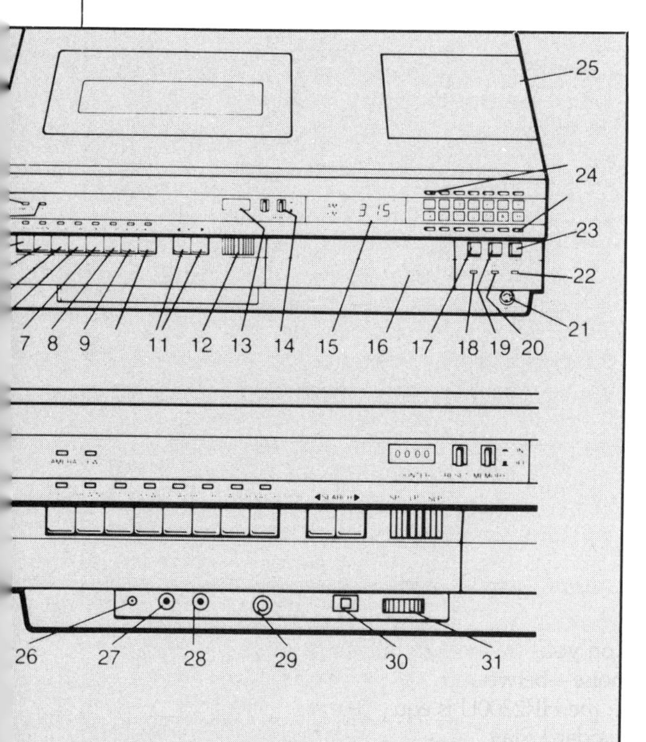

Typical VCR Features

Camera Signal	17 *VCR/TV Switch*
Dew Warning Light	18 *VCR Indicator Light*
Eject	19 *Timer Switch*
Rewind	20 *Timer Indicator Light*
Stop	21 *Remote Control Jack*
Fast Forward	22 *Power Light*
Play	23 *Power Switch*
Record	24 *Channel Selector Switches*
Audio Dub	25 *Tuner Access Door*
Pause	26 *Camera Remote Pause Jack*
Speed Scan	27 *Video In*
Tape Speed Select Switch	28 *Audio In*
Tape Counter	29 *Microphone In*
Memory Switch	30 *Camera/Tuner Selector Switch*
Digital Clock/Timer	31 *Tracking Control*
Timer Control Door	

Illustration: RCA, Inc.

Just how much playing time you actually get at the Beta II or SP speeds depends on the length of the videotape in your cassette. With a standard Beta L-500 or VHS T-120 videocassette you'll get two hours. A Beta L-830 or VHS T-150 will give you well over three hours of quality recording time.

At Beta III (X3) or VHS Long Play (LP) speeds, you will detect a significant loss of image quality (evident as an increase of video noise, or "snow," on your screen) from the original broadcast version of the show you are recording. But the extra recording time—over five hours in either format using the Beta L-830 or VHS T-150 videocassettes—is great for long sports shows, movies, collecting episodes of your favorite TV series, or recording programs while you are away.

And what of the VHS Super Long Play (SLP) mode? True, with this speed and a T-150 videocassette you will be able to cram over seven hours on one tape. But on most VHS units, the image *and* the sound quality of the programs you recorded on SLP can be so poor that you may not want to spend seven hours watching them.

As for tape economics, once again neither format holds any real advantage. The quality of the tape you buy (and how much of a discount you get) will be the major factors in determining the cost per hour of operating a videocassette recorder—not the format.

Some VHS proponents will claim the design of their loading system reduces wear and tear on your (expensive) heads and tape. But this is a claim which few experts have been able to verify at the lab bench.

Which format is best for you? Once again, our advice: See what your friends have. If your best buddy owns a Betamax and a collection of 250 videocassettes, you're crazy to buy a VHS system, even if your video dealer is "giving it away." Unless you get a Beta format system, too, you'll never be able to swap tapes with your friend (which, we might add, is half the fun of home video). You can, of course, always take your machine to your friend's place and laboriously duplicate his tape collection. But not only will you lose image quality this way, you'll probably also lose a good friend.

For what it's worth, one testing organization—Consumer Reports—rates a Beta format machine (Sony's SL-5600, list price $1,350) as its choice for the best VCR console. Statistically, however, the number of VHS format videocassette recorders outnumber Beta units by more than two to one. And, in case you're wondering, most of us at THE SOURCEBOOK own VHS machines. Why? Because our friends do.

Portable Or Console?

The next branch on your "decision tree" is the choice between a console VCR that stays in the house, or a portable unit you can take on the road.

The criteria are very simple. Both console and mobile units can be used with a video camera, but you can't take a console to the kid's graduation or on an offshore sailboat race. So, if you plan any serious live videotape production, invest the extra money in a portable unit now since you'll probably end up buying one sooner or later.

If the idea of portability appeals to you, here are the basic facts you should know:

Every videocassette recorder has two principal components. One is the Playback/Record system. The other is a tuner/timer.

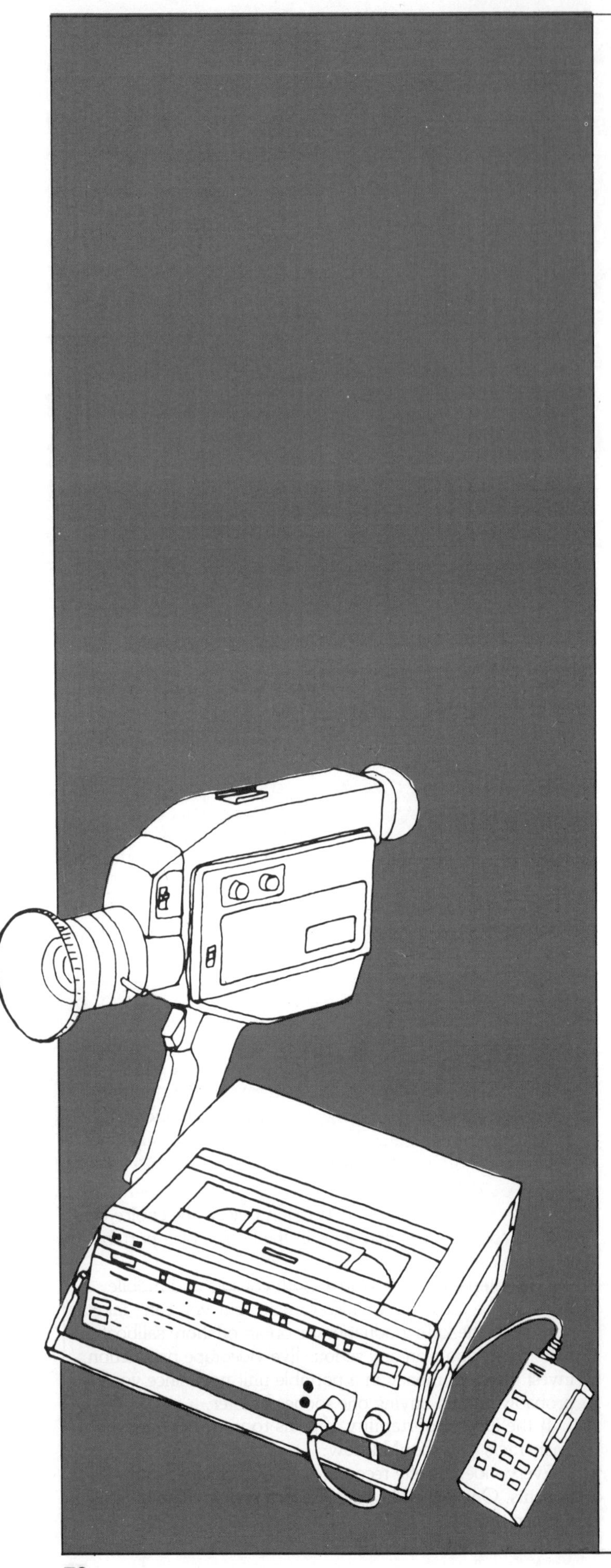

The VCR tuner is actually a duplicate of the tuner in your TV, which is why you can record one channel on the VCR at the same time you are watching another on the television screen. The timer can turn the VCR on when you're not home. (Today's microprocessor-controlled "programmable" timers allow you to pre-set up to 7 or 8 "events" over a 14-day period.)

The first VCRs were strictly living room editions, designed for off-the-air TV recording. As with today's console units, they were heavy—30 to 45 pounds—and required a wall outlet for electrical power. But with the development of an affordable color video camera, the demand for VCR portability skyrocketed.

The electronics industry responded with a portable VCR that had no tuner/timer but could be carried over your shoulder and operated with a rechargeable battery. Unfortunately, this meant owning *two* VCRs if you wanted both portability *and* the ability to record broadcast TV programs.

A second generation of portables that eliminated the need for this costly duplication began appearing in the late 1970s. Units like Panasonic's PV-2100 weighed about 19 pounds, featured rugged "solenoid" controls, and, most important, were sold with a separate tuner/timer. Not only could the portable take to the streets, but back in the living room it could be used for off-the-air recording.

Yet a third generation of portable VCRs is now available. Not only are the new units ultralight—under 14 pounds—but they have many of the sophisticated features found only on the most expensive console models. RCA's Model VEP-170 is perhaps the most versatile of this group. The VEP-170 offers a programmable tuner/timer with a 8-event/14-day memory, along with a 12-function remote control that gives you such special effects as variable speed slow motion, freeze frame, frame-by-frame advance, and even speed scan for locating your favorite scenes. List price is about $1,425, including the tuner/timer.

A unique new model is JVC's HR2200U. This portable VCR has a full-feature remote control, *plus* it is the first VHS videocassette recorder—either portable or console—that does not unload the tape from the audio and video heads when you hit the "Stop" button. The result is less drain on your battery, and fewer "glitches"—or bursts of video noise—between shots. One potentially serious drawback: the HR2200U is equipped with only one speed—Standard Play.

Before investing in a portable videocassette recorder, use "The VCR Buyer's Guide" to examine the alternatives. There are major differences between portables in terms of what features you get for your money.

A final word about VCR portables. If filming the pilot for next season's smash network comedy is your goal, beware. Home video equipment is just that, *home* video. The new video cameras and VCR portables are good, but they are intended to compete with Super 8, not a $50,000 broadcast-quality video production unit. If your goal is to learn the basics of video film-making, however, there's no better hands-on experience than working with a portable VCR.

©1982 THE HOME VIDEO SOURCEBOOK

A WORD ON TAPE SPEEDS

All About Tape Speeds & Recording Times

If you've ever operated a reel-to-reel audio tape recorder, the idea of tape speed should be child's play. If not, take heart. Video tape speed is one of those things that sounds terribly complicated—but isn't. Especially if you think about the evolution of the record player.

The early records were called 78s—namely 78 rpm. And at that speed the crude nature of the recording equipment—not to mention the scratches and nicks in the record itself—wasn't so obtrusive. As recording technology improved, record speeds dropped to 45 rpm and ultimately to 33 rpm, while the length climbed from just a few minutes to close to 1 hour for some modern stereo LP records.

So it has also gone with videotape. The magnetic tape in the first 1975 Betamax whizzed past the video heads at a speed of over 1½ inches per second, or almost 8 feet of tape each minute (while the video heads themselves rotated at 1,800 rpm). At this speed minor imperfections in the tape—analogous to scratches on the surface of a record—moved quickly past the video heads, causing minimal interference. Also, at 1½ inches per second (1.57 ips, to be exact) the video heads had a total of over 20 square feet on which to paint the electronic image of a 1-hour program.

But soon it was clear that the public wanted more than 1 hour of recording time. One answer, of course, would have been to put a longer tape in the videocassette. But Sony's engineers had done such a good job miniaturizing the Beta cassette that this was not possible using mid-1970s tape technology. The only other alternative was to slow down the playing speed—cut it in half, actually, to 0.787 inches per second, or only about 4 feet of video tape each minute. The new speed became known as Beta II (while the original, and soon to be abandoned, recording speed of 1.57 ips was called Beta I). Beta II gave VCR owners what they wanted, the ability to record longer shows, but the tradeoff was a slight reduction in image quality.

Beta II, however, was not the end of it. Once again, goaded by arch-rival JVC, Sony slowed their tape speed even further—to a mere 0.523 inches per second. Now you could get 3 hours on a single Beta cassette. But since you were still using the same 500-foot videotape cassette at this new Beta III speed, your video heads had less than 7 square feet of magnetic tape on which to store the image that they had once painted onto a full 20.5 square feet of tape. Obviously, image quality suffered at Beta III, while "dropouts" caused by tape imperfections became more noticeable.

About this time videotape technology had advanced to the point where Sony was squeezing 750 and, ultimately, 830 feet of tape into their Beta cassettes. The result: recording time could be extended even further—up to 5 hours at Beta III.

VHS has always had a leg up on Beta when it comes to recording times. JVC started the battle by introducing a VCR that could record 2 hours, as opposed to the Betamax's 1 hour. At this speed—1.31 inches per second—JVC could get more recording time because their videocassette was larger, and thus could hold a longer tape. When Sony introduced its own 2-hour speed, JVC countered by slowing *its* machines down, too. The original 1.31 inch per second speed became known as Standard Play—or SP—while the new speed (you guessed it, only half as fast at 0.656 inches per second) was called Long Play, or LP.

Then, to add insult to injury, the VHS crowd added yet another speed of 0.437 inches per second. Now VHS owners could get 6 hours on a tape. Some brands called this speed Super Long Play, or SLP, while others, such as JVC, decided to use the designation EP, or Extended Play. Since these abbreviations are confusing, many videofolk have taken to referring to the 2, 4, or 6-hour modes. Which is fine—except for one thing. The VHS tape manufacturers are now finding that they, too, can play Sony's game and stuff a little more tape into the VHS cassette (usually by using a thinner tape). With the advent of the T-150 videocassette, VHS owners are now getting up to 7½ hours of recording time at SLP. Should we now call it the 7½-hour mode?

Since most video publications have come to more or less automatically refer to the VHS speeds in terms of their recording times—the 2-hour mode, the 4-hour mode, etc.—we won't fight the practice. Just remember that these are the recording times you get with a standard T-120 VHS videocassette. A longer cassette, such as the T-150, will also give you longer recording times.

Tape Speed & Recording Time Comparison Chart

Designated Speed	Speed in Inches per second	Recording Time (Standard Cassette)	Recording Time (Extended Cassette*)
BETA I**	1.57 ips	1 hour	1.6 hours
BETA II	.787 ips	2 hours	3.3 hours
BETA III	0.523 ips	3 hours	5 hours
SP	1.31 ips	2 hours	2.5 hours
LP	0.656 ips	4 hours	5 hours
SLP (EP)	0.437 ips	6 hours	7.5 hours

*Approximate recording times based on the Beta L-830 and VHS T-150 extended length cassettes.

**Beta I is no longer available as a *recording* speed. Some Beta format machines still include Beta I, but as a playback speed only.

NOTE: ALL COMMERCIALLY AVAILABLE PRERECORDED TAPES PLAY AT EITHER BETA II OR VHS SP.

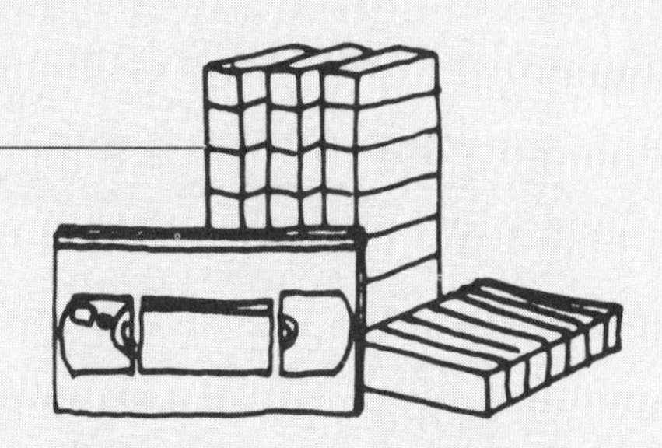

The Standard Features

Here are the basic features you DEFINITELY need on a VCR, and with which virtually every VCR on the market is equipped:

- Play/Stop/Record
- Fast Forward/Rewind
- Pause
- Audio Dub
- Timer/Tuner

As for the transport features—Play/Stop/Record/Fast Forward/Rewind/Pause—these work in essentially the same way as on an audio tape recorder. The only real difference you'll encounter between models here is in the control system. The older design and no-frills models employ the traditional "piano key" method of mechanical transport control. On the newer—and the more expensive—VCRs you will usually find the transport functions operated by push-button electronic, or solenoid, switches. These are slightly easier to operate and a bit more rugged than their piano key predecessors. Electronic switches also allow the addition of a relatively new convenience item—full-feature remote control—and are a distinct advantage on portable VCRs where operating ease is essential for capturing live-action footage.

"Audio Dub" is an intimidating term with a simple meaning. This feature merely allows you to re-record a soundtrack without erasing your video image. Audio dub, however, won't allow you to "overdub," or add sound to the existing audio track. The Audio dub mode automatically erases any soundtrack that is already on your videocassette—so be careful when you hit the pause button, which is adjacent to the audio dub button on many VCRs.

As explained above, the basic tuner/timer will automatically record one event during any 24-hour period without your having to be present. There is only one model you may still encounter on the market—Magnavox's 8370 Portable—for which no tuner/timer is available.

These standard, or basic, features are all you need to record programs or play back prerecorded videocassettes. Sears, Sanyo, Curtis Mathes, Magnavox, and J.C. Penney each markets a no-frills model that offers these basic features—and little else—at list prices between $695 and $795.

If you're looking strictly for economy, one of these units could be just the thing for you. But before you spend good money on a stripped-down VCR, take note: In today's world of competitive retail pricing, the difference between a no-frills unit at list price in a department store like Sears or J.C. Penney and a feature-laden "middle-of-the-line" VCR at a discount video outlet may be very small indeed.

Extended Play Speeds

Every VCR you'll find on the market is equipped with the Beta II and VHS Standard Play (SP) speeds (with the exception of Technicolor's TEC-212, which has its own unique format and speed). Beta II and VHS Standard Play typically provide a video image with good detail definition and a minimum of video background noise; and, as mentioned above, they are the *industry standard* for all legally available pre-recorded films and programs. What you may NOT be able to play back on these speeds are the videocassettes your friends have recorded on VCRs with Beta III, VHS Long Play, or even VHS Super Long Play.

Unless short-term economy is absolutely essential to you, we suggest skipping the one-speed units. True, you get better picture quality at the standard speed for your format (Beta II or VHS SP). But sooner or later you'll want to swap tapes with someone whose entire collection has been recorded at one of the extended play speeds.

In addition, if you do a lot of recording, you will probably find a one-speed machine to be more expensive to operate in the long run. If you're willing to accept the reduced image quality, the Beta III and VHS LP modes will cut your tape costs up to 50%. Translated into dollars, this means a savings of $6.50 per hour based on a list price of $26 for a top quality VHS videocassette like TDK's VAT-120. If you add just one hour of programming to your videocassette library each week, the Beta III or LP mode could save you over $300 a year. Of course, nobody pays full list price for videocassette tapes these days. But if you're an avid program collector, you'll find the extended play speeds will quickly pay for themselves—and a lot of extra features in the bargain.

VCR Operating Costs Per Hour

VHS Format Tapes

Manufacturer/Cassette	List Price	Cost Per Hour at SP	LP	SLP
TDK-VAT-120	$26.00	$13.00	$6.50	$4.34
QUASAR VC-T-120	$20.00	$10.00	$5.00	$3.34
RCA VK-250	$15.00	$ 7.50	$3.75	$2.50

Beta Format Tapes

Manufacturer/Cassette	List Price	Cost Per Hour at Beta II	Beta III
Fuji L-500	$21.50	$10.75	$6.14
Sony L-500 T	$17.00	$ 8.50	$4.86
Zenith L-500	$15.00	$ 7.50	$4.27

Note: These prices are for the "standard" length Beta L-500 and VHS T-120 videocassettes. The cost of using shorter tapes such as Beta L-250 or VHS T-60 will be considerably more *per hour*. This is because a large part of the expense of any videocassette is in the precision-made cassette housing. Longer tapes, like the Beta L-750 or VHS T-150, will usually cost *less* per hour. For a complete rundown on videocassette, and their costs see page 149 "Video Software & Accessories."

Frills For Thrills—The Special Effects

Speed scan, freeze frame, slow motion, and frame-by-frame advance are the principal special effects available on today's VCRs. Deluxe models generally have all these effects, often with interesting twists, like *variable* slow motion, or speed scan in rewind as well as fast forward. None of these features is essential. But, depending on your viewing habits and your reasons for owning a VCR, some special effects may prove extremely useful.

Probably the most convenient of these features for general viewing is speed scan—especially if you plan to do a lot of recording at any of the extended play speeds. True, almost all VCRs are equipped with a counter to assist you in locating a specific segment of a videotape. But the counter, at best, is a crude device, and most of us simply don't want to go through the annoyance of having to "log in" every program we record.

IT'S ALL IN YOUR HEAD

Picture quality. It's the foremost consideration when buying a VCR. But what factor is most responsible for producing a high quality video image? The answer is in your head—your video head.

The video head drum is the critical point of contact between your VCR and the tapes you play on it. When recording a program the video head "paints" a magnetic pattern onto the tape. In playback, another head "reads" the magnetic pattern and passes it along to the decoding and amplification circuitry in the VCR.

Not surprisingly, video heads are one of the most technically sophisticated components ever assembled in a consumer product. Not only do they operate at a relatively high speed—1,800 rpm (your stereo turntable rotates at a mere 33 rpm)—but they are precision machined to within a fraction of a micron. And a micron, in case you've forgotten, is approximately 1/25,000th of an inch. The video heads in a VCR must handle *200 times* more information than the audio heads in your home or car stereo tape deck.

The video head handles this enormous volume of information with the help of a technique known by the imposing name of "helical scanning." A helix (should you have missed Bio 101) is a spiral. In the instance of helical scanning, the helix refers to the path of the magnetic tape around the video head. On VCRs the tape is not only wrapped partially around the video head—it is wrapped in a helical, or spiral pattern.

Also, unlike audio heads, the magnetic signal is not applied to the tape in a straight line, or longitudinal, pattern.

Instead, the video track is painted on the tape by the rapidly revolving video head at a diagonal angle. The original helical scan recording method left wide gaps, or guardbands, between the video tracks to prevent one track from "crossing over" into another.

The basic helical scan method produced a high-quality recorded image. It also required

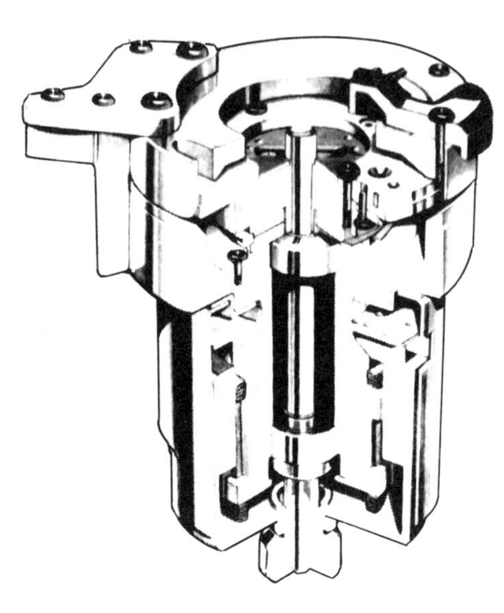

At the heart of today's VCR—the rotating video head—one of the most technically sophisticated components ever assembled into a consumer product.

750 square feet of magnetic tape per hour—or about $1,200 worth for a two hour movie.

The breakthrough, for economical home video, was discovered by the honorable Professor Okamura and subsequently perfected by Sony and Matsushita Electric. The actual process involved reversing the "phase" of alternate video tracks, and then butting them together so there would be no wasted space for guardbands. Professor Okamura called the process "azimuth" helical scan recording. With the new process the number of square feet of videotape required to hold a 1-hour show plunged from 750 to 20. Consumer video had become economically practical.

The earliest VCRs, both Beta and VHS, had one playing speed, and there was not much problem designing video heads to give optimum performance. Today's machines offer 2 and even 3 recording speeds. The designers face a tough choice. Do they use 1 set of heads, compromised to give adequate picture quality at every speed—or do they install 2 separate and expensive sets of video heads, each designed to give optimum results at just 1 speed?

Most brands have done both. They offer lower-priced VCRs with multiple recording speeds, but just one set of video heads. The picture quality is usually good, but nothing to brag about, especially at the longer playing times.

Meanwhile, most deluxe VCRs do, indeed, use two separate sets of video heads. Our advice? If you think you are going to do an extensive amount of recording, get a model with two sets of heads.

Many manufacturers recommend replacing your heads every 1,000 to 1,500 hours—so obviously it will be a while before you start thinking about new heads as a matter of routine maintenance. But should you damage your video heads (for instance, by trying to play a videotape that has been spliced), you'll find the cost of parts and labor to replace them can run as high as $300.

Video heads, just like their audio cousins, require periodic cleaning. This can be done by your local video dealer. Or, you can do it yourself with any of a number of head cleaning-videocassette units that are available. Just one word of caution. The head cleaners are abrasive. Don't exceed the recommend cleaning times, otherwise you're likely to find yourself an early candidate for a video head overhaul. Best bet—a head-cleaning cassette like the "Allsop 3" which shuts off automatically when the cleaning cycle is finished.

In a VCR equipped with speed scan, the videotape stays in contact with the tape heads in fast forward (and sometimes even in rewind). The result: you get a viewable image moving across the screen at speeds up to 40 times faster than normal. You can "scan" an entire 2-hour movie for your favorite scene in just 3 minutes. Or you can race through 3 minutes of commercials in just 4.5 seconds. Only a few deluxe models offer the 40X speed scan, but even with a 9X scan you can get through the same 3-minute commercial break in just 20 seconds.

If you watch a lot of sports programs, or plan to use the VCR for analyzing your own performance in almost any physical activity from ballet to basketball, you'll probably want a unit equipped with slow motion, freeze frame, and frame-by-frame advance. On some VCRs slow motion can be varied, on others the speed will be fixed at ½ or ¼ normal. Freeze frame will allow you to "stop action" by literally freezing a single image on the screen. Beware, however, that both the freeze frame and pause modes (on many models the pause button automatically freezes your video image) put enormous tension on the videotape, and can actually damage or break a tape if overused. Most manufacturers have installed an override on the pause control. After three to five minutes the pause button will automatically unlock. You'll probably discover this the first time you leave your VCR on pause to go make a "quick" phone call.

Programmability

Once you've decided on the format and made up your mind whether to buy a console or a portable, the next step is to consider programmability.

The vanguard videocassette recorders of the mid-1970s could be equipped with a "1 event/1day" timer. You selected channel 2 on your VCR tuner, set the timer for 6:00 and *voila*, you could still catch Eyewitness News when you finally got home from the office at 8:25. Most manufacturers still sell a no-frills VCR with a digital 1 event/1 day timer. It's cheap, and if you don't travel much, it's effective.

With the advent of microprocessors in the late-1970s, VCR timers became truly programmable. Most of today's consoles, and even many of the newer protables, are equipped with a microprocessor-assisted timer/tuner that allows you to automatically record as many as eight different shows, off any combination of channels, over a 2-week period.

This kind of sophisticated automatic programming is a great convenience—especially for the frequent traveler—as long as you are willing to pay an extra $100 to $150 for it. Like many of the space-age innovations available on today's videocassette recorders, just how badly you need them depends entirely on your life-style and the thickness of your wallet.

Remote Control

As mentioned earlier, one advantage of the new solenoid switches is that they allow full-feature remote control. You'll find that on almost all deluxe VCRs not only the basic transport functions—play/stop/record, etc.—but also the special effects, can be operated by a hand-held remote controller. This is a great advantage in recording live action on a portable VCR, as well as a convenience for easy chair viewing.

Electronic Indexing

If you can't get speed scan, the next best thing is "electronic indexing." This feature, available on most programmable VCRs, automatically places an electronic signal on the videotape when the record button is activated. If you then hit the "memory" switch during fast forward, a sensor will read the electronic cues and stop the tape at the beginning of each program. Electronic indexing, or cue-search as it is sometimes called, is a particularly good feature for those addicted to collecting TV episodes like "M*A*S*H" or "The Odd Couple" on 6-hour cassettes.

Sleep Switch

If you're in the habit of dozing off before the show is over, this one's for you. It automatically turns off the VCR at the end of the tape.

Dew Indicator

No, it won't save you from an atomic attack. But it will save your $99 copy of *Alien* from a potentially lethal overdose of moisture that has accumulated on the video head. If you live in an area where moisture is a serious problem, you might be interested in Akai's VP-7300 which has a built-in heater to dry a dewy video head.

Tape Remaining Indicator

Does just what it says—tells you how much tape you have left.

Warranties

The industry standard is 90 days on labor, 1 year on parts. If you're nervous about repairs consider Curtis Mathes. Their F-735-9 comes with an incredible 4-year warranty on parts and labor.

Use "The VCR Buyer's Guide"

Now that you have the basics—and before you rush out to drop $600 to $1,400 on a VCR—take an hour to study "The VCR Buyer's Guide." The Questionnaire on page 83 will help you narrow your selection to those models that suit your needs and your pocketbook. Then fine tune your choice by consulting the brand listings starting on page 86 for a detailed discussion of the best—and the worst—features of each model.

A QUESTIONNAIRE

Defining Your Needs—A Questionnaire

Realistically, how much can you afford to spend on a Home Video System?	$600 to $1,000 $1,000 to $1,500 $1,500 to $2,500
Do you have friends with VCRs? Which format do the majority of them own?	Beta VHS
If you were to spend up to $2,000 for a portable VCR & color camera, do you honestly think you'd use it enough to justify the expense? If your answer is NO, circle CONSOLE, if yes, circle PORTABLE.	Console Portable
Do you plan to use your VCR mainly to watch prerecorded programs? Or do you intend to record off TV as well as buy or rent prerecorded videocassettes? If you don't plan to do much recording, circle NO-FRILLS.	No-Frills
Do you travel extensively, or want to record a series that comes on the air several times a week? If YES, circle programmability.	Programmability
Do you think you'll be actively swapping tapes with your friends, or building your own library of movies & shows? If yes, circle EXTENDED PLAYBACK SPEEDS, SPEED SCAN, and ELECTRONIC INDEXING.	Extended Playback Speeds Speed Scan Electronic Indexing
Do you watch a lot of sports or plan to use your portable for improving physical skills? Are you a dedicated film buff or trivia fanatic? If YES, circle SLOW MOTION, FREEZE FRAME, and FRAME-BY-FRAME ADVANCE.	Slow Motion Freeze Frame Frame-By-Frame Advance
Do you hate to get out of your chair to change stations or alter volume, or do you plan on buying a portable. If YES, circle REMOTE CONTROL.	Remote Control
Do you live in an area where high humidity is a problem? If YES, circle DEW INDICATOR.	Dew Indicator

You can now make a speed scan of your own along the right-hand column. Your only problem will be finding all those extra features you circled for a price you can afford. For some help, turn to page 84 and "The VCR Buyer's Guide."

THE VCR BUYER'S GUIDE

Americans may be piloting the Space Shuttle, but back on earth one thing we are *not* doing is building VCRs. It may come as a surprise, but every videocassette recorder being sold in the United States was designed and manufactured in Japan.

Like virtually all inventions of the Industrial Age—from synthetic penicillin to the electric can opener—the technology of videocassette recording is patented. The majority of VCR patents—covering everything from the design of the actual cassette to the basic technique of "azimuth helical scan" recording (see "It's All in Your Head" on page 81)—are held by just three companies, all of them in Japan. They are Sony, JVC, and Matsushita Electric.

If only three companies can build VCRs, how is it that there are over 20 brands available?

Part of the answer is licensing. Sony, for instance, licenses Sanyo to build Beta format VCRs. Using Sony's patents, Sanyo designs and manufactures its own line of models, paying Sony a small royalty on each unit.

But licensing is only part of the story. It seems that once the video boom really took off in the 1970s many companies got into the act, even though they had no VCR manufacturing facilities of their own. Sears, for example, does not build VCRs. Instead, the Sears' Betavision videocassette recorders are actually produced by Sanyo (who in turn is licensing the patents from Sony). Many of the brands you are most familiar with—Zenith, GE, RCA, Sylvania, Magnavox, Philco, J.C. Penney, Quasar—do not manufacture the videocassette recorders they sell.

About half of all VCRs sold in the United States, including many of the most popular brands like RCA, GE, Quasar, Magnavox, and Panasonic, are built by just one Japanese company—Matsushita Electric. Perhaps you've noticed a certain remarkable similarity between features, and even styling, on some of these competing brands. Your eyes haven't been deceiving you. The only difference between many models is the name on the package. Even the technical experts can't find much to differentiate a unit like Panasonic's PV-1750 from its near-identical siblings—the Quasar VH-5160 and the RCA VET-650. However, you shouldn't necessarily assume equivalent models belonging to the same family tree are *always* identical in every way. Sometimes an original equipment manufacturer like Matsushita Electric will make minor alterations for a big customer like Magnavox or RCA.

Although this system is a bit confusing, we think it serves the consumer well. For one thing, all three major VCR patent holders—Sony, JVC, and Matsushita—set high standards for both their own products *and* the products of the manufacturers they license. As a result, unlike the television or stereo component industries, you won't find any real junk—or even marginal equipment—in the VCR showroom (although, alas, it's still possible that you may be unlucky enough to wind up with a particular unit that happens to be a real lemon).

In addition, the intense competition between brands has produced two substantial benefits for consumers. While the prices on most appliances have been soaring and the level of quality declining, just the opposite has been happening with VCRs. Prices, if not actually falling, are merely keeping pace with inflation, while at the same time features like slow motion, speed scan, solenoid-activated controls, and electronic tuners are making today's generation of machines more reliable, more compact, and far, far more convenient to operate than their forebears.

The VCR Family Tree

Original Equipment Manufacturers	Sold In The U.S. Brand Names
Matsushita Electric	Panasonic (a division of Matsushita) Quasar (a division of Matsushita) RCA Magnavox Sylvania Philco J.C. Penney Curtis Mathes
Hitachi	Hitachi General Electric RCA
JVC (partly owned by Matsushita)	JVC Magnavox
Sanyo	Sanyo Sears
Sony	Sony Zenith

Note: *Toshiba, Sharp, Mitsubishi, and Akai also manufacture original equipment by licensing agreement and market these units exclusively under their own imprint.*

How To Use "The VCR Buyer's Guide"

If you've already read "Selecting Your Videocassette Recorder" and used the questionnaire on page 83, you probably have a good idea which features you do—and don't—want in a VCR. In fact, chances are you've got a better idea about what's going on with VCRs than many TV-stereo salespeople. Beginning on page 86 you'll find a model-by-model rundown on almost every major make and model of videocassette recorder. Use the "Buyer's Guide" to find the specific models that come closest to matching the features you want, at a price you can afford. Be sure to pay close attention to the "Best features" and "Drawbacks." For some thoughts on which VCRs THE SOURCEBOOK likes best see "Home Video's Top 20" on page 116.

When You're Ready To Buy

If you're willing to order your VCR through the mail, you will find that many of the models listed in "The VCR Buyer's Guide" are available at discounts of up to 35% off the list price. We've quoted two prices in our reviews—the suggested retail price, and the lowest mail order discount price we could find.

On page 173, in "The Do's and Don'ts of Discount Shopping," you'll find a full discussion of the advantages and disadvantages of buying through the mail or from a mass-volume discount outlet. We also list some of the better-known discount houses with large video inventories. You can save a lot of money buying your VCR this way. You can also create some big headaches for yourself further on down the road. See page 173 before you buy.

A Word About Specifications

If you have a background in electronics, read no further. You already know what a gold mine the manufacturer's specification sheet can be. Otherwise, it's likely that this fine-print jumble of jargon and numbers will look like so much Greek to you. (In fact, some of the symbols you see on spec sheets are, indeed, in Greek.)

Before you buy, get a spec sheet. Then look for the following 8 categories. If any of this information is missing—and it often is—you might ask yourself why.

- "Video Recording System." You'll see a lot of technical terms here. But the main thing you're after is how many video heads does the unit have—2, 3, or 4. Generally, the more video heads, the better the picture quality—and the more expensive the machine.
- "Tape Speed." If it's a VHS-format unit, does it have SP, LP, and SLP? Beta units can have Beta II and Beta III. Some deluxe models will also play back Beta I (the speed on the original Betamax), but you can ignore Beta I unless you happen to have a library of old Beta videocassette recordings.
- "Maximum Recording Time." This isn't essential, since if you know what speeds you have you can always figure it out. But it never hurts to see it spelled out in print.
- "Operating and Storage Temperature." The reason for this information many seem a little obscure, unless you're buying a portable VCR. A unit with an operating range of 32°F to 104° (which is the typical range of most portables) won't do you much good if you're intending to shoot the Winter Olympics, or even picking up some footage on your next vacation to Vail.
- "Fast Forward/Rewind Time." This is another spec many manufacturers omit, and really it's no big deal. If you want to know how long it takes to rewind you can always time it in the showroom. Typically a 6-hour VHS tape takes 5 to 7 minutes to rewind.
- "Video Signal-to-Noise Ratio," Yet one more spec many manufacturers don't—or won't—publish. This one, however, *is* important. The signal, of course, is the good stuff—Linda Lovelace, or whatever. The noise is the bad stuff—tape hiss and other junk that shows up on the screen as snow. The ratio between the two is given in decibels (dB). 40 dB or above is considered very good for home viewing—the higher the number, the clearer the video image. If the specs for the unit you're interested in DON'T give the signal-to-noise ratio, go ahead and be a little suspicious. We are.
- "Horizontal Resolution." Again, the higher the better. 230 lines is the minimum considered adequate for good viewing.
- "Dimensions and Weight." Not terribly important for consoles, unless you're trying to fit a VCR into a tight space (which isn't a very good idea unless it's well ventilated). But portable buyers will want to take a long hard look at these numbers, particularly considering that the camera can add another 5 to 10 pounds to your load.

The Typical VCR Spec Sheet—What To Look For

SPECIFICATIONS

Temperature	
Operating	: 0° to 40°C (32° to 104°F)
Storage	: –20° to 60°C (–4° to 140°F)
Format	: VHS NTSC standard
Video signal system	: NTSC-type color and monochrome signal, 525 lines
Video recording system	: Rotary, slant azimuth, two-head helical scanning system
Tape width	: 12.7 mm (1/2")
Tape speed	: 33.35 mm/sec. (1.31"/sec.)
Maximum recording time	: 120 min. (with JVC T-120 cassette)
Rewind and fast forward time	: Within 5 min. (with JVC T-120 cassette)
Video signal	
Input	: 0.5 to 2.0 Vp-p, 75 ohms unbalanced
Output	: 1.0 Vp-p, 75 ohms unbalanced
Horizontal resolution	: More than 240 lines (color mode)
Signal-to-noise ratio	: More than 45 dB (Rohde & Schwarz noise meter)
Audio signal	
Microphone input	: –67dBs, high impedance, unbalanced
Line input	: –20dBs, 50k ohms, unbalanced
Line output	: –6dBs, less than 1k ohms, unbalanced
Earphone output	: 0dBs, 10 ohms
Frequency response	: 100 to 10,000 Hz
Signal-to-noise ratio	: More than 40dB
Power requirement	: 12V DC
Power consumption	: 9.6 watts
Dimensions (W × H × D)	: 28.6 × 10.3 × 26.7 cm (11-5/16" × 4-1/16" × 10-9/16")
Weight	: 5.2 kg (11.4 lbs.) including battery pack
Accessories provided	: Remote control unit, cable, earphone, shoulder strap, carrying handle

To help you out, we've included spec sheets for the basic models offered by the major original equipment manufacturers—Panasonic (Matsushita), Sony, JVC, Sanyo, Hitachi. In most cases you'll find the specs of equivalent models sold under different brand names are very similar—but before you spend all that hard-earned money, take 10 minutes and study the specs for the model you are about to buy. You just might find an interesting—and perhaps unpleasant—surprise lurking in the fine print.

The Rating System

Buying a VCR is like ordering your dinner. The choices can range from a burger and fries to chateaubriand and escargot.

In comparing the VCRs, we've adopted a rating system familiar to gourmets and theater goers alike—the 4-star system. And like most such systems, our has its share of subjective, arbitrary, and capricious aspects.

To begin, we've been forced to assume that price is no object. After all, does the gourmet care what the meal costs? Does a movie reviewer knock a film because the theater charges a $7.50 admission? And do you care how much a VCR costs? Of course you do. Price is probably the most important consideration in making your buying decision.

But the problem with writing price into the comparative equation is this: When it comes to price the VCR market place is totally bonkers. List prices mean nothing. We've seen last year's top-of-the line model reduced from $1,425 to $899 while around the corner a no-frills, no-features economy model is actually selling for a few dollars more.

Having ignored price, we've concentrated entirely on three criteria.

● **Quality:** What is the level of quality of the chassis, the various components, and, most important, the video picture produced by the machine?

● **Functions:** Does the machine have all the basic functions (including 3 tape speeds in VHS, 2 in Beta) we expect in a VCR? Or have some basic functions been "traded-off" to make way for fancy features?

● **Features:** Does the unit offer the maximum convenience and flexibility?

With these criteria as a basis, we've given every VCR a rating. Here's what they mean:

★This model has some flaw which, in our opinion, prevents us from recommending it. See the comments for details.

★★Does the basic job, but that's about all. Most no-frills and economy models fall into this category.

★★★It's not quite perfect, but has our approval. Picture quality is good to excellent, it has all the basic VCR functions, and at least some of the most important convenience features and special effects.

★★★★Our highest recommendation. This unit has it all—top-quality craftsmanship and performance—as well as the full range of convenience features and special effects. Generally for reserved the best top-of-the line models.

Note: *Virtually every manufacturer and distributor of home video equipment reserves the right to change product specifications without notice. Changes often result in an upgrading of features, but not always.*

Before you buy a VCR, we urge you to obtain the latest specifications from your dealer. If he doesn't have a spec sheet—and he should—call or write the manufacturer. We've included addresses and phone numbers for exactly this purpose.

It's not a bad idea to check out all the features on a newly purchased VCR before you leave the store. If this is impractical, do it as soon as you get home. Make certain the machine you just bought has ALL the features you were expecting. Then make sure these features work the way they're supposed to work.

In preparing our equipment reports on certain new models we have occasionally had to rely upon the manufacturer's "Preliminary Specifications," which are particularly susceptible to change. Whenever this has happened you'll find a footnote indicating the report is "based on preliminary specifications."

VHS Format Videocassette Recorders

Akai America, Ltd.

For additional product information and a list of Akai dealers in your area call or write the Customer Service Department at:
Akai America, Ltd.
800 West Artesia Boulevard
Box 6010
Compton, CA 90024
213-527-3800

Since Akai is best known in the United States for its quality audio components, it is not surprising to find this company introducing the first stereo VCR to the U.S. market. Akai's videocassette recorders are extremely lightweight, well-built, but generally priced higher than other brands with comparable features. Akai is one of a handful of smaller Japanese electronics companies that actually manufacture their own video equipment under license from JVC and Matsushita.

Akai VS-1U VHS Console ★★★

List price: N/A
Manufacturer: Akai
Best features: 4-head recording system, SP, LP, and SLP playback, SP and SLP recording, 8-event/14-day programmability, auto rewind, solenoid-activated transport controls, memory rewind, remote control, speed scan in forward and reverse.
Drawbacks: No slow motion*, no recording in LP mode.

This is Akai's first console VCR, and at 22 pounds it isn't much heavier than the first and second generation *portable* VCRs. If home video is also business video or if for any other reason you're looking for a one-piece console VCR that is also convenient to tote—this is a good bet.

Like many other top-of-the-line VHS videocassette recorders, the VS-1U offers 4 rotary video heads. One set of heads is optimized for SP, one set for SLP. The stepchild turns out to be LP, the 4-hour VHS speed. You can play back at LP (so Uncle Zeke can bring his tapes next time he visits), but you can't record. The speed scan, auto

Preliminary Specifications

Video recording system	4 rotary heads, helical scanning system
Tape format	VHS, NTSC
Tape speed	SP, LP. SLP (playback), SP, SLP (record)
Power requirements	120V/60Hz
Power consumption	50 W
Operating temperature	50°C to 40°C
Video input	0.5 to 2V p-p, 75 ohms
Video output	1.0V p-p, 75 ohms
Signal-to-Noise Ratio	Better than 45 dB
Frequency Response	240 lines
Dimensions	17.3" (W) x 5.5" (H) x 12.9" (D)
Weight	22 lbs. (10 kg)

Note: For improvement purposes, specifications and design are subject to change without notice.

rewind, and microprocessor-assisted timer/tuner are all features we've come to expect in deluxe model VCRs. One special effect is missing, however. And if you're a sports fan you may regret the lack of slow motion*.

Based on Preliminary Specifications

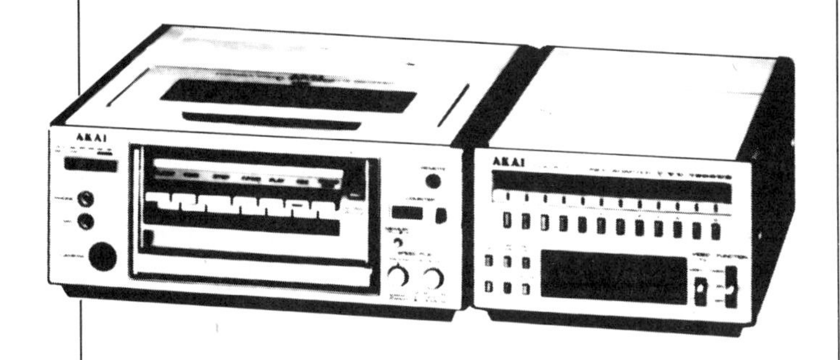

Akai VPS-7300 VHS Portable★

List price:	$1,500
Best features:	4x speed scan, variable slow motion, 2-event/7-day programming.
Drawbacks:	No remote, no LP or SLP speeds, requires special adapters to use videocameras other than those built by Akai.

Akai has designed and built a quality machine in the VPS-7300 portable, but in our view the rather spartan features—there's no remote control or extended playing speeds—don't justify the hefty price tag. As with the other one-speed VHS portable—JVC's HR-2200—you'll also find yourself at a big disadvantage when it comes to recording off network or cable TV. Namely, many movies and most sports events just won't fit on a 2-hour cassette.

If you're looking for a portable and are impressed with Akai, we suggest you spend a few dollars more and go with the VP-735OU. It has much more to offer.

Akai VP-735OU VHS Portable★ ★ ★

List price:	$1,700
Manufacturer:	Akai
Best features:	2-channel stereo with Dolby B noise-reduction system, lock-up security, 6-event/7-day programming, IPLS (Instant Program Location System), special effects including freeze frame, slow motion, and variable speed scan.
Drawbacks:	No LP mode, piano key controls, remote control costs extra as an option, special adapters needed to use another manufacturer's videocamera.

Many video dealers started out in the audio business, and sound is still their first love. So don't be surprised if you hear a lot about this model in your quest for a perfect portable. The Akai VP-735OU happens to have a stereo sound.

In Japan, where all VCRs are made and TV has been broadcast with 2-channel sound (often with English on the second channel) for almost half a decade, it's no big deal to find stereo in a VCR. But in the U.S., where the closest thing to stereo TV has been a few FM radio simulcasts, there are no stereo VCRs. Correction, there *were* no stereo VCRs, because the Akai VP-735OU is the first.

If you're interested in the Akai VP-735OU because you want to record stereo programs off television, you may be in for a long wait. The first stereo TV broadcasts aren't expected until after the middle of the decade. But we still like the idea of stereo in a *portable* VCR, since the 2-channel sound capability vastly increases your potential for creative and soundtrack editing.

As for prerecorded films in stereo—yes, there are a couple on the market (ironically, the first stereo videocassette happens to be a porno flick) and more are certain to be released. But don't expect an avalanche.

Our biggest quarrel with the VP-735OU is the unit's lack of an LP playback and recording speed. Once again we repeat our argument: What happens when your uncle Zeke shows up with a trunk full of great tapes made at LP? And LP is, arguably, the most popular VHS recording speed for owners of mid-priced VCRs. It also seems downright insulting to pay extra for a remote controller on a unit which, if price is any criterion, should be the Rolls Royce of home video. And, alas, you probably won't find any grand discounts on that sky-high list price of $1,695 (which includes the VP-735OU and its separate programmable timer/tuner). The best we could do was $1,300—a mere 24% off list.

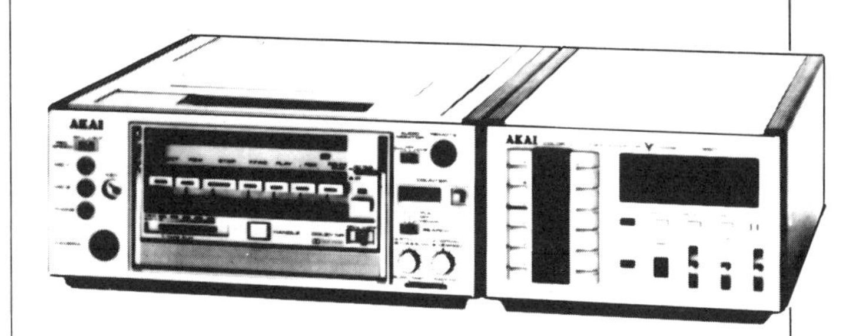

Specifications

Format	VHS System
Recording system	Rotary, slant azimuth two-head helical scan system. Maximum 260 min. using T-120 video cassette (at super long play)
Video, input	0.5 to 2V p-p, 75 ohms, unbalanced
Output	1V p-p, 75 ohms, unbalanced
Horizontal resolution	Color, more than 240 lines (in standard play)*
Signal-to-noise ratio	Better than 45 dB (in standard play)*
Audio, input	MIC, 70 dB, 600 ohms, unbalanced (left, right)
Output	LINE, 6 dB (400 mV)
Track	2 Tracks
Frequency response	100 Hz to 10 kHz (in standard play)*
Signal-to-noise ratio	Better than 50 dB at 5kHz (in standard play w/Dolby on)
Earphone	150 mV/8 ohms
Tape speed	33.35 mm/sec (in standard play)* 11.12 mm/sec (in super long play)
F. forward/rewind time	Within 240 seconds using T-120 video cassette
Operating temperature	0°C to 40°C (32° to 104°F)
Power requirements	12V DC
Power consumption	12.5W (in standard play)*
Battery operating fime	60 min continuous (using video camera)
Dimensions	292(W) x 123 (H) x 307 (D) mm (11.5 x 4.8 x 12.1″)
Weight	6.6 kg (without battery)

Curtis Mathes

For additional product information and a list of Curtis Mathes dealers call or write the Customer Service Department at:
Curtis Mathes
One Curtis Mathes Parkway
Athens, TX 75751
800-527-7646

Curtis Mathes video products are distributed primarily through their dealer network, which is currently best established in the south and southwest. Curtis Mathes VCRs and video cameras—like those marketed by Panasonic, Quasar, Magnavox, RCA, and others—are actually manufactured by Matsushita Electric.

Curtis Mathes, however, offers by far the best warranty in home video—*4 years on parts and labor.* Such a warranty, purchased on a year-by-year basis from an independent video specialty store would cost in excess of $400, if you could get it at all. Repairs are done at the Athens, Texas, Service Center, and some Curtis Mathes dealers will charge a $15 to $20 shipping fee for forwarding a VCR or camera for warranty work.

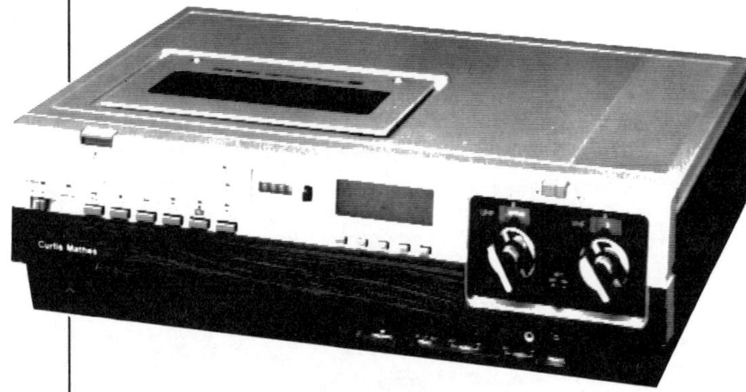

Curtis Mathes G-748★ ★

List price: $879.95
Manufacturer: Matsushita
Best features: 4-year warranty, solenoid transport controls, remote pause, SL, LP, and SLP speeds.
Disadvantages: No special effects or speed scan, limited programmability.

Here's Matsushita's basic VCR updated. Instead of piano key controls it has solenoid-assisted push buttons. These are definitely easier to use and should be less prone to breakdown as well. There is also a remote pause, which means that if you happen to be watching a show at the same time you are recording it, you can edit out the commercials without leaving your chair.

The Panasonic PV-1270, RCA VFT-190, Quasar VH-5011, and Magnavox 8315 are virtually identical models at lower prices. But you won't get the Curtis Mathes warranty.

Curtis Mathes F-755 VHS Portable★ ★ ★

List price: $1,499.95
Manufacturer: Matsushita
Best features: 4-year service warranty, SP, LP, and SLP speeds, electronic controls, camera-mounted remote control when used with Curtis Mathes G-760 video camera (list price: $1,099.95), cue and review (speed search), and 8-event/14-day programmability.
Drawbacks: No slow motion. Freeze frame and frame-by-frame advance work only in the SLP mode.

This is the Panasonic PV-4100 portable with its very handy addition—the cue and review (speed scan). For the Curtis Mathes version, there is an optional camera with a built-in remote control (most video cameras have a pause control only). We like almost everything about the F-735, except the lack of slow motion and the limiting of the special effects to the SLP mode where image quality is almost always marginal in a 2-video head machine—our same complaints about the Panasonic prototype. As long as special effects aren't a big deal with you, this is good portable to investigate.

Curtis Mathes G-750 VHS Console★ ★ ★ ★

List price: $1,599.95
Manufacturer: Matsushita
Best features: 4-year warranty, full special effects, electronic transport controls, wireless remote control, 4-video-head recording system, speed scan, 8-event/14-day programmability.
Drawbacks: No serious disadvantages.

Except for the name on the package and the addition of the wireless remote control, this unit is virtually indentical to Panasonic's top-of-the-line PV-1750. As with its Panasonic prototype, the Curtis Mathes G-750 offers just about every frill available on a VCR, from speed scan and microprocessor-assisted programming to the feature we like best of all—4 video heads. Again, you probably won't find Curtis Mathes discounted as much as Panasonic, RCA, Quasar, or Magnavox—but don't forget that those brands can't match the Curtis Mathes warranty.

General Electric

For additional product information and a list of General Electric retail outlets call or write the Customer Service Department at:
General Electric
Television Division
Portsmouth, VA 23705
804-483-5600

General Electric is one of the best known and widely established domestic electrical-appliance manufacturers. GE has thousands of local retail outlets in the U.S., including many major department-store chains, and also maintains numerous regional repair facilities capable of servicing GE products. The primary components of General Electric's VCRs are manufactured in Japan by Hitachi under licensing agreements with JVC and Matsushita.

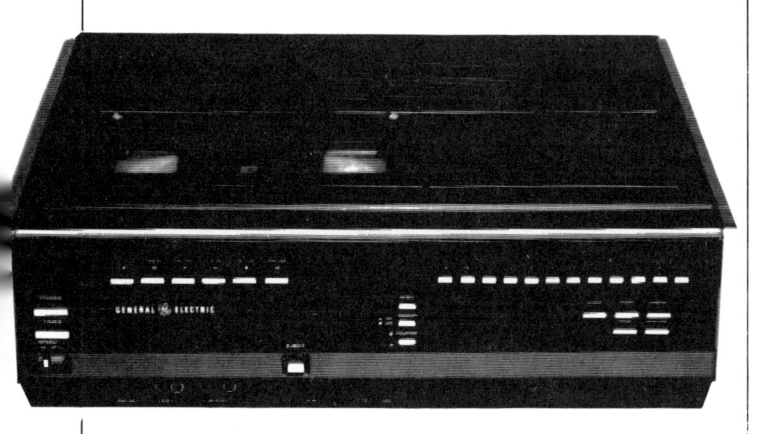

General Electric 1VCR-2002X VHS Console Portable★

List price: $989
Manufacturer: Hitachi
Best features: Solenoid controls, designer styling, remote pause, light weight.
Drawbacks: No LP speed, no special effects, limited programmability.

This is GE's entry into the no-frills, economy-price VCR market. The 1VCR-2002X is light weight, handsomely styled, and has convenient solenoid controls. Unfortunately, one thing it lacks is an LP speed—and in our opinion LP is the most practical tape speed on a low-priced machine. True, Super Long Play (SLP) is more economical, but after you've spent a couple of hours watching blurred scenes in colors that bleed all over the screen, we think you, too, will have your doubts about SLP. And, yes, SLP is certainly the best speed for image quality. But take a look at your "TV Guide." How many of the movies that you want to record will fit on a 2-hour cassette?

As for the 1VCR-2002X, we're glad to see a new economy VCR, especially one with solenoid controls. But we also think dropping the LP speed was a mistake. Magnavox, Quasar, RCA, Panasonic, Curtis Mathes, and others market economy models that *do* have the Long Play mode.

General Electric 1CVP-2020X VHS Portable★ ★ ★

List price: $1,399
Manufacturer: Matsushita
Best features: SP, LP, SLP speeds, speed scan, remote control, special effects (freeze frame and frame-by-frame advance), excellent 8-event/14-day programmability.
Drawbacks: No slow motion.

This is GE's first portable VCR and we like what we see in their preliminary specs. Because of the full range of recording and play back speeds, an 8-event/14-day microprocessor-assisted tuner/time, speed scan (GE calls it "video scan"), 4-function remote control, and special effects including freeze frame and frame-by-frame advance, the 1CVP-2020X ranks with other deluxe portables in terms of versatility on location and in the living room. At 13 pounds with battery, the 1CVP-2020X is only marginally heavier (less than 2 pounds) than the ultra-light weight units like RCA's VFP-170 and JVC's HR-2200U. The only drawback we can find is the lack of slow motion, which is a fault shared by many if not most portable VCRs. And if your only use for slow motion is to get a closer look at those controversial close plays, the freeze frame and frame-by-frame advance will be a more than satisfactory substitute. If the purpose of your portable is to help your daughter analyze her figure skating form, or to assist your Little Leaguer in getting a handle on why he's missing those low curve balls, then you may want to investigate the RCA VFP-170 or Quasar 5300, which *do* offer slow motion.*

**Based on Preliminary Specifications.*

General Electric 1VCR-2014W VHS Console★ ★ ★ ★

List price: $1439
Manufacturer: Hitachi
Best features: 4-video-head recording system, solenoid "soft touch" transport controls, 8-event/14-day programmability, speed scan, full special effects, wireless remote control. SP, LP, and SLP (GE uses the designation EP) speeds.
Drawbacks: None

This is GE's first 4-video-head recorder, and in our view it's a winner. We recommend this unit right along side the Panasonic PV-1750 if you're shopping for a deluxe VCR that offers top image quality and the full spectrum of state-of-the-art VCR features.

What we like best is the 4-video-head recording system. The 2-video-head basic and middle-of-the line VCRs are, indeed, kinder on your pocketbook. But it's not long until most owners abandon the SLP speed on these units. The reason? One set of video heads simply cannot do a good job at all 3 speeds. Usually it's the 6-hour SLP speed that suffers the worst—faces and details blur, colors shift and bleed, and the entire screen sometimes appears awash in a sea of video noise.

With GE's 1VCR-2014W we found image quality in the 6-hour mode that rivals anything we've seen at LP in mid-priced unit. Investing in this model—or other 4-video head units like JVC's HR-7300, Panasonic's PV-1750, or RCA's VET 650—will cost you more initially. But if you record just 3 movies a week for your video library at SLP, you'll surely end up saving money in the long run.

Hitachi Sales Corporation of America

For additional product information and a list of Hitachi dealers in your area call or write the Customer Service Department at:
Hitachi Sales Corporation of America
1200 Wall Street
Lyndhurst, NJ 07071
201-935-8980

Hitachi Sales Corporation of America
401 West Artesia Boulevard
Compton, CA 90220
213-537-8383

Hitachi is one of 4 Japanese electronics manufacturers licensed by JVC and Matsushita to build VHS videocassette recorders. Hitachi supplies VCR components to General Electric and also sells and services its own line of video equipment in the U.S. through the Hitachi Sales Corporation of America with regional offices in New Jersey, Illinois, Georgia, and California. Hitachi products are available through video specialty dealers, many department stores, and mail-order discount houses such as Video Wholesalers, Inc., in Miami.

Hitachi VT-6500A VHS Portable ★ ★ ★

List price: $1,150 for the VCR deck.
$450 for the VT-TU65A tuner/timer

Manufacturer: Hitachi

Best features: Ultra-light weight (11 pounds with battery), low power consumption (5.3 Watts), automatic "power saving function," sound-on-sound audio dubbing, full special effects (slow motion, freeze frame, frame-by-frame advance, *and* speed scan), 13-function remote controller, SL, LP, and EP (same as SLP) recording and playback speeds, solenoid controls, 8-event/14-day microcomputer programmability, timer back-up circuit, good picture quality.

Drawbacks: Just one—special effects work in 6-hour EP mode only.

Just a few years ago, when the first 20-plus pound, one-speed-only VCR portables were coming onto the market, a mobile VCR as ultra-light and as feature-packed as the VT-6500A was a pie-in-the-sky fantasy. But no more.

The fact that we could only find one flaw in this model that justified complaint is in itself remarkable and ought to give a clue as to our attitude toward the VT-6500. For quality, convenience, and all-round excellence, the Hitachi VT-6500A gets our highest recommendation.

This portable has virtually every feature we could have asked for on a VCR—portable or console—as well as a few features that maybe we should have thought of, but didn't. Anyway, Hitachi did.

Suffice it to say that all the basics are covered—all 3 tape speeds in both playback and record, solenoid transport controls, full-feature remote control, great programmability, and very high signal-to-noise ratios. The only thing we can really take Hitachi to task for is engineering this unit with the special effects (particularly the speed scan) on the 6-hour EP mode only. Hitachi obviously feels that EP is the speed that people will use the most. We don't agree. Image and sound quality in the 6-hour mode is so marginal on 2-video head recorders like the VT-6500A that we think most people end up using the LP 4-hour speed, even though they get ⅓ less on a cassette. This is doubly true with a portable VCR where you are limited to a couple of hours shooting time by your batteries. When you're making your own video recordings with a $1,000 portable and a $900 camera, your prime concern is image quality, not saving $3.50 on tape costs. Whether it's baby's first step or a nude picnic at Zuma Beach, you're going to want the sharpest, most vibrant image you can get. Which means shooting at SP, or at the slowest LP. But when you get home and want to play that favorite sequence in slo mo it's going to be no go with the VT-6500A.

What about those extra features we should have though about, but didn't? There's 3 we think are particularly useful:

● **Sound-on-sound recording.** This is the first VCR we've seen that *does* permit overdubbing the sound track (you may recall that in "Selecting Your VCR" we said you couldn't overdub, which is true—unless you own this Hitachi portable).

● **Timer back-up circuit.** A few VCRs have this feature, but it's rare. The back-up circuit keeps the timer ticking away through a momentary voltage interruption (which happens almost daily in some homes) or even an hour-long power failure. And the trouble with voltage interruptions, even very short ones, is that they cause your timer to reset to 12:00 AM. Which means that unless the interruption happens at mid-night, your entire timer/tuner program is effectively destroyed. You'll still record the channels you programmed, but at different times than you had in mind. Instead of *High Noon* off the "Late Show," for instance, you may get the last 27 minutes of the 6:00 AM "Farm Report" and the first 1 hour and 33 minutes of "Good Morning America." The timer back-up circuit is a feature we'd like to see on more deluxe VCRs with multi-program, multi-day tuner/timers.

● **Power saver.** The VT-6500A has one of the lowest power consumption figures we've ever seen for a portable VCR. Just 5.3 Watts compared to 9.6 Watts for Panasonic's PV-3200 and 9 Watts for JVC's HR-2200U. The advantage of low power consumption? The less power your portable uses, the longer your batteries will last. In addition, there's a "power saver" circuit that kicks in automatically if the unit has been left in pause for more than 5 minutes, reducing battery drain by 80%.

List price for the VT-6500A with its separate tuner/timer is $1,600.

Specifications

VT-6500A Portable Video Deck	
Recording	Rotary two head helical scan, azimuth recording
Cassette:	VHS-type video cassette tape 33.35 mm/sec (SP), 16.67 mm/sec (LP), 11.12 mm/sec (EP)
RF input	VHF 75 ohms UHF 300 ohms
Video input	0.5—2 Vp-p 75 ohms unbalanced
Video output	1 Vp-p 75 ohms unbalanced
Video S/N ratio	Better than 46 dB (SP), better than 43 dB (LP), better than 40 dB (EP)
Horizontal resolution	240 lines (SP), 230 lines (LP), 230 lines (EP)
Audio Response	50 Hz—10 kHz (SP), 50 Hz—7 kHz (LP), 100 Hz—6 kHz (EP)
Audio S/N ratio	Better than 43 dB (SP), better than 40 dB (LP), better than 40 dB (EP)
Power input	DC 12 V
Power consumption	5.3 W nominal (rec mode at DC 12 V)
Dimensions (WxHxD)	10-3/8″ x 4-5/16″ x 10-1/8″ (263 x 108 x 257 mm)
Weight	11 lbs/5 kg (approx.) (including Battery Pack VT-BP65A)

Hitachi VT-8500A VHS Console★ ★ ★

List price: $1,395
Manufacturer: Hitachi
Best features: "Video enhancer" sharpness control, automatic tape electronic tape indexing, auto rewind, full special effects including speed scan, 13-function remote control, LP, SP, and EP (or SLP) mode, 8-event/14-day programming, solenoid transport controls, lightweight.
Drawbacks: 2-video-head recording system, special effects in EP only.

Among the features we like about the VT-8500A are Hitachi's "Picture Sharpness Control" which allows fine tuning each channel for maximum sharpness, automatic tape indexing which electronically "codes" the start of each new recorded segment for easy location, and the 13-feature hand-held remote control. The most significant drawbacks? As with its sister VCR, the VT-6500A, this Hitachi's special effects—including that all-important speed scan—only work in the 6-hour mode.

This limitation wouldn't be so bad if the VT-8500A were a 4-video head VCR with a separate set of playback and recording heads optimized for the 6-hour EP mode. But it isn't. And while the picture quality of the VT-8500A is very good, it still can't compare to a 4-head unit in EP. What's more, for only a few dollars more you can buy a 4-video head, top-of-the-line model like Panasonic's PV-1750.

VT-8500A Specifications

Recording	Rotary two head helical scanning azimuth recording
Tape speed	33.35 mm/sec. (SP), 16.67 mm/sec. (LP), 11.12 mm/sec. (EP)
Cassette tape	VHS-type video cassette tape
Record/playback time	6 hours max. with T-120
RF input	VHF: 75 ohms, UFH: 300 ohms
RF output	Channel 3 or 4 (selectable) 75 ohms
Horizontal resolution	SP mode: 240 lines, LP mode: 230 lines, EP mode: 230 lines
Video s/n radio	SP mode: 46 dB, LP mode: 43 dB, EP mode: 40 dB
Audio s/n ratio	SP mode: 43 dB, LP mode: 40 dB, EP mode: 40 dB
Power requirement	120 V/60 Hz
Power consumption	38 W
Dimensions (WxHxD)	17-1/8" x 5-9/16" x 13" (435 x 145 x 330 mm)
Weight	24.2 lbs. (11 kg)

*Specifications are subject to change without notice due to product improvement.

Hitachi VT-8000A VHS Console★ ★

List price: $1,295
Manufacturer: Hitachi
Best features: "Feather-touch" solenoid controls, 3-speed recording and playback visual scan and special effects, auto rewind, 7-function remote control, light weight.
Drawbacks: No special effects or speed scan in SP and LP modes, 1-event programmability.

This is essentially the same VCR as the VT-8500A, but without 8-event programmability, electronic indexing, and a couple of other features. Our opinion? Go for the VT-8500A instead. The extra features are more than worth the $100 difference in list prices.

Victor Company of Japan, Ltd. (JVC)

For additional product information or a list of JVC video dealers in your area write or call the Customer Service Department at:
JVC
58-75 Queens Midtown Expressway
Maspeth, New York 11378
212-476-8300

JVC is a partially owned subsidiary of Matsushita Electric, but maintains its own research and manufacturing facilities. JVC is credited with pioneering the development of the VHS (Video Home System) format. JVC's original Vidstar VHS videocassette recorder was the first model to seriously challenge Sony's domination of the infant VCR market. Its advantage was a 2-hour recording time, compared to the Betamax's one-hour, and within a year over 15 companies had adopted JVC's VHS (Video Home System) format. Today the VHS format is outselling Beta in most areas of the country by a margin of 3 to 1. Ironically, in some regions consumers have come to view the Beta format—home video's first successful design—as an "offbeat" system.

JVC video equipment is widely distributed through video specialty shops, department stores, and is often available from mail-order houses and discount outlets.

JVC HR-6700U VHS Console★ ★ ★

List price: $1,350
Manufacturer: JVC
Best features: Exceptional picture quality at both SP and SLP speeds, full special effects (including 2X speed scan in SP, 3X speed scan in SLP), and 6-event/7-day electronic programming.
Drawbacks: No LP speed, no solenoid-activated transport controls.

If picture quality ranks at the top of your list of priorities, you owe it to yourself to take this model for a test drive. JVC, along with their parent, Matsushita Electric, pioneered the basic research on the VHS format, so it should come as no surprise that the HR-6700U and HR-7200U are among the most advanced VCRs on the market.

JVC's advertising proclaims "Four heads are better than two." In the case of the HR-6700U and HR-7300, we think they may be right.

Both these models use *two separate sets of recording/playback heads.* One pair is designed for optimum performance at the SP speed, while the second pair is intended specifically for use at the SLP (JVC uses the designation EP) mode. It works: the horizontal resolution, color purity, and signal-to-noise ratio for this unit are almost as good in the Standard Play mode as at the Super Long Play speed. Which is a complicated way of saying that even at SLP you get a damn fine picture (and cut your tape costs by ⅓ as well).

Special effects on the HR-6700U include the speed scan (although we wouldn't mind seeing something a little faster than 3X to really zip through those commercial breaks), variable slow motion, freeze frame, frame-by-frame advance, and a remote controller that operates the transport functions but not the tuner. There is a nifty electronic cue that halts the fast forward at the beginning of each new recording on a tape. You'll also find a host of other "automatic switching" features (such as a sensor that automatically selects the speed at which your cassette was recorded) that are so sensible was JVC's decision not to offer the LP speed on this model. Yes, we can understand their reasoning: With such excellent image quality in the SLP mode with its 6-hour recording time, who would want to bother with LP? Which is fine if you happen to be a hermit, or all your friends own HR-6700Us. But if neither of these descriptions fits you, then chances are someday you're going to want to swap tapes with a friend who has recorded the show you want to see at the LP speed. Sorry, Charlie, but then you're just out of luck.

Another inconvenience—and this one really falls into the category of nit-picking—is the old fashioned, piano key controls. With a machine that is state-of-the-art in so many ways, it's hard to understand why JVC went with the mechanical transport controls instead of the easy-to-use electric push buttons. But then at least these transport controls do have a memory circuit which lets you press rewind or fast forward immediately after the stop key has been depressed (and before the tape is fully unloaded from the heads).

The list price of $1,350 sounds a little intimidating. But don't take to too seriously. We found the rock-bottom low to be $795 from a NYC mail order house, with our local JVC dealer not far behind at $900.

Preliminary Specifications

Format	VHS standard
Recording system	4 rotary heads, slant azimuth, helical scan system
Video signal system	NTSC-type color signal
Tape width	½ inch
Tape speed	SP mode, 1.31 inches/second, SLP (EP) mode, 0.43 inches/second, At speed scan SP mode, 2.62 inches/second SLP mode, 1.31 inches/second
Maximum recording time	120 min. or 360 min. (with JVC T-120 video cassette)
Temperature	
Operating	41°F to 104°F
Storage	-4°F to 140°F
Antenna	75-ohm external antenna terminal for VHF, 200-ohm external antenna terminals for UHF
Channel coverage	VHF Channels 2 - 13 UHF Channels 14 - 83
VHF output signal	Channel 3 or Channel 4 (switchable; set to Channel 3 when shipped.) 75 ohms, unbalanced
Power requirement	120 V AC, 60 Hz
Power consumption	55 watts
Video	
Input	0.5 to 2.0 Vp-p, 75 ohms unbalanced
Output	1.0 Vp-p, 75 ohms unbalanced
Signal-to-noise ratio	More than 45 dB (Rohde & Schwarz noise meter)
Horizontal resolution	More than 240 lines
Audio	
Signal-to-noise ratio	More than 40 dB
Timer	24-hour digital indication, 6-way programmable
Dimensions	18-1/2" x 5-13/16" x 13-3/4"
Weight	31 lbs.

*Specifications subject to change without notice.

JVC HR-7300U VHS Console★ ★ ★

List price:	
Manufacturer:	JVC
Best features:	4-video-head recording system, SP, LP (play back only), and EP (JVC's designation for the 6-hour mode), 8-event/14-day programming, high-speed "shuttle search," electronic transport controls, wireless remote control, auto rewind, excellent picture quality.
Drawbacks:	No special effects (including freeze frame, frame-by-frame advance, and slow motion).

In reporting on the HR-6700U we grumbled about the piano-key controls, the lack of an LP speed, and even suggested that JVC speed up the shuttle search a little. And they did all these things. Unfortunately, for some reason we cannot fathom, they also dropped the special effects features—slow motion, frame-by-frame advance, and freeze frame. And what a shame. This could have been the one VCR with everything. Now it misses by so little.

If the special effects are irrelevant to you (most people don't use them much once the novelty wears off), you will find almost everything else you could ask for in the HR-7300U. The transport controls are solenoid-activated with a wireless remote controller, there's an LP mode for playing back Uncle Zeke's homemade tapes, and JVC has even upped the shuttle search speed from 3 times normal on the HR-6700U to 21 times normal (in the EP mode) on the HR-7300. That means you'll be able to shuttle through the typical 2-minute commercial break in less than 6 seconds. And, most important, the 4-video-head recording system that was pioneered by JVC will give you a picture in the EP mode that we think you'll find compares very favorably to LP and, even, SP recordings on 2-head VCRs.

If you're looking for tops in picture quality and convenience, and are willing to overlook the missing special effects, the JVC HR-7300U has our recommendation.*

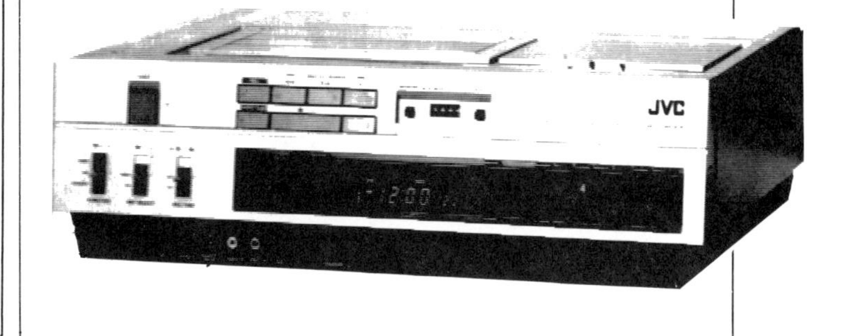

Preliminary Specifications

Format	VHS standard
Recording system:	Rotary, slant azimuth two-head helical scan system with two pairs of video heads, one pair exclusively for the SP mode and one pair for the EP mode.
Tape width	12.7 mm (½ inch)
Tape speed	(SP): 33.35 mm/s (1-5/16 ips) (1.31 ips) (LP): 16.67 mm/s (21/32 ips) (0.65 ips) (EP): 11.12 mm/s (7/16 ips) (0.43 ips)
Maximum recording time (SP)	120 min. with JVC T-120 videocassette (EP): 360 min. with JVC T-120 videocassette
Temperature	
Operating	5°C to 40°C (41°F to 104°F)
Storage	-20°C to 60°C (-4°F to 140°F)
Antenna	(VHF): 75 ohms, unbalanced, (UHF): 200 ohms, unbalanced
Channel coverage	(VHF): Channels 2 - 13 (UHF): Channels 14 - 83
Power requirement	120 V AC, 50 Hz
Video	
Input	0.5 to 2.0 Vp-p, 75 ohms, unbalanced
Output	1.0 Vp-p, 75 ohms, unbalanced
Signal-to-noise-ratio	More than 45 dB (Rodhe & Schwarz noise meter)
Horizontal resolution	More than 240 lines
Audio	
Signal-to-noise ratio	More than 40 dB
Timer	14-day programmable timer (8 programs)
Dimensions	440 mm(W) x 140 mm(H) x 330 mm(D) (17-3/8″x5-9/16″x13″)
Weight	10 kg (22 lbs)
Provided accessories	Channel number film, anntenna cable, F-type matching transformer

*based on the manufacturer's Preliminary Specifications

JVC HR-2200U VHS Portable★ ★

List price: $1,420 (including timer/tuner)
Manufacturer: JVC
Best features: Excellent picture quality, (Edit Start Control feature reduces video noise between "takes," ultra-lightweight and compact, electronic remote control, push-button solenoid transport, and special effects controls.
Disadvantages: One speed (Standard Play) only, programmability limited to 1 event.

Like one of its big brothers, JVC's HR-6700U console, the HR-2200U portable is state-of-the-art in just about every way, save one.

JVC has a reputation for getting high-quality images from its equipment, and the HR-2200U is no exception. The picture is as vibrant, sharp, and crisp as we've seen on any consumer portable. JVC claims a signal-to-noise ratio of more than 45 dB, and a horizontal resolution of over 240 lines, which puts this little 11.4-pound unit right in the same ball park as the top-of-the-line consoles when it comes to image quality.

The HR-2200U boasts one advantage we've never seen in a VCR—console or portable—before. Called ESC—for Edit Start Control—this feature promises to "eliminate noise at the intersegment gap." The ESC, in other words, gets rid of the glitches between "takes" for smooth transitions from scene to scene when shooting your own video footage.

Weighing less than an unabridged dictionary, and under 1 foot on its longest dimension, the HR-2200U compact and lightweight as any member of the current generation of portables. Part of the credit goes to a material known as FRP—fiber reinforced plastic—from which the chassis is molded.

Despite its diminutive dimensions, the HR-2200U is big on deluxe features and special effects. One we are particularly happy to see is the use of electronic push-button controls, which greatly enhance your ability to record live action sequences. Another plus is the remote control, which allows you to wear the unit on a backpack or shoulder cart.

Other noteworthy features: variable slow motion (1/6 to 1/30 normal speed), freeze frame, frame-by-frame advance, dew indicator, and a 10X speed scan (JVC calls it "shuttle search") for rapid viewing in forward and reverse. The Ni-Cad battery provides one hour of recording time, and can be recharged in 90 minutes. Additional batteries are available, if you plan to shoot more than one hour at a session. There is also an optional car-battery cord which allows the HR-2200U to be powered from any cigarette-lighter jack.

As for our one reservation about JVC's portable? If you are seriously considering this unit—and it's a good one if *you're* serious about amateur videography—understand that it has ONE SPEED ONLY. Yes, it will record off your TV, and do a magnificent job of it, but only for two hours with a standard T-120 cassette. Which means in many cases your tape will run out before the movie or sports event your trying to record is finished. So if your thing is recording long shows off the air-beware!

List price on the HR-2200U is $1,420 for the portable VCR and its separate tuner timer. We found them discounted as low as $945.

Preliminary Specifications

Power consumption	9.6 watts
Weight	5.2 kg (11.4 lbs.) including battery pack
Dimensions	11-5/16° x 4-1/16° x 10-9/16°
Temperature	
Operating	32° to 104°F
Storage	-4° to 140°F
Video recording system	Rotary, slant azimuth, two-head helical scanning system
Maximum recording time	120 min. (with JVC T-120 cassettee)
Rewind and fast forward time	Within 5 minutes (with JVC T-120 cassette)
Video signal	
Input	0.5 to 2.0 Vp-p, 75 ohms unbalanced
Output	1.0 Vp-p, 75 ohms unbalanced
Horizontal resolution	More than 240 lines (color mode)
Signal-to-noise ratio	More than 45 dB (Rohde & Schwarz noise meter)
Audio signal	
Signal-to-noise ratio	More than 40 dB
Accessories provided	Remote control unit, cable, earphone

Magnavox Consumer Electronics Company

For additional product information or a list of Magnavox dealers and service centers in your area call or write the Customer Service Department at:
Magnavox Video Systems
1700 Magnavox Way
Fort Wayne, IN 46804
219-482-4411

Over half of the VCRs sold in the United States are manufactured by Matsushita Electric—the world's largest consumer electronics company. Magnavox—which was acquired in 1981 by another multinational electronics giant, the $17 billion-a-year Dutch-based Philips Company—is one of the 8 brand names under which the Matsushita VCRs are sold in the U.S.

But while Magnavox offers just about the same product as Panasonic, Quasar, RCA, Curtis Mathes, J.C. Penney, Sylvania, and Philco, this brand offers two slight advantages to the consumer. First, Magnavox's list prices are on the low side—considerably lower than the other North American Philips subsidiaries, Sylvania and Philco. And second, Magnavox's service support system is among the best in consumer electronics.

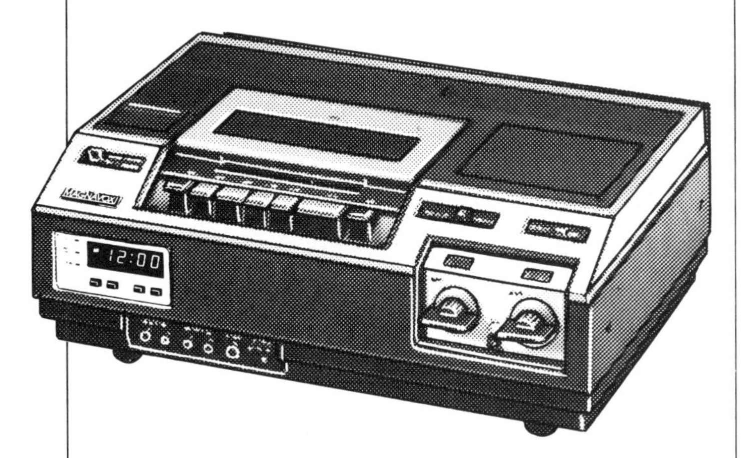

Magnavox 8310 VHS Console★

List price: $795
Manufacturer: Matsushita
Best features: SP, LP, and SLP recording and playback tape speeds, 1-event/24-hour tuner/timer, remote pause, DEW indicator, auto shut off.
Drawbacks: No special effects, mechanical-transport controls and tuner.

This is essentially the Panasonic PV-1210 with a remote pause feature. Quasar's VH-5020, RCA's VCT-500, and J.C. Penney's 5012 are also equivalent models. All are solid, economy-minded VCRs, but in today's world of "push button" controls the piano keys are a little passe. Our recommendation is to pass this model by in favor of the Magnavox 8315 or an equivalent. The price is the same, but the electronic controls with their logic memory circuit are a big improvement. Our survey found $625 to be the lowest discount price for 8310.

Magnavox 8315 VHS Console★ ★

List price: $795
Manufacturer: Matsushita
Best features: Same as the 8310, but has updated styling and electronic transport controls.
Drawbacks: No special effects, limited programmability.

An improved version of the 8310, and a good bet if price is your biggest consideration.

The electronic solenoid-assisted controls are easier to use, and are more trouble-free to boot. Features are very basic—but you do get all 3 speeds, a simple digital 24-hour timer that will record one program (or several consecutive programs on the same channel), and a remote pause that lets you edit out TV commercials. The unit is rugged, reliable, and the price is right. We'll repeat our single reservation, applicable to all no-frills VCRs—buy a basic model and it may not be long until you start wishing you had a couple of special effects and a little more flexible programming ability.

Equivalent versions of this basic VCR with electronic controls include the Quasar VH-5011 and RCA's VFT-190. If you are service-conscious, Curtis Mathes offers a near-identical model—the G-748—that is covered by its incredible 4-year warranty.

We expect discount prices on the 8315 to fall in about the same range as the 8310—about $625 to $675.*

*Report based on manufacturer's Preliminary Specifications.

Magnavox 8376 and 8377 VHS Portables★ ★ ★

List price: 8376 with 1-event/24-hour tuner/timer—$1,325
8377 with 8-event/14-day programmability—$1,400
Manufacturer: Matsushita

Here is the 8371 portable deck upgraded with the addition of 2 very important features—speed scan and remote control.

The 8376 is limited to automatically recording just 1 event in a 24-hour period. The 8377 has full 8-event/14-day programmability. A feature we think more the justifies the $75 difference between the two models.

The speed scan and remote control greatly enhance the convenience and usefulness of these mobile units, both in the field and back at home. We strongly recommend them over the Magnavox 8371, especially since the difference in price is almost negligible.

At discount you should expect to find these units available in the $925 to $1,025 range.*

*Report based on manufacturer's Preliminary Specifications.

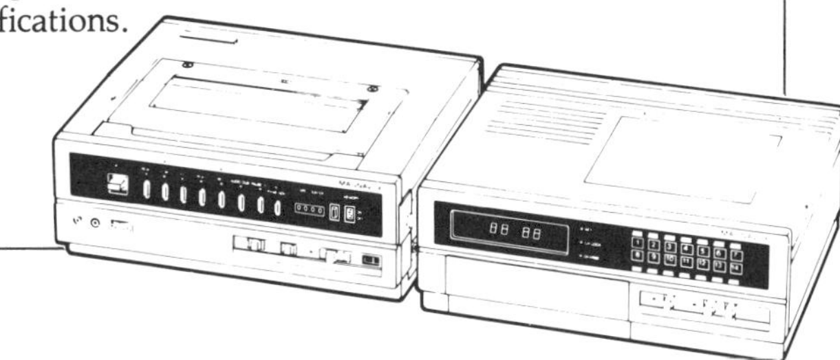

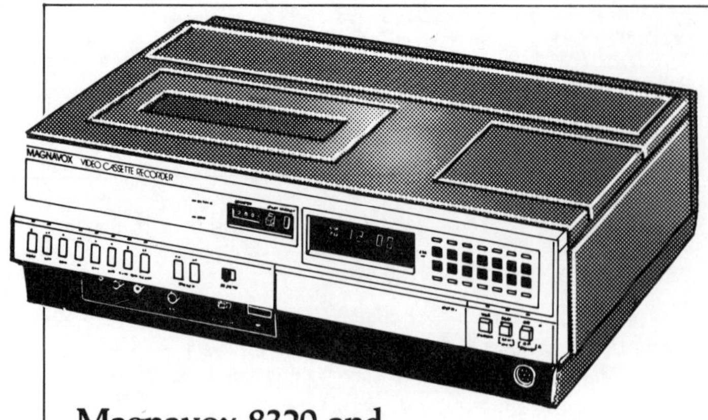

Magnavox 8320 and 8330 VHS Consoles★ ★ ★

List price: $995 for the 8320 and a 1-event/24-hour digital tuner/timer.
$1,145 for the 8330 with a 8-event/14- day progammable tuner/timer.
Manufacturer: Matsushita
Best features: See the Panasonic PV-1300 and PV-1400—they're virtually identical to the 8320 and 8330.

Here's the Matsushita mid-priced console with a Magnavox logo on it. The 8320 has a standard 1-event/24-hour tuner/timer, while the 8330 sports an 8-event/14-day programmable tuner/timer that will appeal to frequent travelers. Both units have the Matsushita speed scan—9 times normal speed in forward and reverse at the LP and SP modes—and hand-held remote controllers with 3 functions (channel selection, speed search, and pause). Transport functions are all operated by electronic pushbutton controls, and the random-access electronic tuner can be programmed for any 14 channels (UHF, VHF, and cable) and has Automatic Fine Tuning.

Equivalent models, in addition to the Panasonic prototypes, include the RCA VET-250, Quasar VH-5030, J.C. Penney 5013, RCA VET-450, Philco V 1550, and Sylvania VC 3100.

The biggest drawback? Same as the Panasonic models—no slow motion, or frame-by- frame advance.

Look for a model number change on these units in mid-1982. The 8320 will become the 8325 and the 8330 will be renumbered 8335. The new versions will offer a freeze frame mode and slightly higher list prices. Magnavox's 8325 will sell for $1,200, the 8335 will list for $1,325.

Magnavox 8340 VHS Console★ ★ ★ ★

List price: $1,395
Manufacturer: Matsushita
Best features: See Panasonic PV-1750

This is Matsushita's deluxe VCR in a Magnavox wrapper. And regardless of the name on the box—Panasonic, Magnavox, Curtis Mathes, Quasar, RCA, Sylvania, or Philco—we think this 4-video head recorder is one of the best there is.

The die-cast aluminum chassis, the direct-drive motors, the signal processing circuitry, and even the special effects are virtually identical to the Panasonic PV-1750 prototype. So is the price.

In the special-effects department there is speed scan (9X normal speed) in all three modes, and there is double speed, variable slow motion, freeze frame, and frame-by-frame advance in SP and SLP. The tuner is random-access electronic with Automatic Fine Tuning. There is 7-event/14-day programmability, with a backup circuit to protect your timer from resetting to 12:00 AM in the event of a power interruption.

As for picture quality, the 4-head system produces a crisp, stable image in SP and LP. Even at SLP the picture is quite good—comparable, we think, to the image quality you'll find at the LP mode on many of the 2-head models.

During 1982 the Magnavox 8340 will be renumbered. The new model—Magnavox 8345—will have similar features with a suggested price of $1,525.

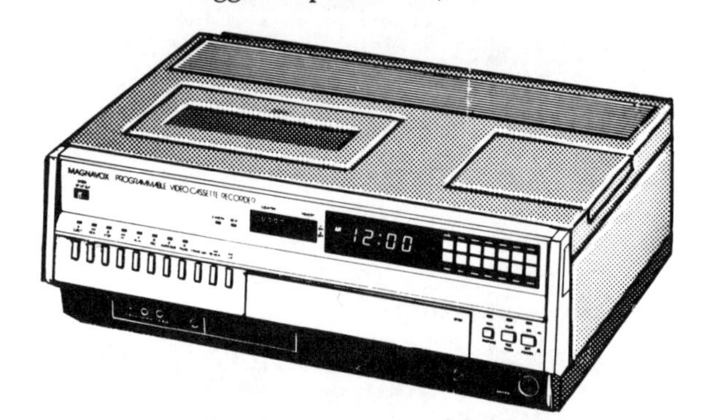

Magnavox 8371 VHS Portable★ ★

List price: $1,295
Manufacturer: Matsushita
Best features: See Panasonic PV-3200

Matsushita's standard portable with a separate 1-event/24-hour tuner/timer. We regret the lack of speed scan, slow motion, and remote control. All important in a portable. Nor are we wild about the fact that the freeze frame and frame-by-frame advance work only in the SLP mode, with its marginal picture quality.

If you do decide to buy this unit, don't order the Magnavox 8370 by mistake. True, these are identical portable decks, but the 8370 has no tuner/timer. You can play back what you've recorded with a camera, but you won't be able to tape shows off television.

Our discount shopping turned up a low price of $895 for the Magnavox 8371 portable VHS recorder.

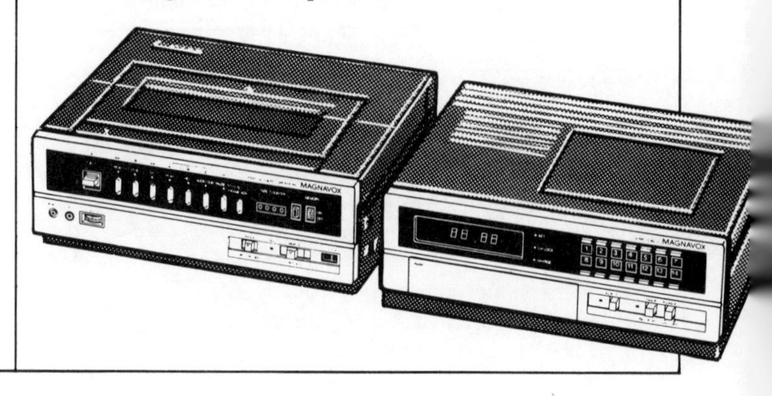

Mitsubishi Electric Sales America, Ltd.

For additional product information and a list of Mitsubishi dealers in your area write or call the Customer Services Department at:
Mitsubishi Electric Sales America, Ltd.
3030 East Victoria Street
Compton, CA 90221
213-537-7132

Mitsubishi Electric (not to be confused with Matsushita Electric) offers a line videocassette recorders that are noteworthy for their emphasis on styling and special-effects features. Mitsubishi manufacturers its own video products under license from Matsushita and JVC.

Mitsubishi HS-300U VHS Console★

List price: $1,450
Manufacturer: Mitsubishi
Best features: 5 direct-drive motors, full special effects (including 2 slow-motion settings, 15X speed scan, freeze-frame, and frame-by-frame advance), memory counter, DEW indicator, 6-event/6-day programmability, 15-function wireless remote control.
Drawbacks: No LP speed.

Although this model has been superseded by the HS-310U, you still may encounter it in some showrooms. And it's very tempting. The HS-300U is as loaded with extra goodies as any VCR on the market. Unfortunately, it also lacks one of the most fundamental features of all—the LP mode. Our opinion? It's an inconvenience, but you can live without LP in a 4-video head VCR with its superb image quality in the 6-hour mode. But in a 2-video head machine like the HS-300, the LP, or 4-hour mode, is absolutely essential. Unless you find this model at a fire-sale price you just can't refuse, we suggest you go for the HR-310U instead. It's got all the features—and all the speeds.

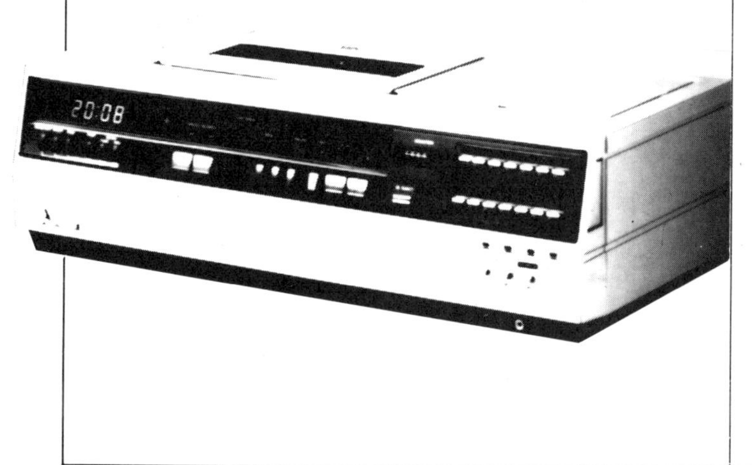

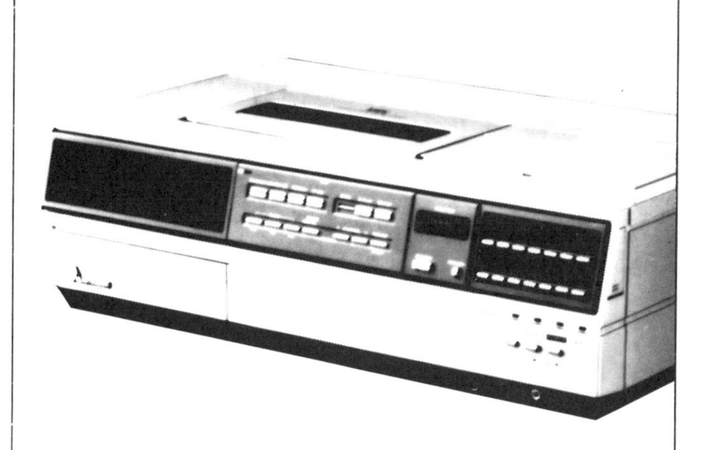

Mitsubishi HS-310U VHS Console★ ★ ★ ★

List price: $1,350
Manufacturer: Mitsubishi
Best features: 5 direct-drive motors, full special effects (including 2 slow-motion settings, 15X speed scan, frame-by-frame advance, and freeze frame), memory counter, DEW indicator, 8-event/14-day programmability, 14-function wireless remote control, and full range of VHS speeds.
Drawbacks: None.

The Mitsubishi HS-310U is one of the most feature-packed 2-video-head VCRs on the market.

We are particularly impressed by the picture stability and image quality on the HS-310U, which are exceptionally good for a 2-video-head unit. The SLP in mode, in particular, while not without some increased video noise, loss of resolution, and color bleeding, was still among the sharpest and the most stable we've seen in the 2-video head class. Part of the reason is Mitsubishi's system of 5 direct-drive motors. It's one of the most sophisticated in home video.

The convenience and special effects features on this model are also outstanding. There's slow motion at 1/3rd and 1/10th normal speed, a blazing 15X speed scan, freeze frame, frame-by-frame advance, and a 14-function wireless remote control to operate them all from your easychair. Even more important Mitsubishi has overcome a serious flaw in the previous version of this model by adding the LP speed. The microprocessor-assisted tuner/timer has also been upgraded to allow 8-event/14-day programmability.

Our opinion of the HR-310U? It's one of the best engineered and equipped 2-video-head VCRs around. One suggestion, however. Before you buy, compare prices with a full-feature, 4-video-head unit like Panasonic's PV-1750.

Specifications

Tape format	VHS
Power source	120 V AC: 60Hz
Power consumption	48 watts
Video recording system	2 rotary heads, azimuth helical scanning system
Audio track	1 track
Tape speed	1-5/16 ips (standard play), 7/16 ips (extended play)
Record/playback time	120 min. with T-120 cassette, 60 min. with T-60 cassette (timed is tripled in EP mode)
Fast forward/rewind time	Approx. 3 min. with T-120 cassette
Heads	
Video	2 rotary hot press ferrite heads
Audio/control	1 stationary head
Erase	1 full track: 1 for audio dubbing
Video input	0.5 to 2.0 Vp-p, 75 ohm unbalanced RCA pin plug
Video output	1.0 Vp-p, 75 ohm unbalanced RCA pin plug
Audio output	0dB, ik ohm unbalanced RCA pin plug
TV tuner	VHF input, Ch 2 - 13: 75 ohm unbalanced, F connector, UHF input, Ch 14 - 83: 300 ohm unbalanced
Horizontal resolution (monoscope test pattern)	More than 240 lines (SP), 220 (EP)
Signal-to-noise ratio	Video: better than 45dB Audio: better than 35dB
Operating temperature	41°F to 104°F
RF channel output	Channels 3 or 4, switchable
RF input/output	VHF: F connectors/75 ohm UHF: screw terminals/300 ohm
Weight	33 lbs.
Dimensions	19-3/8″ (W), 13-1/2″ (D), 6-1/4″ (H)
Standard accessories	75 ohm VHF output cable, 300 ohm UHF connector cable, dust cover
Clock	Digital clock, 24 hour display format
Channel programmer	6 programs, for any channels in a week/every day, 24 hour digital display
Clock accuracy	Synchronized with power line frequency
TV/video switch	Automatic
SP/EP switch	Playback: automatic, recording: manual

Panasonic Company

For additional production information or a list of Panasonic dealers in your vicinity, write or call the Customer Service Department at:
Panasonic Company
One Panasonic Way
Secaucus, NJ 07094
201-348-7000

Panasonic is a wholly owned subsidiary of Japan's giant Matsushita Electric Industrial Company, which has been involved in the development of videotape recording equipment since the early 1950s. Together with another subsidiary—The Victory Company of Japan (JVC)—Matsushita invented and holds patents on the principal components and signal processing techniques used in the VHS-format videocassette recorder.

As a rule, VCRs marketed by Panasonic are prototypes of similar models sold under 7 different brand names. These include Quasar (another Matsushita subsidiary), RCA, Magnavox, Sylvania, Philco, J.C. Penney, and Curtis Mathes. Despite the enormous demands placed upon Matsushita's VCR manufacturing facilities by the worldwide video explosion of the past several years, we've detected no systematic decline in the quality of any of these brands.

As for the differences between Panasonic and its "siblings": While the chassis and basic components seldom vary between equivalent models, there are often significant differences in styling, features (especially in programmability and special effects), and price.

Panasonic video products are among the most heavily advertised and widely distributed in the United States, and Panasonic maintains an extensive dealer network as well as a number of regional service centers. Panasonic VCRs and video cameras are also available from many mail-order houses and mass-merchandise discount outlets.

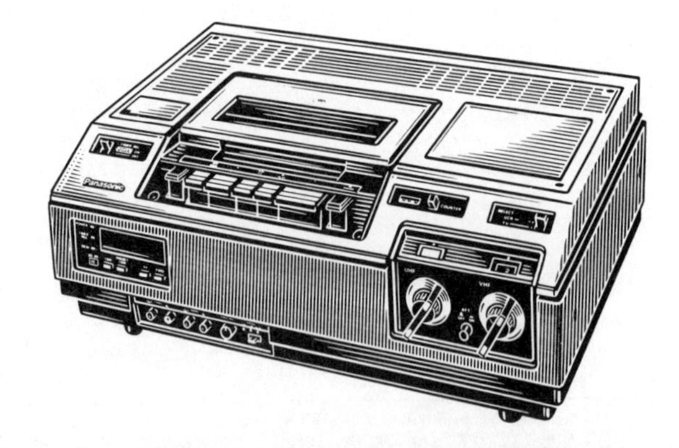

Panasonic PV-1210 VHS Console★ ★

List price: $995
Manufacturer: Matsushita
Best features: SP, LP, and SLP speeds, DEW indicator, auto shut off.
Drawbacks: No special features, special effects, or remote control. Marginal picture quality at SLP speed.

This is your basic, no-frills economy model. It lists for $995, but you can find it for as low as $625 in the discount outlets.

The picture quality at SP and LP is good, but we weren't too impressed by the SLP mode on the machines we've seen. Transport functions are operated by manual piano keys, there's no remote control, and channel selection is with the old-fashioned manual dials. Still, the PV-1210 is as good a way as any to get into home video on a tight budget—but you may soon outgrow its limited features. If the extra bucks can somehow be found, we recommend the PV-1400 or PV-1750 instead.

The PV-1210 is Matsushita's prototype for equivalent models including the Quasar VH-5020, RCA's VCT-500, the Magnavox 8310, and J.C. Penney's 5012.

Panasonic PV-1300 and PV-1400 VHS Consoles★ ★ ★

List price:	PV-1300 (with 1-event/24-hour digital tuner/timer)—$1,095 PV-1400 (with 8-event/14-day programmability)—$1,295
Manufacturer:	Matsushita
Best features:	9X speed scan (forward and reverse), good picture quality, streamlined styling, electronic controls and tuning, auto stop, remote control.
Drawbacks:	Limited special effects.

The PV-1300 and PV-1400 are actually the same model, sold with two different tuner/timers. The chassis and signal processing circuitry of this model represent Matsushita's prototype for no less than 10 other mid-priced VCRs, including units sold by Magnavox, RCA, Quasar, J.C. Penney, Curtis Mathes, Philco, and Sylvania.

The playback and recording system uses a 2-video-head system, and the image quality in the SP mode is among the best we've seen. Picture quality in the LP and SLP modes is also quite good, although the SLP mode is not as sharp or as noise-free as can be found in higher-priced, 4-video-head models like the Panasonic PV-1750. Since the PV-1300 and PV-1400 offer the full range of recording and play back speeds—SP, LP, and SLP—there is no problem viewing videocassettes recorded on any VHS format machine. A built-in sensor will automatically select the correct playback speed. There's also an auto-stop feature that will turn the VCR off when the end of the tape is reached.

Two of the biggest frustrations experienced by owners of the first and second generation of VCRs were trying to locate specific program segments and having to sit through all those ads in programs recorded off commercial TV. Why couldn't the VCR manufacturers give us a visible fast forward? Then we could scan through our tapes for the scenes we wanted to see. And we could use the same feature to make those long commercial breaks zip past in just a couple of seconds.

Enter the "Omni-Search." This is Panasonic's name for a speed scan that gives you a visual image at 9 times the normal viewing speed. True, you'll find a few stripes of brilliant white video noise across the screen when scanning, but who cares? You'll see enough to know what's going on. Commercials will whiz by like mini Chaplinesque comedies and you'll find yourself able to zero in on the part of the tape you want to watch, without spending half the evening hunting around for it. Once you've used speed scan, it's doubtful you'll ever want a machine without it.

Transport controls on the PV-1300 and PV-1400 are solenoid-activated and have a logic circuit that allows you to go directly from one function to another without having to "shift" through stop as with most mechanical piano key operated VCRs. A minor convenience, but nice. Other features include a pause circuit that backs the videotape up several frames to help reduce glitches when editing, a 3-function hand-held remote controller (search, pause, and channel selection), and a random-access electronic tuner with 14 channels and Automatic Fine Tuning.

The tuner/timer on the PV-1300 is a basic 1-channel, 1-day job. The PV-1400 is fully programmable for 8 different channels (or shows) over a two-week period. The added programmability will cost you $200 at list price. But we found the price difference dropped to $80 in the discount marketplace.

For a mid-priced unit there's really just one disadvantage to this machine. There are no special effects—slow motion, freeze frame, or frame-by-frame advance. For most folks these are just gimmicks anyway, played with a few times and soon forgotten. But if your purpose is more than just home entertainment, and you're looking for a VCR to use in sales presentations, analysis of sports events, and the like, the lack of special effects might prove troublesome. Also take note: The speed scan does not operate at the SP speed.

Look for a model number and price changes on these units during 1982. The PV-1300 will become the PV-1370 with a list price of $1,150. The PV-1400 will become PV-1470 with a suggested retail of $1,200. New features will include freeze frame and frame-by-frame advance.

Specifications

Power source	AC 120V 60 Hz
Power consumption	Approx. PV-1300; 53W, PV-1400; 55W
Video recording system	2 rotary heads, helical scanning system
Audio track	1 track
Tape speed	1-5/16 ips (33.35mm/s) standard play 6565 ips (16.675mm/s) long play 4375 ips (11.2mm/s) super long play
Record/playback time	360 min. with NV-T120
FF/rew time	Less than 6 min. with NV-T120
Heads	Video: 2 rotary heads Audio/control: 1 stationary head Erase: 1 full track erase 1 audio track erase for audio dubbing
Input level	Video: video in: 1.0 Vp-p 75 ohms unbalanced TV tuners: VHF; Ch2 - Ch12 75 ohms unbalanced UHF; Ch14-Ch83 ohms balanced
Output level	Video: video out; 1.0Vp-p 75 ohms unbalanaced
Weight	PV-1300; 27.5 PV-1400; 27.7
Dimensions (H x W x D)	PV-1300/PV-1400; 5-7/16″ x 19-3/16″ x 14-1/16″

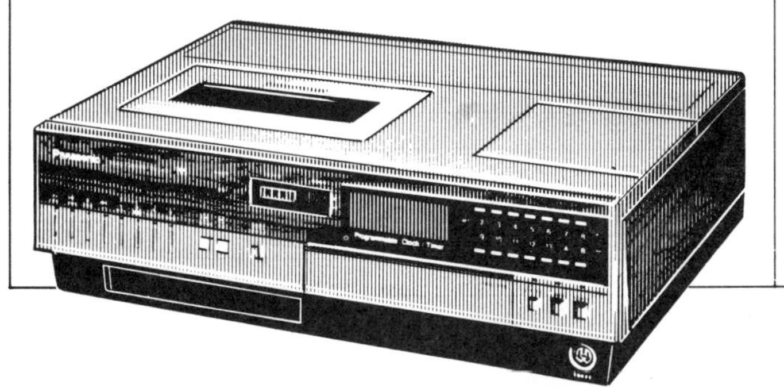

Panasonic PV-1750 VHS Console ★ ★ ★ ★

List price:	$1,440
Manufacturer:	Matsushita
Best features:	4-video-head recording system, aluminum die-cast chassis, long-life direct-drive motors, 8-event/14-day programmability, 14-channel random access electronic tuner with Automatic Fine Tuning, 11-function remote control, speed scan, special effects including slow motion, frame-by-frame advance, freeze frame, 2X normal speed quick advance, add-on recording, pause safety, DEW indicator, playback tape speed sensor, counter memory.
Drawbacks:	None

The PV-1750 gets our unqualified endorsement. It's one of the best-built, most feature-laden VCRs you can find. And its 4 top-of-the line clones—the Magnavox 8340, the RCA VET-650, the Curtis Mathes G-750, and the Quasar VH-5160—aren't bad either.

Our favorite feature is the 4-video-head recording system. The picture quality in all 3 modes—SP, LP, and SLP—is superb.

Next comes speed scan—9 times normal speed, and it works in every mode no less. There's also a separate "double speed" control. Other special effects include a 1/4th to 1/30th normal-speed variable slow motion, freeze frame, and frame-by-frame advance. While Panasonic may have been a little stingy with the special effects in some of their other models, they've been generous indeed with the PV-1750. Except for the speed scan, the special effects will work only in the SP and SLP modes, but that's almost irrelevant. Image quality at the economical SLP 6-hour mode is so good, you'll probably never use LP anyway, except to record music-oriented programs (the extra video heads still don't do much for the mediocre audio quality of the SLP speed).

During 1982 you'll find most dealers replacing the PV-1750 with the PV-1770. Features on the PV-1770 will be about the same, but the list price will go up to $1,600. As for the PV-1750, our discount shopping turned up a rock-bottom low of $1,050.

Specifications

Power source	AC 120V 60Hz
Power consumption	Approx. 62W
Video recording system	4 rotary heads, helical scanning system
Audio track	1 track
Tape speed	1-5/16 ips (33.35mm/s) standard play 21/32 ips (16.675mm/s) long play 7/16 ips (11.2mm/s) super long play
Record/playback time	360 min with NV-T120
FF/rew time	Less than 6 min. with NV-T120
Heads	Video: 4 rotary heads Audio/control: 1 stationary head Erase: 1 full track erase 1 audio track erase for audio dubbing
Input level	Video: video in; 1.0 Vp-p 75 ohms unbalanced Line in; -20dB 100k ohms unbalanced TV tuners: VHF; Ch 2-Ch 13 75 ohms unbalanced UHF: Ch 14-Ch 83 300 ohms balanced
Output level	Video: video out; 1.0Vp-p 75 ohms unbalanced
Weight (lbs)	33
Dimensions (H x W xc D)	6-3/8" x 18-15/16" x 14-5/8"

Panasonic PV-3200 VHS Portable ★ ★

List price:	$1,295 (including 1-event/24-hour digital tuner/timer).
Manufacturer:	Matsushita
Best features:	Good picture quality, light weight, SP, LP, and SLP speeds, electronic transport controls, memory rewind, add-on recording in pause, freeze frame and frame-by-frame advance in SLP.
Drawbacks	No slow motion, no speed scan, no AUDIO IN for dubbing sound or copying tapes, special effects in SLP mode only, no remote control.

The Panasonic portable PV-3200 is the prototype for the Magnavox 8371, the Quasar VH-5300, and the J.C. Penney 5503. It offers three strong advantages for those looking to buy a portable VCR: a full set of recording and play back speeds, easy-to-operate solenoid-activated transport controls, and picture quality that matches most 2-video head console models. You can operate the PV-3200 with AC current in the living room, from your car battery using an optional adapter, or with a rechargeable battery pack. Recording time with the battery pack is approximately 1 hour, depending on the power consumption of your video camera. A special circuit rewinds the tape a few frames in pause to help reduce glitches between recordings, or "takes," when you are using a video camera.

The PV-3200 is a little heavier (but only by about 2 pounds) than the ultra-lightweight models like RCA's VEP-170 or JVC's HR-2200U. But this portable's biggest drawback is not the extra couple of pounds, but the lack of special effects.

What we lament the most is the lack of slow motion. We can understand leaving slow motion off a console VCR. But on a portable it doesn't make sense. One of the big reasons for buying a portable video taping system is for things like analyzing a weak backhand, or trying to show a budding Tom Seaver why his fastball always ends up in the left field bleachers. (At least so we rationalize.) And one of the best tools for such video clinic enterprises is slow motion.

True, the PV-3200 does have a freeze frame and frame-by-frame advance—both good substitutes for a real slow motion. But we've got a bone to pick here, too. These features only operate in the SLP mode on the PV-3200. Which is where your color, detail resolution, and general image quality is the worst. Another gripe: There's no remote controller. It's sure a heck of a lot easier to use a portable in the field when there is a remote controller available. One more thing. Speed scan is a feature we love on a console, but it's probably even more important on a portable. The reason? Unless you're a cinematographic genius, we doubt even you are going to want to watch *everything* you shot last spring at the Renaissance Fair. Alas, you won't find a speed scan on the PV-3200 either.

The PV-3200 package, which includes a standard 1-event/24-hour tuner/timer, lists for $1,295. We found it for as low as $945. You can also buy the deck separately, list $1,150 and about $795 discount. A programmable 8-event/14-day tuner/timer lists for $395. We found the PV-3200 portable deck and the PV-A35P programmable tuner being sold together at one Miami video discount house for $1,090. A very good buy—if you don't need the special effects.

Specifications

Power source	Deck; DC 12V Tuner (PV-A32E); AC 120V 60Hz Adaptor (PV-A30); AC 120V 60 Hz
Power consumption	Deck; 9W at play mode, Tuner; 51W, Adaptor; 44W
Input level	Video; video in 1.0Vp-p 75 ohms unbalanced, TV tuners; VHF Ch 2-Ch 13 75 ohms unbalanced UHF Ch 14-Ch 83 300 ohms unbalanced
Output level	Video: video out 1.0Vp-p 75ohms unbalanced
Weight (lbs)	Deck; 13.2 (with battery), Tuner; 10.1, Adaptor; 5.0
Dimensions (H x W x D)	Deck; 4-1/2″ x 12-1/8″ x 9-4/5″, Tuner; 4-2/5″ x 11-1/2″ x 9-4/5″ Adaptor; 4-2/5″ x 4-1/5″ x 9-4/5″

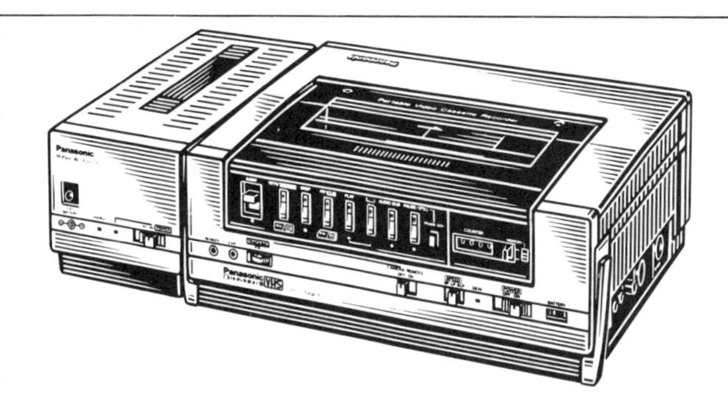

Panasonic PV-4100 VHS Portable★ ★ ★

List price: $1,195

This is the PV-3200 deck with the addition of that most convenient of the special effects features—speed scan. The PV-4100 comes with an AC adapter for living room playback, but not tumer/timer. An 8-event/14-day progammable tuner timer is available separately at a retail list price of $395.

Panasonic PV-1270 VHS Console★ ★

List price: No recommended list on this unit.
Manufacturer: Matsushita
Best features: Electronic transport controls, remote pause, DEW indicator, auto shut off, SP, LP, and SLP modes, 1-event/24-hour digital tuner/timer.
Drawbacks: No special effects, limited programmability, mechanical tuning.

This is the PV-1210 chassis updated with electronic "push button" transport controls and a remote pause. If economy demands you stick with a no-frills VCR this is not a bad package to consider. Equivalent models worth comparing include the Curtis Mathes G-748, Magnavox 8315, Quasar 5011, and RCA VFT-190. We think the electronic controls and the remote pause give this unit a significant edge over the PV-1210 and its equivalents.

As expected when flying economy class, the features are very basic—a simple digital tuner/timer that lets you record one event within 24 hours, a remote pause that's handy for editing out TV commercials if you are watching while you record, and an automatic shut off.

The 2-video-head recording system gives good results in SP and LP, and rather marginal picture quality in the SLP mode—which is typical of most VHS format machines with 2 video heads. There are no special effects to play with, which is probably no great loss. But there is also no speed scan on this VCR, which is a definite inconvenience.

The bottom line? A good choice if you're on a budget (depending on what kind of deal you can get). But beware. Buy an economy model now, and you may wish you owned something with more features later.

Specifications

Power source	AC 120V 60Hz
Power consumption	Approx. 34W
Video recording system	2 rotary heads, helical scanning system
Audio track	1 track
Tape speed	1-15/16 ips (33.35mm/s) standard play 21/32 ips (16.675mm/s) long play 7/16 ips (11.2mm/s) super long play
Record/playback time	360 min. with NV-T120
Heads	Video: 2 rotary heads Audio/control: 1 stationary head Erase: 1 full track erase 1 audio track erase
Input level	Video: video in; 1.0 Vp-p 75 ohms unbalanced, TV tuners: VHF; Ch 2-Ch 3 75 ohms unbalanced, UHF, Ch 14-Ch 8 75 ohms balanced
Output level	Video: video out; 1.0 Vp-p 75 ohms unbalanced, Rf modulated: Ch 3 or Ch 4 72 dB (open voltage) 75 ohms unbalanced
Weight (lbs)	30.8
Dimensions	6-7/8″ x 19-1/8″ x 15-1/2″ (H x W x D)
Accessories supplied	1 pc. remote control, 1 pc. 75 ohms one touch F-F coaxial cable, 1 pc. 300 ohms twn lead cable, 1 pc. 75 - 300 ohms VHF matching box, 1 pc. 300 -75 ohms VHF antenna adaptor

J.C. Penney Company

The VCRs you find in the TV department of the nation's third largest retailer may say Penney on the outside, but on the inside what you'll find is pure Panasonic. While Penney's prices don't match a mass volume discounter, they are at least lower than Panasonic's list prices. You'll also find Penney's VCRs are subject to frequent "price breaks" averaging about $150.

Here's a list of J.C. Penney's VCRs and their equivalent Panasonic models. For additional information on features and performance see the Panasonic listings.

J.C. Penney 5012 VHS Console★ ★

List price: $750
Panasonic equivalent: PV-1210

J.C. Penney 5013 VHS Console★ ★ ★

List price: $1,095.95
Panasonic equivalent: PV-1400

J.C. Penney 5503 VHS Portable★ ★

List price: $1,095
Panasonic equivalent: PV-3200

J.C. Penney 5507 VHS Portable★ ★ ★

List price: $1,250
Panasonic equivalent: PV-4100 (with tuner/timer)

Philco and Sylvania (N.A.P. Consumer Electronics Corp.)

For additional product information and a list of Philco or Sylvania dealers in your area call or write the customer service department at:
Sylvania, N.A.P. Consumer Electronics Corp.
700 Ellicott St.
Batavia, NY 14020
716-344-5200

Philco and Sylvania are both owned by North American Philips, and the VCRs sold by these companies are nearly identical to each other—as well as to their Panasonic prototypes. Pricing on the Philco/Sylvania models is also about the same as Panasonic, although discount prices are more difficult to find.

Here's a list of Philco and Sylvania models along with their Panasonic equivalents. For additional information on what to look for in the way of features and performance, see the Panasonic listings.

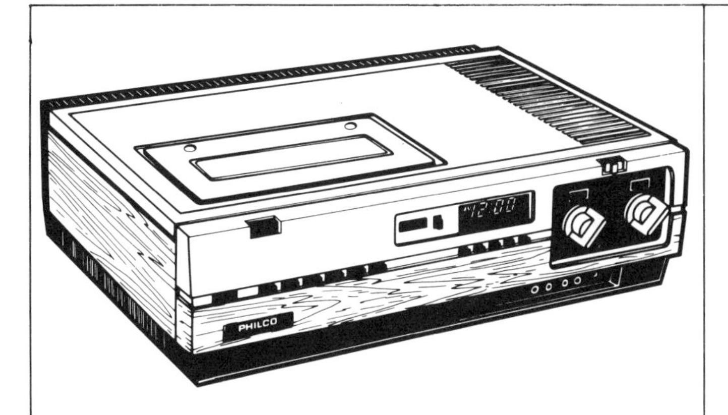

Philco V-1011 VHS Console★ ★
Sylvania VC-2210 VHS Console★ ★

List price: Approximately $850
Panasonic equivalent: PV-1270
Features: SP, LP, and SLP speeds, electronic function controls, mechanical 1-event/24-hour tuner/timer, wired remote pause.

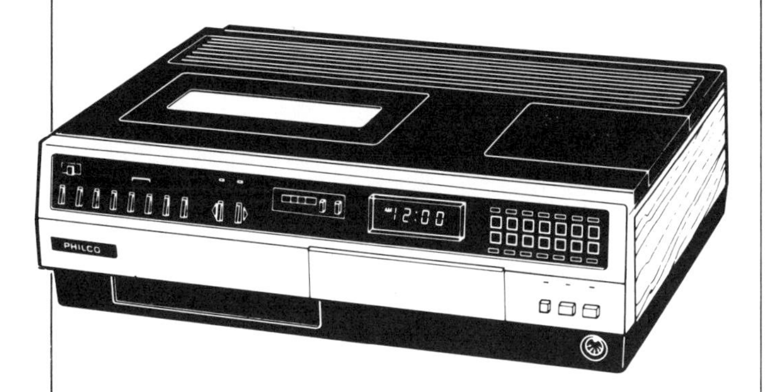

Philco V-1441 VHS Console★ ★ ★
Sylvania VC-2910 VHS Console★ ★

List price: $1,050
Panasonic equivalent: PV-1300
Features: SP, LP, and SLP speeds, electronic function controls, electronic 1-event/24-hour tuner/timer, speed scan, freeze frame, wired remote control. Infra-red wireless remote available as an option.

One small but significant advantage over the Panasonic PV-1300, the Philco and Sylvania models are both equipped with freeze frame.

Philco V-1151 VHS Console★ ★ ★
Sylvania VC-3110 VHS Console★ ★ ★

List price: $1,200
Panasonic equivalent: PV-1400
Features: SP, LP, and SLP needs, electronic function controls, electronic 8-event/14-day programmable tuner/timer, speed scan, freeze frame, wired remote control. Infra-red remote control available.

Again, the Philco and Sylvania models have freeze frame, the PV-1400 does not.

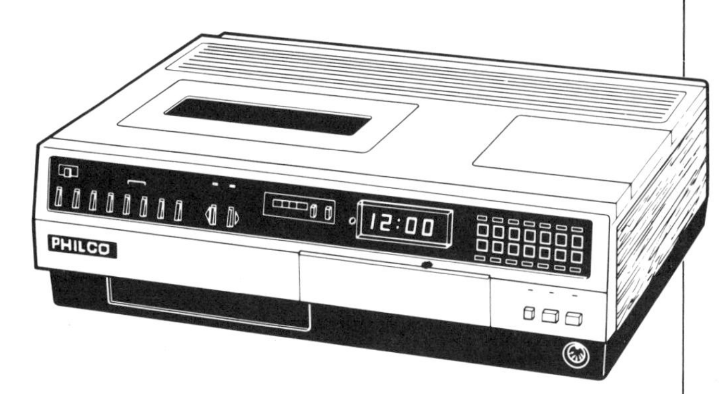

Sylvania VC-3610 Console★ ★ ★ ★

List price: $1,500
Panasonic equivalent: PV-1750
Features: SP, LP, and SLP needs, electronic function controls, electronic 8-event/14-day programmable tuner/timer, speed scan, freeze frame, frame-by-frame advance, variable slow motion, double speed forward, 4 video heads, built-in infra-red remote control.

A top-of-the-line unit with a price tag to match. The built-in frame infra-red remote control is a feature you won't find on the PV-1750, although it is available as an option of the Quasar 5610.

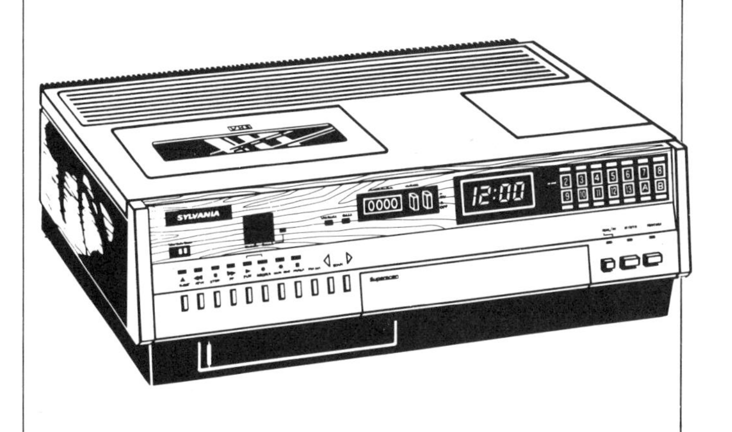

Quasar Electronics Corp.

For additional product information and a list of Quasar dealers in your vicinity call or write the Customer Service Department at:
Quasar Electronics Corp.
9401 W. Grand Avenue
Franklin Park, IL 60131

Quasar, like its sister company, Panasonic, is a subsidiary of Matsushita. With the exception of a few minor features such as the wireless remote control on Quasar's deluxe VCR, the Quasar models are virtually identical to their Panasonic prototypes. Quasar prices are also in the same ball park as Panasonic. For a complete discussion of features and performance please turn to the Panasonic listings.

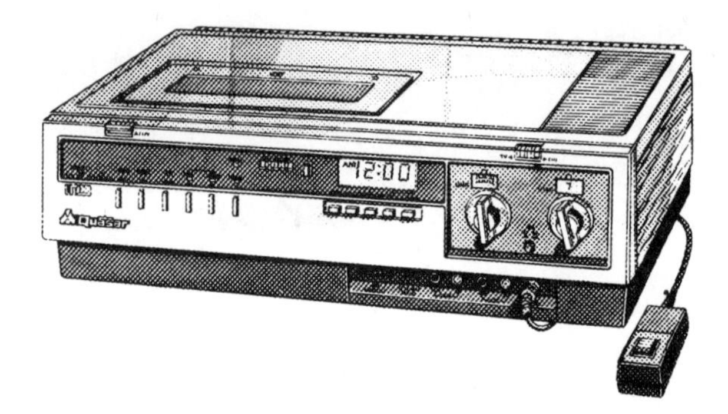

Quasar 5011 VHS Console★ ★

List price:	Approximately $900
Manufacturer:	Matsushita
Features:	SP, LP, and SLP speeds, electronic function controls, manual tuning, 1-event/24-hour tuner/timer, DEW indicator, remote pause. See Panasonic PV-1270.

Quasar 5210 VHS Console★ ★

List price:	$1,100
Manufacturer:	Matsushita
Features:	SP, LP, and SLP speeds, electronic function controls and 1-event/24-hour tuner/timer, speed scan, optional multi-function remote controls, standard remote control with speed scan, pause, and channel change, auto tape rewind. See Panasonic PV-1300.

Nearly identical to the PV-1300 in styling, features, and performance. One minor difference—the Quasar 5210 has an automatic tape rewind feature not found on the Panasonic model. Whether this is any advantage is debatable, since the PV-1300 is equipped with an auto shut-off instead. If, like some of the folks we know, you have a propensity to nod off before the program's over, the auto shut-off will save you both electricity and wear and tear on your machine.

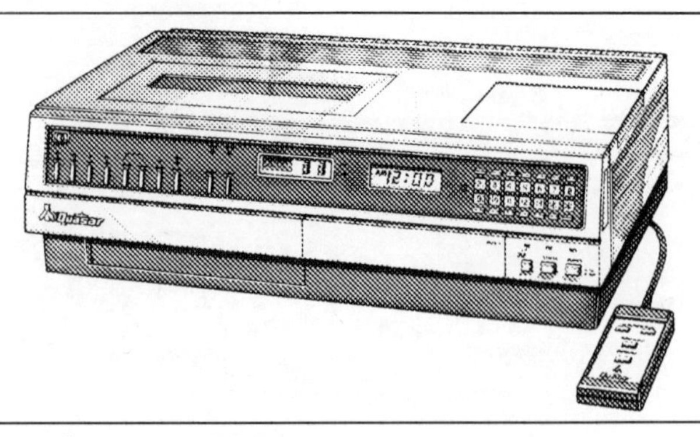

Specifications VH-5011

Video recording system	2 rotary heads, azimuth, helical scanning system
Audio track	1 track
Tape speed	$1\frac{5}{16}$ i.p.s. (standard play)
Record/playback time	1, 2 or 3 hours with VC-T60 cassette; 2, 4 or6 hours with VC-T120 cassette
Fast forward/rewind time	Less than 6 min. with VC-T120
Heads: Video	2 rotary hot press ferrite heads
Audio control	1 stationary head
Erase	1 full track, 1 for audio erasing
Video input	1.0V p-p, 75 ohm unbalanced RCA pin plug
Audio input	Line: -20 db, 100K ohm unbalanced, RCA pin plug
TV tuners	VHF input, Ch2-13; 75 ohm unbalanced "F" connector, UHF input, Ch14-83; 300 ohm balanced
Video output	1.0V p-p, 75 ohm unbalanced, RCA pin plug
Audio output	Line: -6 db, 600 ohm unbalanced, RCA pin plug
Horizontal resolution (monoscope test pattern)	Color and b/w: more than 230 lines
Audio frequency response	SP: 100-8000 Hz (-10db) LP: 100-6000 Hz (-10db) SLP: 150-5000 Hz (-10db)
Signal-to-noise ratio	Better than 40 db (Luminance by Rohde & Schwarz Noise Meter) Audio: better than 43 db (SP mode)
Operating temperature	41°F to 104°F
RF channel output	Channels 3 or 4, switchable
RF input/output	VHF: F connectors/75 ohm, UHF: screw terminals/300 ohm
Weight	21 lbs.
Dimensions	19" (W), $14\frac{1}{2}$" (D), $5\frac{3}{8}$" (H)
Standard accessories	Remote pause control with 20 ft. cable; 5-foot, 75 ohm VHF output cable; 5-foot, 300 ohm UHF connector cable; 75/300 ohms VHF matching transformer; 300/75 ohm VHF antenna adaptor

Quasar 5310 VHS Console★ ★ ★

List price:	$1,200
Manufacturer:	Matsushita
Features:	SP, LP, and SLP speeds, electronic function control and 8-event/14-day programmable tuner/timer, speed scan, freeze frame, auto rewind, standard 3-function remote controller, optional 9-function wired and wireless remote control. See Panasonic PV-1400.

If you're a sports fan, or plan to use your VCR for business or educational presentation purposes, you'll want to note that the 5310 is equipped with freeze frame. The Panasonic PV-1400 is not.

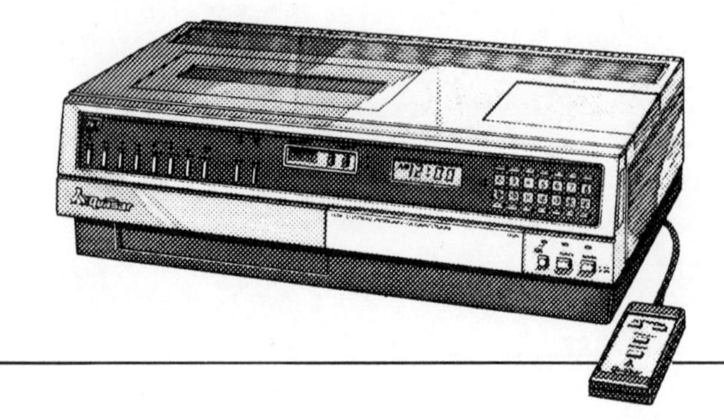

Quasar 5610 VHS Console ★ ★ ★

List price: $1,500
Manufacturer: Matsushita
Features: Identical to the PV-1750, with the addition of a multi-function wireless remote control.

Regardless of the brand name on the case—Panasonic, RCA, Sylvania, Curtis Mathes, Quasar, or Magnavox—inside you'll find Matsushita's top-of-the-line VCR. We think it's among the best money can buy.

One suggestion. Before you buy the 5610 be certain to do a little comparison shopping. The Magnavox 8340, which is identical except for the wireless remote control, can be purchased at discount outlets like J&R Music World in Manhattan for $990. The RCA 650, also identical save for the wireless remote control, discounts for $1,050, as does the Panasonic PV-1750.

Specifications VH-5610

Power consumption	Approx. 62 watts
Video recording system	4 rotary heads, azimuth, helical scanning system
Audio track	1 track
Tape speed	1⁵⁄₁₆ i.p.s. (standard play)
Record/playback time	1, 2 or 3 hours with VC-T60 cassette; 2, 4 or 6 hours with VC-T120 cassette
Fast forward/rewind time	Less than 5 min. with VC-T120
Heads: Video	4 rotary hot press ferrite heads
Audio/control	1 stationary head.
Erase	1 full track; 1 for audio dubbing
Video input	1.0V p-p, 75 ohm unbalanced, RCA pin plug
Audio input	Line: -20db, 100K ohm, unbalanced, RCA pin plug; Mic: -70db, 600 ohm unbalanced, ¼" phone jack
TV tuners	VHF input, Ch 2-13/mid/super band; 75 ohm unbalanced "F" connector; UFH input, Ch 14-83: 300 ohm balanced
Video output	1.0V p-p, 75 ohm unbalanced, RCA pin plug
Audio output	Line: -6 db, 600 ohm unbalanced, RCA pin plug
Horizontal resolution (monoscope test pattern)	B/W: more than 270 lines; Color: more than 230 lines
Audio frequency response	SP: 100-8000 Hz (-10 db) LP: 100-6000 Hz (-10 db) SLP: 100-5000 Hz (-10 db)
Signal-to-noise ratio	B/W better than 40 db (Rohde & Schwartz Noise Meter); Audio: better than 42 db (SP mode)
Operating temperature	41°F to 104°F
RF channel output	Channels 3 or 4, switchable
RF input/output	VHF: F connectors/75 ohms UHF: screw terminals/300 ohm
Weight	33 lbs.
Dimensions	18⅞"W, 14⅝"D, 6⅜"H
Standard accessories	VC-T60 1/2/3 hr. video cassette special effects Infra-Red remote control; 5-foot, 75 ohm VHF output cable; 5-foot 300 ohm UHF connector cable; 75/300 ohm VHF matching transformer; 300/75 ohm VHF antenna adaptor

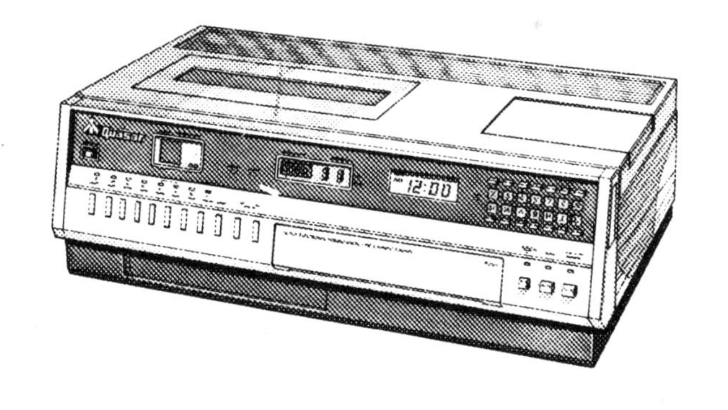

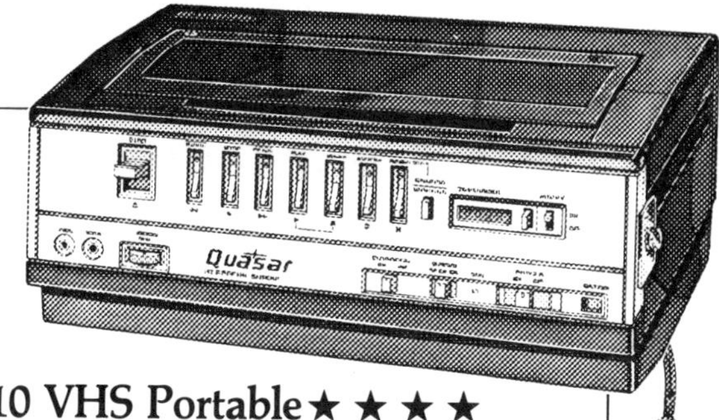

Quasar 5410 VHS Portable ★ ★ ★

List price: $1,000 + tuner.
Manufacturer: Matsushita
Best features: Lightweight, slow motion, speed scan, freeze frame, SP, LP, and SLP speeds, remote control, electronic function control, battery meter.
Drawbacks: No serious disadvantages

The Quasar 5410 is similar in design and appearance to the Panasonic PV-4500 portable (without tuner). But the Quasar model offers one big advantage over the Panasonic version—slow motion. In our opinion the Quasar 5410, along with the RCA VFP-170, is one of the best portables around.

For off-the-air recording two optional tuner/timers are available for the 5410. The Quasar VA-512 is a standard 1-event/24-hour digital unit and lists for $250. With the VA-520 tuner/timer you can program the Quasar 5410 portable to record up to 8 events during the 14-day period. The price is $350.

Except for the lack of 4-video-head recording quality, the 5410, when coupled with its VA-520 random access electronic tuner/timer, is a close match for any top-of-the-line VCR console when it comes to special effects and features. Not bad, considering you also have the added versatility of battery-operated location recording. If you're searching for a portable VCR which also offers deluxe console features, this one deserves your undivided attention.

Specifications VP-5410

Power source	12V DC or built-in battery
Power consumption	Approximately 9.4 watts (playback mode)
Video recording system	2 rotary heads, azimuth, helical scanning system
Audio track	1 track
Tape speed	1⁵⁄₁₆" ips (standard play)
Record/playback time	1, 2 or 3 hours with VC-T60 cassette; 2, 4 or 6 hours with VC-T120 cassette
Fast forward/rewind time	Less than 5 min. with VC-T120 cassette
Heads: Video	2 rotary hot press ferrite heads
Audio/control	1 stationary head
Erase	1 full track; 1 for audio dubbing
Video input	1.0V p-p, 75 ohm unbalanced, RCA pin plug
Audio input: camera	-20 db, 10K ohm unbalanced (included into 10 pin camera jack); Mic: -70 db, 4K ohm unbalanced Mini Jack; Note: Use attached attenuator cable for input audio signal when recording from audio equipment.
Video output	1.0V p-p, 75 ohm terminated RCA pin plug
Audio output	Line: -6 db, 600 ohm unbalanced, RCA pin plug
Horizontal resolution	B/W/Color: more than 230 lines
Audio frequency response	SP: 100-8000 Hz (-10 db) LP: 100-6000 Hz (-10 db) SLP: 100-5000 Hz (-10 db)
Signal-to-noise ratio	More than 40 db (Luminance by Rohde & Schwarz Noise Meter) Audio: better than 43 db (SP mode)
Operating temperature	32°F to 104°F
Relative Humidity	10% to 75%
RF channel output	Channels 3 or 4, switchable
RF input/output	VHF: F connectors, 75 ohm
Weight	Approx. 12 lbs. (approx. 13½ lbs. with battery)
Dimensions	4¼" high, 11½° wide, 9¾° deep
Standard accessories (included)	VC-T60 cassette, battery pack, shoulder strap, earphone, special effects remote control, battery connector cord, 75 ohm VHF output cable, 75/300 ohm VHF matching transformer, mic attenuator, mic plug matching adaptor

RCA Corporation

For additional product information or a list of RCA dealers and service centers in your area call or write the RCA Customer Service Department at:
RCA Consumer Electronics
600 North Sherman Drive
Indianapolis, IN 46201
317-277-5000

A couple of lean years at NBC, RCA's television network, combined with a big corporate commitment to the videodisc, has resulted in some financial lumps for RCA, including the lowering of RCA Corp.'s bond, commercial paper, and preferred stock ratings by Standard & Poor. But for the VCR consumer, RCA's line of Matsushita-built machines offer lower than average list prices with healthy discounting, and above-average customer service on VCR repairs.

RCA VFT-190 VHS Console★ ★

List price: Approximately $900
Manufacturer: Matsushita
Best features: Identical to the Panasonic PV-1270 (as well as the Magnavox 8315, Quasar 5011, Curtis Mathes G-748, J.C. Penney 5012, Philco V-1011, and the Sylvania VC-2210

The VFT-190 represents Matsushita's latest generation basic VCR. The mechanical piano key controls found on older no-frills models such as the RCA 500, Quasar 5020, and Magnavox 5310 have been replaced with state-of-the-art "soft touch" electronic switches. The result is increased reliability and ease of operation. Otherwise, the VFT-190 offers the basic VCR functions at a basic price.

But in VCRs, basic does not necessarily mean better. Video retailers report the most frequently requested VCR feature is speed scan. Once you've used this feature to locate the beginning of a movie buried in the middle of a 6-hour tape, or to zip through 3 minutes of commercials in 20 seconds, you'll understand why. But speed scan isn't available on a basic unit such as the VFT-190 or Panasonic PV-1270.

If economy is your biggest concern, then by all means check out the VFT-190 and its identical twins. But take heed. For just a few dollars more you can get a VCR that will provide far greater convenience and versatility.

RCA VFT-250 VHS Console★ ★ ★

List price: $995
Manufacturer: Matsushita
Best features: Same as Panasonic PV-1300, except for freeze frame and auto rewind.

You'll find the RCA 250 in discount stores for about $750—a few dollars less than the Panasonic PV-1300 and Magnavox 8320. Not a bad buy either, considering the RCA model has auto rewind and freeze frame capability, which its near-identical Panasonic and Magnavox twins do not.

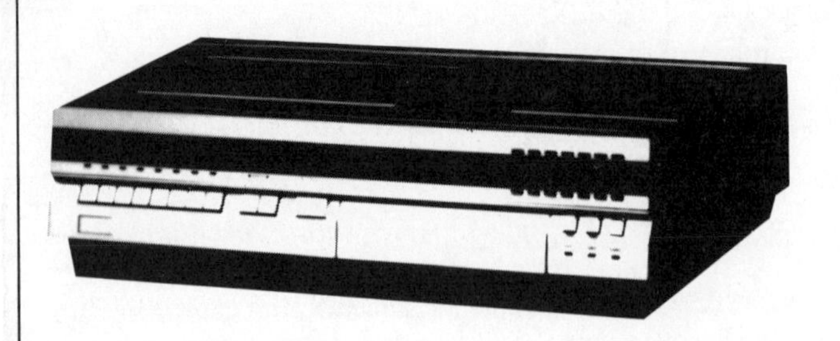

The picture quality of the RCA 250 and 450 models is excellent in the SP and LP modes. As with most 2-video head VCRs, image quality in the SLP mode is erratic. Sometimes it's quite good, sometimes it isn't.

The only serious disadvantages to the RCA 250 is the lack of a microprocessor-assisted programmable tuner/timer. For that feature you will have to move up to the RCA 450.

RCA VFT 450 VHS Console★ ★ ★

List price: $1,100
Manufacturer: Matsushita
Best features: Same as the Panasonic PV-1400, with the addition of freeze frame and automatic rewind.

The RCA 450 can be found in discount outlets for as low as $845, slightly less than its Panasonic and Magnavox equivalents. And once again, as with the RCA 250, this mid-priced programmable Matsushita-built VCR is equipped with a freeze-frame mode and automatic rewind while its Panasonic and Magnavox siblings are not. In our opinion the RCA 450 is a good buy if you're looking for a maximum in quality, features, performance, and service at a minimum price.

RCA VET 650 VHS Console★ ★ ★ ★

List price: $1,350
Manufacturer: Matsushita
Best features: Virtually identical to the Panasonic PV-1750

Whatever the wrapper—Panasonic, Magnavox, Quasar, Sylvania, Curtis Mathes, or in this case, RCA's 650 VHS Console—we think Matsushita's 4-video-head machine is just about the best VHS videocassette recorder on the market today. It's hard to imagine how Matsushita can improve on the quality and features of this model, although someday we suspect it will. Nevertheless, here's a VCR that should give you many years of superb viewing performance and convenience.

In addition to the Panasonic PV-1750, other equivalent versions of this unit include the Magnavox 8340, the Curtis Mathes G-750, the Quasar VH-5160, and the Sylvania VC-3610 (see Philco/Sylvania). As it happens, with a list price of $1,350 the RCA VET 650 is the least expensive of the lot. Discount shoppers will find the RCA 650 available for $1,050, which puts it within a few dollars of the Panasonic PV-1750 ($1,040) and Magnavox 8340 ($990) at the mass volume outlets.

SelectaVision 650

6-Hour Video Cassette Recorder with Special Effects and Picture Search

Adoring prose is typical of most specification sheets. But you'll also find a wealth of useful information about a VCR's features and how they work.

High-Speed Picture Search by remote control

This new SelectaVision remote control feature lets you review recorded material in a fraction of the time it would normally take. You can scan in either direction—forward or reverse.

Special multi-speed playback effects

SelectaVision 650 gives you a wide choice of special effects, in either the Standard Play or Super Long Play mode, all of which are activated by remote control. Remote hand unit comes with 20-ft. cord.

Double speed. Lets you review taped material in half the normal time, skimming through unwanted segments.

Slow motion. Slide control varies speed from 1/4 to 1/30 of normal.

Stop action. Freezes the action so you can examine details easily missed in a moving picture.

Single frame advance. Moves the image—one frame at a time—until you find exactly the one you want.

Other remote control functions:

Play, Record, Stop, Fast Forward and Rewind.

Pause/Still. Lets you stop and start the tape at the touch of a button during recording or playback.

Channel Change. Enables you to change channels remotely—during normal viewing as well as when recording—even on a manually tuned TV. Safety lock prevents accidental channel change during off-the-air recording.

Model VET650

14-day electronic programmer

A microprocessor circuit and special timer display let you program the shows you want to watch *up to two weeks in advance.*

You can preset the VET650 to record as many as eight different programs even on different channels. Or you can record the same program automatically every day for as long as the tape lasts. And programming is simple. Just feed the memory circuits these four pieces of information: 1) day of program; 2) time program starts; 3) time program ends; 4) program channel.

Electronic touch-button tuning

Deluxe tuner is electronic for smooth, quiet, instantaneous operation. Selector positions are easily set up to receive any 14 of the 82 VHF and UHF channels, in any sequence desired. Indicator light tells which channel is being recorded.

Automatic tape rewind

SelectaVision rewinds your video cassettes automatically when end of tape is reached. (Automatic rewind functions in all modes except timer mode.)

Tracking Control

Fine tunes your SelectaVision unit for the best playback of tapes recorded on other VHS machines.

Audio dub

Lets you use an optional microphone to put your own soundtrack on a previously recorded cassette. Dubbing automatically erases any previous soundtrack.

Camera and microphone input jacks

Both are located on the front of the unit. When not in use, a small panel slides into place to hide them from view.

Tape counter with memory

Four-digit counter makes it easy to find the start of a program. Flip on the memory switch, set the counter to zero before you record, and tape will stop at that point automatically when you rewind.

Soft-touch function controls

The new SelectaVision VET650 features soft-touch electronic controls that let you smoothly activate all primary VCR functions. Operation is convenient, reliable and virtually instantaneous. Most controls feature lighted indicators.

Automatic TV/VCR switching

When VCR is turned off, unit goes to the TV mode automatically to prevent interference with normal TV viewing.

DEW moisture control

Prevents operation when excessive moisture might cause the tape to stick to the headwheel. Indicator light signals when control is operative.

Direct-drive tape transport system

Two direct-drive motors provide precise speed control to maintain a steady picture during recording and playback. The headwheel motor is deactivated during "Fast Forward" and "Rewind" to reduce wear. The capstan motor keeps the tape speed steady, minimizing audio wow and flutter.

Scene transition stabilizer

Minimizes picture break-up between scenes edited with VCR's pause control or recorded with video camera, so you get smooth transitions between individual shots.

SPECIFICATIONS:

Video Recording System:
Rotary four-head helical scan

Video Signal:
EIA standard: NTSC color

Antennas:
VHF—75 ohm ext. terminals
UHF—300 ohm ext. terminals

Fast Forward/Rewind Time:
4 minutes with VK250 cassette

Power Requirements:
110 to 130 volts, 60Hz, AC

Power Consumption: 62 watts

Weight: 33 lbs.

Dimensions: H-6⅜", W-19", D-14¾"

Optional Cameras

- **CC007 color camera** with professional "mini-cam" look and lightweight design, flip-over electronic viewfinder, 6:1 power zoom lens, fade in/out button, boom microphone, and easy-grip handle.
- **CC005 color camera** with electronic viewfinder, 3:1 zoom lens, built-in microphone, trigger-grip handle, and recording pause control.

Advertisement Courtesy ©RCA

RCA VFP 170 VHS Portable★ ★ ★ ★

List price:	$1,400
Manufacturer:	Matsushita
Best features:	SP, LP, and SLP speeds, electronic function controls, separate 8-event/14-day tuner timer, multi-function remote control, speed scan, slow motion, freeze frame, frame-by-frame advance, battery meter, ultra-lightweight.
Drawbacks:	None.

The most advanced portable in home video. It weighs a mere 11 pounds, but when coupled with its 8-event/14-day tuner/timer the VFP 170 offers virtually every feature you'll find on a deluxe console like RCA 650, save the 4-video-head recording system. If you want the maximum in features, plus the versatility of a portable, the VFP 170 is a hard act to beat. It gets our vote.

Specifications

Video recording system	VHS; rotary two-head helical scan
Video signal	EIA standard; NTSC color
Antennas	VHF—75 ohm ext. terminals UHF—300 ohm ext. terminals
Power requirements	12V nominal, +15%, -10% (recorder), 120V AC, 60 Hz (tuner/timer)
Approx. weight	11 lbs. (recorder), 13 lbs. (tuner/timer)
Dimensions	H-4¼" W-10⅝", D-10⅛" (recorder) H-4¼", W-9", D-10⅜" (tuner/timer)

Beta Format Videocassette Recorders

Sanyo Electric Inc.

For additional product information and a list of Sanyo dealers in your area, call or write the Customer Service Department at:
Sanyo Electric, Inc.
1200 West Artesia Boulevard
Compton, CA 90220
213-537-5830

It's worth taking note that Sanyo is one of the only VCR manufacturers willing to back their models with a full one-year warranty on both parts *and* labor.

Sanyo 9100A Beta Console★

List price:	$695
Manufacturer:	Sanyo
Best features:	Good image quality, remote pause, 1 year warranty on parts and labor.
Drawbacks:	Beta II speed only (3 hours of recording time), mechanical controls, 1-event/24-hour tuner/timer, no speed scan or special effects. No audio dub.

The Sanyo 9100A has one of the lowest list prices of any VCR on the market. It also has among the fewest features. Both image and sound quality are good. But that's partly because the 9100A has only one tape speed, Beta II, which delivers just 3 hours of recording time on a L-750 videocassette. This is a serious limitation because it increases your recording costs, and limits your ability to swap tapes with other Beta format VCR owners.

The mechanical controls and tuner are old-fashioned and a bit cumbersome when compared to the latest generation of all-electronic VCRs, but are perfectly serviceable. The remote pause lets you edit out commercials from your easy chair (of course you have to watch them while recording, but not on playback). Sanyo's "timed phase editing" helps to eliminate some of the picture instability created by the editing process.

Since there's no speed scan, you'll have to rely on the counter to locate programs when you've recorded more than one show on a tape. And, even more annoying in our opinion, you'll have to sit patiently through all those commercials when watching shows you've automatically taped off the air. Another sacrifice to economy: no slow motion, freeze frame, or frame-by-frame advance. Ironically, despite the lack of features the 9100A is, at 44 pounds, one of the heaviest VCRs around. If you intend to occasionally move your VCR from room to room, we recommend a thorough physical exam first.

The bottom line? This is a bare-bones VCR at a bare-bones price. If you can possibly afford it, find yourself a unit with both Beta speeds.

Specifications Sanyo 9100A

Recording system	Rotary 2-head helical scan with azimuth recording
Tape speed	20mm (0.79 inches) per second
Cassette length	1, 3, or 3 hours
Cassette dimensions	6⅛"L x 3¾"W c 1"H
Rewind/fast forward time	Less than 4 min. for 2 hour cassette
Load/unload time	Less than 3 sec.
Video output	1V p-p, 75 ohms, negative sync.
Video S/N ratio	Luminance better than 43dB, Chrominance better than 35dB
Horizontal resolution	Color, 240 lines minimum, B&W; 250 lines minimum
Microphone input	-60dB (600 ohms unbalanced)
Audio output	560mV (10k ohms, unbalanced); phono jack
Audio distortion	Less than 2%
Audio S/N ratio	Better than 40dB
RF adaptor	Switchable channel 3 or 4
Dimensions	19.5"W x 14.6"D x 7.7"H
Weight	44 lbs.

(Specifications subject to change without notice)

Sanyo 4200 Beta Console★ ★

List price:	N/A
Manufacturer:	Sanyo
Best features:	Electronic transport controls, electronic tuning, Beta II and Beta III speeds, auto rewind, auto shut off, remote pause.
Drawbacks:	No speed scan or special effects, 1-event/3-day programmability.

This all-electronic console VCR looks like the higher-priced deluxe models, but isn't. Its one big advantage over the Sanyo 9100A is the addition of the all-important long play Beta III speed. The "feather touch" controls are nice, and give this unit the same "look" as its feature-laden big brothers. But you won't find any speed scan on this unit, nor special effects such as slow motion and freeze frame. Also missing are any remote transport functions. The Sanyo 4200 does have a remote controller, but it's a pause-only unit and can't be used for fast forward, rewind, or channel selection.

Our opinion? The Sanyo 4200 is a well-built and reliable no-frills Beta VCR. But we'd rather hold out for a unit with speed scan.

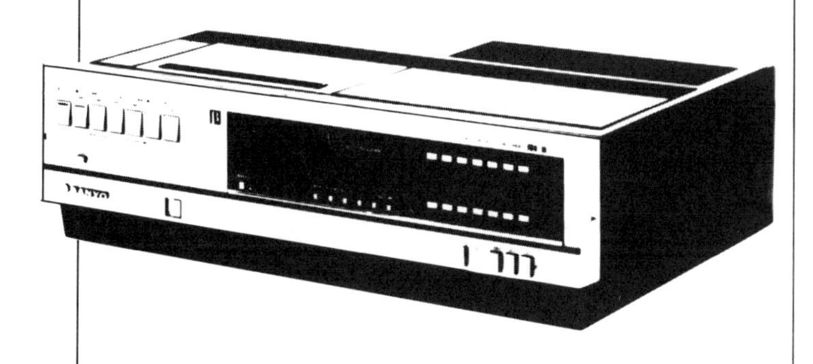

Specifications Sanyo 4200

Video recording system	Rotary, 2-head helical scan system, BII/BIII switchable
TV system	NTSC standard (525 lines, 30 frames, 60 fields/sec.)
Mechanism	3 DC motors
Tuner input	VHF: 2-13; UHF: 14-83
Video S/N ratio	43dB luminance; 35dB chrominance
Audio S/N ratio	43dB
Audio wow & flutter	0.2% WRMS (BII)
Power	120VAC, 60Hz, 40 watts
Dimensions	18⅞"W x 5¼"H x 13¾"D
Weight	22 lbs.

Sanyo 4300 Beta Console★ ★ ★

List price:	N/A
Manufacturer:	Sanyo
Best features:	Electronic transport and tuning controls, Beta II and Beta III speeds, speed scan, freeze frame, auto rewind, auto stop, lightweight, full remote control.
Drawbacks:	No slow motion, limited 1-event/7-day tuner/timer.

Here's the basic Sanyo chassis dressed up with some of the features that we found wanting in the 4200 model. Speed scan—9 times normal speed—is our favorite. There is a freeze frame—which can be used as a passable substitute for slow motion on those rare occasions when you want to see for yourself whether your team's wide receiver caught the ball inbounds like you thought, or out-of-bounds like the referee claims. The full-function remote lets you control it all from your chair.

Beware, however, of the so-called "7-Day Programmable Timer" advertised on the Sanyo 4300. By our definition this really isn't a "programmable" timer at all, since it can't be set to record *multiple events* off *multiple channels* over an extended period.

Instead, the Sanyo 4300 is equipped with a *single* program timer that can be set for 7 days rather than the usual 24 hours. This does, indeed, increase your capability for automatic recording over most economy model VCRs. With the Sanyo 4300 you can automatically tape any *single* event within the upcoming week. Not bad if you're leaving for a California business trip but can't bear the thought of missing the Red Sox versus the Yankees at Fenway Park on Thursday night.

The Sanyo 4300 can also be set to turn on and off on a fixed channel at the same time every day. A godsend for soap opera fanatics. But while the Sanyo 4300 will let you tape "As The World Turns" every day at 1 PM off KNXT it won't let you catch Danny Kaye in *The Inspector General* on Channel 13 Sunday at 4:30, *Moonraker* on Select-TV Monday at 2 PM, and *Dracula's Castle* with John Carradine (in the declining years of his career) on the Channel 40 "Late Show" at 3 AM on Friday. All of which would be duck soup for a true microprocessor-assisted tuner/timer with multiple event, multiple channel capability.

As for image quality, we found the Sanyo 4300 on a par with most other 2-video head Beta VCRs. And the 1-event/7-day tuner/timer is not a bad compromise for the discriminating viewer who travels frequently, but is interested in recording one show while away.

Specifications Sanyo 4300

Video recording system	Rotary, 2-head helical scan system, BII/BIII switchable
TV system	NTSC standard (525 lines, 30 frames, 60 fields/sec.)
Mechanism	3 DC motors
Tuner input	VHF: 2-13; UHF: 14-83
Video S/N ratio	43dB luminance; 35dB chrominance
Audio S/N ratio	43dB
Audio wow & flutter	0.2% WRMS (BII)
Power	120VAC, 60Hz, 33 watts
Dimensions	18⅞"W x 5¼"H x 13¾"D
Weight	22 lbs.

Sanyo 4800 Beta Portable★ ★ ★

List price: N/A
Manufacturer: Sanyo
Best features: Speed scan, freeze frame, electronic transport controls, Beta II and Beta II speeds.
Drawbacks: No slow motion, remote controller operates pause function only.

This is a rugged, though ultra-lightweight (just 8.75 pounds), Beta portable. On the plus side it does offer both Beta speeds, as well as freeze frame and speed scan. In the minus column: No slow motion. Also the remote controller operates only the pause function. This makes backpacking the 4800 extremely difficult, since you'll have to take off the recorder every time you want to operate the transport controls—unless you also buy Sanyo's VSC 450 color camera. The advantage of the Sanyo camera? It has set of transport controls built into it.

VTT481 is the designation Sanyo has chosen for the separate tuner/timer that allows the 4800 portable to record from your television. This is essentially the same 1-event/7-day unit that is described in the report on the Sanyo 4300 Console. It beats the basic 1-event/24-hour tuner/timer hands down, but is no contest for a "real" multi-event, multi-channel programmable unit. Unfortunately, Sanyo doesn't offer a programmable, multi-event tuner/timer. At least not yet.

Specifications Sanyo 4800

Recorder	
Video recording system	Rotary, 2-head helical scan system, BII/BIII switchable
Mechanism	3 DC motors
Power	120 VAC, 60Hz; 9.5 watts
Dimensions	10¾"W x 4"H x 10½"D
Weight	8¾ lbs.
Tuner/Timer	
Tuner input	VHF: 2-13; UHF: 14-83
Channel selection	Electronic, 12-position tuner
Antenna in/out	F-type
Power	120VAC, 60Hz, 60 watts
Dimensions	10¾"W x 4"H x 10½"D
Weight	16½lbs.

Sears, Roebuck and Company

For additional product information contact the TV/Stereo Department at your nearest Sears outlet.

Sanyo is the primary supplier for Sears line of videocassette recorders. However, you will find considerable differences in features between seemingly identical Sears and Sanyo models. In addition, Sears offers two models that you won't find in any Sanyo catalog. One is Sears deluxe console—model 5322. The other is Sears 5360 Beta portable. Both of these units offer outsanding 4-video-head performance at generally competitive prices.

Sears 53055 Beta Console★

List price: N/A
Manufacturer: Sanyo
Features: See Sanyo 9100A

This is Sears cheapest model VCR. It's built on the Sanyo 9100A chassis, and offers much the same features (or absence of features, to be precise) or its Sanyo prototype. In fact, the Sears version really has even less to recommend it, lacking even the Sanyo 9100A's "clean edit" circuitry and auto shut-off capability. These super stripped-down VCRs are also just about the only videocassette recorders, in either VHS or BETA, that don't even offer audio dub. But the real clincher, as far as we are concerned, is the lack of the Beta III speed.

Our opinion? Steer clear of the Sears 53055.

Sears 5310 Beta Console★ ★

List price: $685
Manufacturer: Sanyo
Best features: Electronic "micro touch" transport and tuning controls, Beta II and Beta III speeds, remote pause.
Drawbacks: No speed scan, no special effect, 1-event/24-hour tuner/timer WITHOUT shut off.

Here's the Sanyo 4200 with some slight variations. The low list price makes this unit an attractive buy, especially if you can find it marked down during one of Sears periodic video equipment sales.

Don't be deceived, however, by the electronic controls and fancy styling. As with its Sanyo counterpart this model looks like a deluxe VCR, but isn't. Witness the 1-event/24-hour tuner/timer that does not have a "time off preset." The Sears 5310 can be set to start recording automatically, but not to stop. Instead the machine merely runs until all the tape on your cassette has been exhausted. Which means you can't insert a show on a tape that already has some programming on it without being physically present to turn the VCR off when the show you are recording is finished.

Our advice? Unless the price is irresistible, go with the Sanyo 4200 instead. At least it has auto shut off.

Sears 5314 Beta Console★ ★

List price: $785
Manufacturer: Sanyo
Features: Identical to the Sears 5310 with the following exceptions: This model DOES have auto shut-off, the tuner/timer is a 1-event/3-day version with an "everyday" feature, and, most important, the Sears 5314 is equipped with speed scan.

The primary difference between the Sears 5314 and the Sanyo 4300 is that the Sanyo prototype has full remote control, the remote controller on the Sears model is a pause-only version. The Sears also lacks freeze frame capability, which you will find on the Sanyo 4300. Both models, however, do have speed scan. This is not a bad machine for the price, especially if you can catch it on sale.

Sears 5318 Beta Console★ ★ ★

List price: Approximately $900
Manufacturer: Sanyo
Features: Virtually identical to the Sanyo 4300. Has all the features of the Sears 5314 PLUS freeze frame and a full-feature remote control.

Sears 5322 Beta Console★ ★ ★ ★

List price: $995
Best features: 4-video head recording system, electronic controls, 8-event/14-day tuner/timer, speed scan, special effects including freeze frame, frame-by-frame advance, and slow motion, full-feature remote control.
Drawbacks: No serious drawbacks to this machine.

Although we're not wild about some of their lower-priced models, we think Sears deserves congratulations for coming up with a first-class Beta format VCR at a very reasonable price.

The image quality of this machine is excellent, as well it should be with *two* sets of video heads. In the case of the Sears 5322, one set of heads is used for normal speed recording and playback, while the second set is dedicated exclusively to the freeze frame and slow motion special effects. The result, this is one of the few VCRs we've seen that give you freeze frame and slow motion WITHOUT any noise bars.

The 8-event/14-day tuner/timer may take a little effort to master, but once you've learned how to program it, the dividends in convenience are enormous. Convenience also describes the full-function wired remote control that not only operates your basic transport operations—it also gives you armchair command of the speed scan (in forward and reverse), variable slow motion from ⅓ to 1/30th normal speed, frame-by-frame advance, freeze frame, and even a 2 times normal speed playback speed.

If you can stretch your VCR budget to $995 we recommend looking into this model. It's head and shoulders above the Sears no-frills and mid-priced units, and compares very favorably to other deluxe Beta format VCR brands.

Sears 5360 Beta Portable★ ★ ★ ★

List price: $1,145
Best features: 4-video-head recording system, electronic controls, 8-event/14-day tuner/timer, speed scan, special effects including slow motion (forward and reverse), freeze frame, and frame by frame advance, as well as an 11-function wired remote controller.
Drawbacks: None.

Another winner. As with the Sears 5322 the picture quality is excellent in both Beta II and Beta III. There's hardly a feature this unit lacks, and yet at 11.7 pounds you'll find it among the lightest portables on the market. If you're hunting for a state-of-the-art Beta portable, don't overlook this one. It gets our enthusiastic endorsement.

Sony Corporation of America

For additional information and a list of Sony dealers in your vicinity call or write the Customer Service Department at:

Sony Corporation of America
9 West 57th Street
New York, NY 10019
213-371-5800

Sony was the first to invent a successful consumer VCR—and for much of the late 1970s the word Betamax actually became a widely used generic term for the videocassette recorder.

Recently, however, the avalanche of 6-hour VHS format VCRs has drastically cut into Sony's one-time domination of the home video industry. Sony has fought back with a series of feature innovations—including the development of speed scan, or "Betascan" as Sony calls its visual scan mode. The end result? Don't let the rather dowdy styling of the current generation of Sony Betamax VCRs deceive you. Despite its plain-Jane exterior, the Betamax is still at the head of the pack when it comes to convenience and quality.

In addition to its line of Betamax consoles, Sony offers the world's smallest and lightest Beta format portable—the Betapak SL-2000. Although it weighs under 10 pounds and is about the size of a typical three-ring notebook, the Betapak is equipped with advanced features found only on top-of-the-line consoles such as speed scan and slow motion. Sony also has several projects in development, including a video camera with its own *built-in* recorder, that have the potential to revolutionize the home video industry by the end of the decade.

Sony SL-5400 Beta Console ★ ★ ★

List price: $1,250
Manufacturer: Sony
Best features: Electronic tuning, speed scan, Beta II and Beta II speeds (also Beta I in playback only), remote control with pause, freeze frame, and speed scan functions.
Drawbacks: No slow motion, 1-event/3-day tuner/timer.

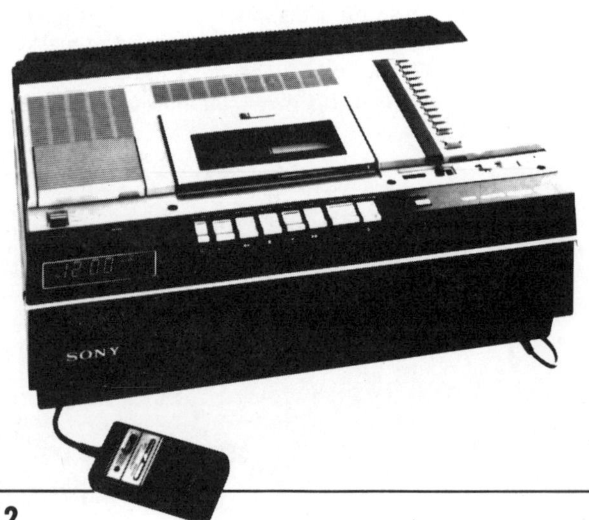

Despite its hefty list price of $1,250, you'll find that the Sony 5400 is available for as low as $775. Picture quality is tops for a 2-video-head recorder with both speed scan and freeze frame, this model has the basic features we think are essential for operating convenience. The biggest drawback is the 1-event/3-day tuner/timer—but then this is a feature that is only likely to cause you grief if you are away from home for extended periods.

Specifications Sony SL-5400

General	
Video recording system	Rotary 2-head helical scanning system
Video signal	EIA standards, NTSC color
Power requirements	120 VAC ± 10%; 60 Hz ± .5%
Power consumption	55 Watts
Weight	33 lb. 1 oz.
Dimensions	19¾"W x 15"D x 6½"H
Video	
Horizontal resolution	Monochrome: more than 280 lines; color: more than 240 lines
Signal-to-noise ratio	Better than 45dB (Monochrome)
Audio	
Frequency response	50 to 10,000 Hz (Beta II); 50 to 7,000 Hz (Beta III)
Signal-to-noise ratio	Better than 40dB
Supplied accessories	Videocassette tape; RM-55W remote control unit; channel indicators; antenna connectors EAC-24W and EAC-25; 75-ohm coaxial cable; 300-ohm twin-lead cable

All specifications are subject to change without notice.

Sony SL-5600 Beta Console ★ ★ ★

List price: $1,350
Manufacturer: Sony
Best features: Electronic transport controls and tuning, speed scan, Beta II and Beta III (also Beta I in playback only), 4-event/14-day programmability, special effects including 3X and 13X speed scan (forward and reverse), freeze frame, electronic indexing, 3-function remote control.
Drawbacks: Lacks slow motion.

Again, don't be deceived by the list price. Discounts on Sony are substantial. We had no trouble finding the SL-5600 at $875, and you shouldn't either.

In appearance as well as performance the SL-5600 is very similar to Sony's SL-5400. The main differences are the addition of full-logic solenoid controls (although these are operated by piano keys, not push buttons as on the other brands), an electronic indexing feature for easy program location (which is a handy feature indeed), and a microprocessor-controlled 4-event/14-day tuner/timer. The image quality, as with all the Sony VCRs, is excellent. Our only criticism—there's no slow motion on this machine. But then for most of us, slow motion is a seldom-used gimmick anyway.

Our opinion: At a discount price in the $900 range, the Sony SL-5600 is one of the best buys available in Beta.

Specifications Sony SL-5600

General	
Video recording system	Rotary 2-head helical scanning system
Video signal	EIA standards, NTSC color
Power requirements	120 VAC ± 10%; 60 Hz ± .5%
Power consumption	55 Watts
Weight	36 lb. 6 oz.
Dimensions	19½"W x 14⅞"D x 6½"H
Video	
Horizontal resolution	Monochrome: more than 280 lines; color: more than 240 lines
Signal-to-noise ratio	Better than 45dB (Monochrome)
Audio	
Frequency response	50 to 10,000 Hz (Beta II); 50 to 7,000 Hz (Beta III)
Signal-to-noise ratio	Better than 40dB
Supplied accessories	Videocassette tape; RM-56W remote control unit; channel indicators; antenna connectors EAC-24W and EAC-25; 75-ohm coaxial cable; 300-ohm twin-lead cable

All specifications are subject to change without notice.

Sony SL-5800 Beta Console ★ ★ ★

List price: $1,450
Manufacturer: Sony
Best features: Same as SL-5600 but also offers variable speed scan (5 to 20 times normal speed), variable slow motion, 10-minute power back-up for timer clock, auto rewind, and 7-function remote control.
Drawbacks: None.

You'll find Sony's deluxe VCR in discount houses for $1,040—virtually the same price as most top-of-the-line VHS format models.

Unlike the Sears top model, as well as many deluxe VHS recorders, the Sony SL-5800 does not have 4 video heads. But, frankly, the picture quality is truly remarkable, even without the extra pair of video heads. Signal-to-noise ratios for this unit have been reported in excesss of 47dB—the best we've seen on any VCR, regardless of format. At the Beta II speed it is almost impossible to see any difference between the SL-5800 recording and the original broadcast. Several consumer testing organizations have singled out the SL-5800 as the best Beta format VCR available, and we can find no good reason to disagree.

Yes, we do have a couple of minor criticisms. One is the rather old-fashioned styling common to all the Sony consoles. The other is the placement of channel select buttons on the *top* of the unit. Should you want to install your SL-5800 (or any Sony console) in a video cabinet, you'll have to leave ample room for access, or find a cabinet with a sliding shelf. But when it comes down to the essentials—picture quality, dependability, features, and convenience of operation—the Sony SL-5800 is the standard to which the other Beta brands should be compared. In fact, unless you own a top-quality TV set, you probably will be unable to realize the Sony SL-5800's full potential.

Specifications Sony SL-5800

General	
Video recording system	Rotary 2-head helical scanning system with a Double-Azimuth Video Head.
Video signal	EIA standards, NTSC color Power requirements:
Power requirements	120 V ± 10%; 60 Hz ± 5%
Power consumption	65 Watts
Weight	36 lb. 6 oz.
Dimensions	19½"W x 14⅞"D x 6½"H
Video	
Horizontal resolution	Monochrome: more than 280 lines; color: more than 240 lines
Signal-to-noise ratio	Better than 45dB
Audio	
Frequency response	50 to 10,000 Hz (Beta II); 50 to 7,000 Hz (Beta III)
Signal-to-noise ratio	Better than 40dB
Supplied accessories	Betamax Videocassette tape; RM-58W Remote Commander, channel indicators; antenna connectors EAC-24W and EAC-25; 75-ohm coaxial cable; 300-ohm twin-lead cable

All specifications are subject to change without notice.

Sony SL-2000 Beta Portable ★ ★ ★

List price: N/A
Manufacturer: Sony
Best features: Electronic transport function and tuner controls, ultra-small, ultra-lightweight, Beta II and Beta III speeds (with playback in Beta I), speed scan, slow motion, freeze frame, smooth edit, and index control.
Drawbacks: None.

At 9.25 pounds and approximately the size of the average telephone book, this feature-packed Beta portable is obviously the harbinger of yet one more generation VCR portables. As of press time we have yet to see the SL-2000, but if this new model lives up to its advance billing it should be the most advanced portable VCR ever.

Sony AG-300 Beta Autochanger

List price: $350
Manufacturer: Sony

20-Hour Recording The Sony BetaStack

Sony's AG-300 BetaStack Programmable Videocassette Autochanger is a simple-to-install accessory designed for use with Betamax models SL-5800, SL-5600, and SL-5400. With it, you can record automatically on up to four

Betamax videocassettes, for a total of 20 consecutive hours (with approximately 15 seconds changeover time between cassettes). Even automatic playback and automatic rewind capability, too.

That's not all you can do with BetaStack. The AG-300 has a built-in micro computer—and when you use it with a Betamax SL-5800 or SL-5600 programmable recorder, the AG-300 can be programmed to match your Betamax programming schedule. That means you can record each program on its own videocassette for replay later. No more searching through a single videocassette to find what you're looking for...and programs you want to erase can easily be separated from those you decide to save.

Specifications

Designed for	Betamax models SL-5800, SL-5600, SL-5400 only
Total cassette capability	4 (1 in deck, 3 in BetaStack)
Maximum cassette length accepted	L-830
Power requirements	AC 120V, 60Hz (AC adaptor supplied)
Power consumption	12 watts (max.)
Automatic changeover mode	Playback, record, rewind
Dimensions	9⅜"W x 5⅛"H x 13½"D
Weight	5 lb. 1 oz. (2.3 kg)
Supplied accessories	Adaptor plate, screwdriver, cassette holder, AC power adaptor, autochanger control cord

All specifications are subject to change without notice.
Sony, Sony Corporation of American and BetaStack are trademarks of Sony Corporation.
**Programmability with SL-5600 and SL-5800 only.*

Toshiba America, Inc.

For additional product information or a list of Toshiba dealers in your vicinity call or write:
Toshiba America, Inc.
82 Totowa Road
Wayne, NJ 07470
201-628-8000

Toshiba 8000 Beta Console★ ★ ★

List price: $1,245
Manufacturer: Toshiba
Best features: Electronic transport and tuner controls, Beta II and Beta III speeds with Beta I playback, special effects including speed scan and slow motion, auto rewind with auto shut-off, and wired remote control.
Drawbacks: 1-event/7-day tuner/timer.

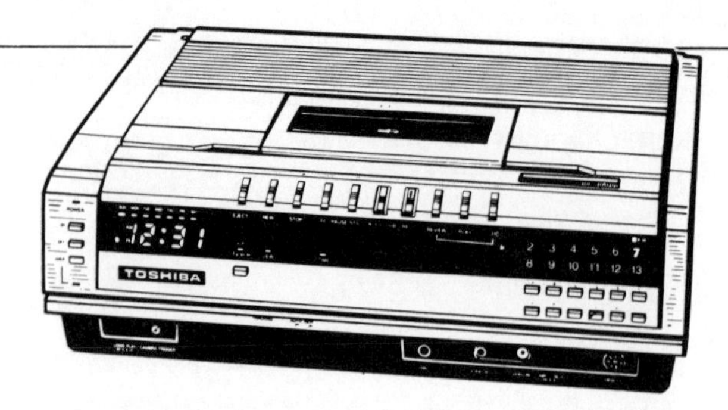

The Toshiba 8000 gets top marks for picture quality, features, and operating convenience. Its only serious drawback: Programmability is limited to one event. Also, you'll find this unit at Toshiba dealers only—so discount prices will be difficult to obtain.

Toshiba 8500 Beta Console★ ★ ★ ★

List price: $1,495
Manufacturer: Toshiba
Best features: Virtually the same as Sears 5322

We can't find much difference between the Toshiba 8500 and the Sears 5322—except the price. The Sears unit lists for $500 less.

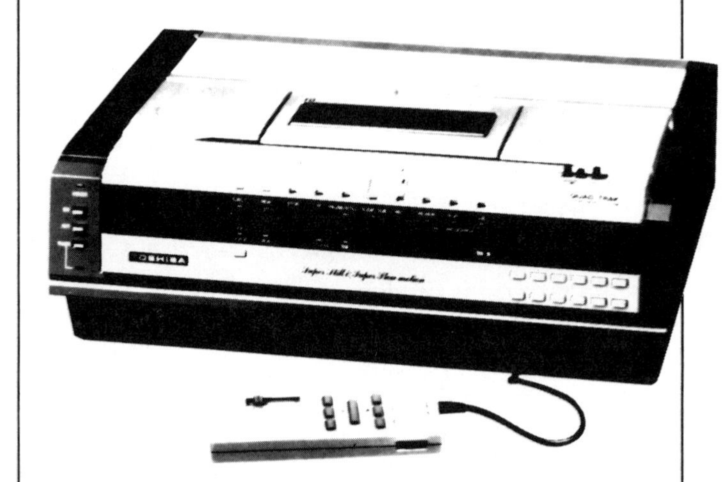

Specifications V-8500

General	
Video recording/playback system	Rotary 4-head helical scanning; (Heads: A, B, B_1, B_2,) A, B, for recording/playback, B_1, B_2, for playback only
Storage temperature	-20°C to +60°C (-4°F to 104°F)
Operating temperature	5°C to 40°C (41°F to 104°F)
Antenna	75-ohm external antenna terminal for VHF; 300-ohm external terminal for UHF
Channel coverage	VHF channels 2-13; UHF channels 14-83
VHF output signal	Channel 3 or 4 (selectable), 75 ohms unbalanced
Power consumption	53W
Weight	13.8 kg (30.4 lbs.)
Dimensions	465(W) x 153(H) x 360(D) mm; (18.3 x 6.02 x 14.17")
Video	
Input	Video line in: phono-type connector 75 ohms unbalanced
Output	Video line out: phono-type connector 75 ohms, unbalanced
Signal-to-noise ratio	Better than 45dB
Audio	
Input	Audio line in: phono-type connector
Output	Audio line out: phono-type connector
Frequency response	SP: 50Hz to 8,000Hz LP: 50Hz to 7,000Hz
Signal-to-noise ratio	Better than 40dB
Audio distortion	Less than 4% at 400Hz
Tape transport	
Tape speed	Beta II, 20.0 mm/sec., Beta III, 13.3 mm/sec.
Maximum recording time	5 hours (by L-830, Beta III mode)
Fast forward time	Within 3.5 min. (L-500)
Rewind time	Within 3.5 min. (L-500)

Zenith Radio Corporation

For additional product information or a list of Zenith dealers in your vicinity call or write the Customer Service Department at:
Zenith Radio Corporation
1900 North Austin Avenue
Chicago, IL 60639
312-745-2000

Zenith 9700 Beta Console★ ★ ★

List price: $1,300
Manufacturer: Sony
Best features: Equivalent to Sony 5600 without Beta I playback.

A fine machine in every respect and comparably priced with its prototype—the Sony 5600. Unfortunately, in the discount stores we were unable to get the same generous price cuts offered on Sony models. Thus, while the Sony 5600 could be had for $875, we couldn't do any better than $945 on the Zenith 9700. Perhaps you'll have better luck.

One fine distinction between the Sony 5600 and Zenith 9700: The Sony is equipped with Beta I playback, the Zenith is not. This won't make much difference to you, however, unless you happen to have some vintage Beta tapes recorded at the old Beta I speed.

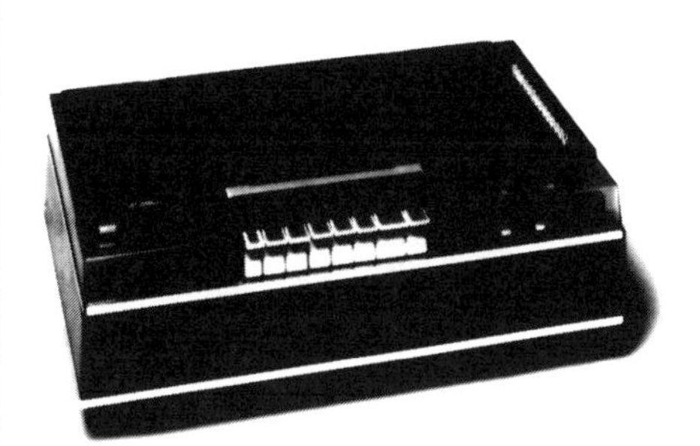

Zenith 9775 Beta Console★ ★ ★

List price: N/A
Manufacturer: Sony
Best features: Ultra-compact, front-loading, Beta II and Beta III speeds, electronic transport controls and tuning, 4-event/14-day programmability, electronic indexing, electronic counter with tape remaining indicator, special effects including speed scan, freeze frame, and slow motion (forward and reverse), full-function wireless remote control.

With the 9775 Zenith is introducing what is essentially the first model in a new generation of ultra-compact console VCRs. If the 9775 wins rapid consumer acceptance, you can expect to see an across-the-board reduction in the size of top-of-the-line console VCRs as other brands—both VHS and Beta—follow Zenith's lead.

In fact, the Zenith 9775 looks like no VCR we've ever seen before. At a shade over 3 inches tall it's less than half the height of other deluxe consoles such as the Sony 5800 or Panasonic PV-1750. It's also *front loading*—a feature that up until now was available only in the Sharp VHS console. But unlike the Sharp VC-7400, Zenith is offering the features to match its space-age design.

You'll find recording and playback in both Beta II and Beta III, as well as playback only in Beta I. All the special effects we've come to expect in a top-of-the-line unit are available, including speed scan and slow motion—in forward and reverse, no less. There is also electronic indexing for the ultimate in easy program location. And since you can't see the tape because of the front loading design, Zenith has added an electronic counter with a digital tape remaining display.

Only one thing puzzles us about this incredible unit. We can't understand why Sony continues to insist on equipping the models it manufactures with a mere 4-event programmability, while its competitor—Matsushita—has been offering 8-event tuners for several years.

As of press-time Zenith had not announced a list price on the 9775, although Zenith's president J.L. McCallister has assured his dealers the price for this new generation deluxe console will be "aggressively competitive." If this means the 9775 ends up in the $1,050 to $1,150 range at discount outlets, we urge you not to overlook this model if you're shopping for the most advanced console available in home video. Also, before you buy, we recommend you touch base with your Sony dealer. Since Sony is building this model for Zenith, we suspect it may not be long before Sony adds an equivalent to its own product line.

Zenith 9800 Beta Portable★ ★ ★ ★

List price: N/A
Manufacturer: Sony
Best features: See Sony SL-2000

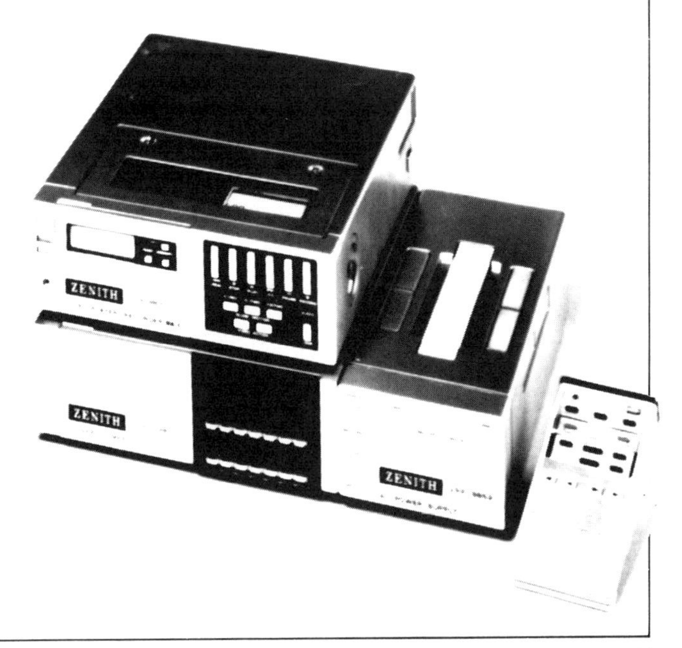

HOME VIDEO'S TOP 20

Brand	Model/Format	List Price Discount Price	Speeds	Transport Controls	Tuner/Timer	Remote Control
AKAI	375 VHS Portable	$1,700 $1,300	SP, SLP (No LP)	Piano Key	Separate Electronic 6-Event/7-Day	Optional
Curtis Mathes.	6-750 VHS Portable	$1,600	SP, LP, SLP	Electronic Push Buttons	Electronic 8-Event/14-Day	Yes/Wireless
General Electric	2014 VHS Console	N/A	SP, LP, SLP	Electronic Push Button	Electronic 8-Event/14-Day	Yes/Wireless
Hitachi	6500 VHS Portable	$1,600	SP, LP, SLP	Electronic Push Buttons	Separate Electronic 8-event/14-day	Yes/ 13-Function Wired
JVC	7300 VHS	N/A	SP, LP*, SLP	Electronic Push Buttons	Electronic 8-Event/14-Day	Yes/Wireless
Magnavox	8340 VHS Console	$1,395 $1,000	SP, LP, SLP	Electronic Push Buttons	Electronic 8-Event/14-Day	Yes/Wired
Mitsubishi	310 VHS Console	$1,350	SP, LP, SLP	Electronic Push Buttons	Electronic 8-Event/14-Day	Yes/14-Function Wireless
Panasonic	1400 VHS Console	$1,295 $ 875	SP, LP, SLP	Electronic Push Buttons	Electronic 8-Event/14-Day	Yes/ 3-Function Wired
Panasonic	1750 VHS Console	$1,440 $1,050	SP, LP, SLP	Electronic Push Buttons	Electronic 8-Event/14-Day	Yes/ 11-Function Wired
Quasar	5610 VHS Console	$1,500 $1,050	SP, LP, SLP	Electronic Push Buttons	Electronic 8-Event/14-Day	Yes/Wireless
RCA	250 VHS Console	$995 $750	SP, LP, SLP	Electronic Push Buttons	Electronic 1-Event/1-Day	Yes/ 3-Function Wired
RCA	650 VHS Console	$1,350 $1,050	SP, LP, SLP	Electronic Push Buttons	Electronic 8-Event/14-Day	Yes/ 11-Function Wired
RCA	170 VHS Portable	$1,400	SP, LP, SLP	Electronic Push Buttons	Separate 8-Event/14-Day	Yes/Wired
Sanyo	4300 Beta Console	$895	Beta II, Beta III	Electronic Push Buttons	Electronic 1-Event/7-Day	Yes/Wired
Sears	5322 Beta Console	$995	Beta II, Beta III	Electronic Push Buttons	Electronic 3-Event/14-Day	Yes/Wired
Sears	5360 Beta Portable	$1,145	Beta II, Beta III	Electronic Push Buttons	Separate 8-Event/14-Day	Yes/ 11-Function Wired
Sony	5600	$1,350 $ 875	Beta I*, Beta II, Beta III	Electronic Piano Keys	4-Event/14-Day	Yes/ 3-Function Wired
Sony	5800	$1,450 $1,040	Beta I*, Beta II, Beta III	Electronic Piano Keys	4-Event/14-Day	Yes/ 7-Function Wired
Sony	2000	N/A	Beta I*, Beta II, Beta III	Electronic Push Buttons	4-Event/14-Day	Yes/Wired
Zenith	9775	N/A	Beta I*, Beta II, Beta III	Electronic Push Buttons	4-Event/14-Day	Yes/Wired

**Playback only.*

Our nominees for the top 20 videocassette recorders—consoles and portables—in quality, features, and value.

Speed Scan	Freeze Frame	Frame-by-Frame Advance	Slow Motion	Electronic Indexing	Comments
✓	✓	✓	✓	✓	High price & no LP speed are serious disadvantages. However, 2-channel stereo with Dolby B noise reduction make this model unique.
✓	✓	✓	✓	—	Incredible 4-year warranty is a big plus. Same basic unit as Panasonic PV-1750, has 4 video heads.
✓	✓	✓	✓	—	4 video heads insure excellent picture quality.
✓	✓	✓	✓	—	Timer back-up circuit and sound-on-sound dubbing, are worthwhile extras.
✓	—	—	—	—	Lack of special effects is a big minus. Otherwise look for tops in quality and performance.
✓	✓	✓	✓	—	Timer back-up circuit, 4 video head image quality.
✓	✓	✓	✓	—	2 video head unit, but otherwise offers deluxe features.
✓	—	—	—	—	Good bet for a mid-range model. The RCA 450 is an equivalent model which has two extra features—freeze frame and auto rewind—at a slightly lower price.
✓	✓	✓	✓	—	4 video heads. A sure bet if you're looking for a top quality full-feature VCR.
✓	✓	✓	✓	—	Identical to Panasonic 1750, except for wireless remote control.
✓	✓	—	—	—	Good buy for a mid priced unit. Biggest disadvantages—limited programmability, no slow motion. Has freeze frame and auto rewind which its Panasonic equivalent—the PV-1300 does not.
✓	✓	✓	✓	—	Virtually identical to Panasonic 1750.
✓	✓	✓	✓	—	Excellent features, ultra-compact, and lightweight.
✓	✓	✓	—	—	Limited programmability and no slow motion are the only serious drawbacks.
✓	✓	✓	✓	—	Best programmability in the Beta format. Super features at a reasonable price.
✓	✓	✓	✓	—	One of the best portables we've seen in Beta.
✓	✓	✓	—	—	A lot of machine for the money.
✓	✓	✓	✓	—	Many experts consider this the best Beta format VCR.
✓	✓	✓	✓	—	Ultra-compact, ultra-light.
✓	✓	✓	✓	—	Preliminary specifications suggest this may be the most advanced VCR available in either format. Front loading, ultra-compact.

SELECTING YOUR VIDEODISC PLAYER

Despite all the hoopla, there's nothing new about the videodisc. True, the latest species (*genus* VHD, for Video High Density, a plastic-shelled creature said to have been discovered at the higher elevations of Mount Fuji by a trusted employee of the Matsushita Electrical Company) did not actually reach most video showrooms in the United States until the spring of 1982. But the disc-like record that could play both sound and pictures has been around for almost half a century.

Credit John Logie Baird—an enigmatic Scotsman—not only for developing a working videodisc player back in 1927, but for also being the first person to televise a moving picture of any sort. Unfortunately for Baird, his videodisc came nearly 5 decades too soon, and his television broadcasting system was passed over by BBC in favor of a competing technique developed by Marconi Electrical and Musical Instruments (EMI). So much for the luck of the Scots.

Even the contemporary videodisc player has also been around longer than you may imagine. The first VLP format disc player (for Video Long Play), Magnavox's "DiscoVision" player, made its very low-key test market debut at an Atlanta shopping mall late in 1978.

Inside the Magnavox player, an optical laser beam decodes billions of bits of information that have been impressed into the surface of a grooveless disc. It was the first consumer product ever to use that darling of High Technology—the laser.

About a year later a second brand of videodisc player, the Pioneer LaserDisc, appeared on the market. Like the Magnavox unit, it, too, employed laser-beam technology developed by N.A. Philips, the giant Dutch electronics firm that owns Magnavox, Sylvania, and Philco.

Despite one significant difference between the Magnavision 8000 and Pioneer LaserDisc VP-1000—the Pioneer unit can be instructed to give immediate random access to any of the 54,000 tracks that are on one side of a disc, the Magnavox model cannot,—Both units play the same MCA DiscoVision discs. In the parlance of the trade—their software is compatible.

Shoppers marvelled at these units but seldom bought them. One reason was price. Both listed for over $700. Another was software. There were very few titles in the MCA catalog (it's improving). In addition, many of the programs that appeared in the catalog never actually made it into production. And of those that finally did reach the consumer, a very high percentage proved defective. Between 1978 and 1980 Magnavox and Pioneer together sold fewer than 50,000 units of this remarkable space-age home video product.

Sluggish laser disc sales did not discourage the marketing wizards at RCA. Their projections showed Americans buying 1 million disc players by 1984, and as many as 5 million a year by 1990. At that rate videodiscs alone could surpass the 7-billion-dollar-a-year color-TV industry before the end of the decade. The only problem. RCA didn't have a videodisc player—yet.

April 1981. After investing over $200 million in research and development—far more than it spent to develop the color television—RCA rolled out the SelectaVision SFT 100, and the Era of the Videodisc entered its second phase.

To the seasoned video buff, phase two—the Age of CED, for Capacitance Electronic Disc—was actually something of a throwback. This is because instead of using a high-tech laser like the Magnavox and Pioneer systems, the CED technique developed by RCA used a rather ordinary diamond-tipped stylus and a grooved disc, not terribly unlike a conventional phonograph record.

To the cognizant, this meant the RCA system could offer fewer features and less flexibility—no slow motion, freeze frame, or random access. But to marketing boys at RCA, all that was irrelevant. Because at a list price of $499—going as low as $399 in some discount houses—they figured they could sell a lot of videodisc players. Especially if RCA could come up with a list of blockbuster films at reasonable prices. Which it did. Films that, incidentally, could not be played on laser machines, or vice versa.

As with videocassette recorders, the consumer was now faced with the choice between two different and technically incompatible versions of what was essentially the same machine. And unlike VCRs, where thousands of prerecorded tapes now existed, as well as the option to record off TV, the owner of a videodisc had only a few score of films to choose among. In fact, at 3.5 viewing hours per day the average family could conceivably go through the entire RCA videodisc catalog in about 3 months.

While two competing and incompatible videodisc systems may have seemed like enough to most American consumers, it didn't to the Japanese. Having a virtual stranglehold on the American VCR market, the Japanese electronics companies were understandably far from thrilled to see a new video product, the CED videodisc player, that was both competitively priced and, of all things, actually manufactured in the United States. As everybody knows, the Americans can only make transistor radios. The result? A crash program to develop a Japanese-designed and manufactured alternative.

Spring of 1982 saw the videodisc enter its third phase—the Age of VHD. The instigator was Matsushita Electric, the world's largest consumer electronic company, which together with its partially owned subsidiary, JVC, had developed yet a third incompatible videodisc format, the Very High Density, or "grooveless capacitance" system.

The VHD machine—marketed under the brand names JVC, Panasonic, Quasar, General Electric, and others—retails for less than the VLP laser systems but more than CED units. It also offers the same special effects as the laser players—slow motion, fast motion, freeze frame, search, and an optional random-access feature. And the VHD technique does all this with a stylus. But unlike the RCA VideoDisc, there are no grooves, just invisible microscopic pits. VHD—its proponents claim—puts less wear on machine and disc than CED, but offers the special effects previously found only in laser units, at a lower price.

Which one is best for you? We suggest that before you rush off to Video Land to buy any new videodisc player, you give it a second thought. We think there are at least 3 good reasons to wait a little longer before spending anywhere from $400 to $800 on a videodisc player.

The Future of VideoDisc

There is no denying that the future belongs to the videodisc—both in the home and industry.

It's certainly no accident that Magnavox's first customer was the CIA. Or that Pioneer sold 11,000 of its industrial laser disc players to just one company—General Motors.

For entertainment, the videodisc player is great. But for administrative, training, record keeping, and educational purposes, these units are nothing short of revolutionary. When coupled with a minicomputer or microprocessor, the videodisc player becomes the most awesome informational tool the world has ever known. "They threaten to change the way that employees are trained, equipment is maintained, students are taught, and products are demonstrated and sold," predicted *Business Week* in 1980.

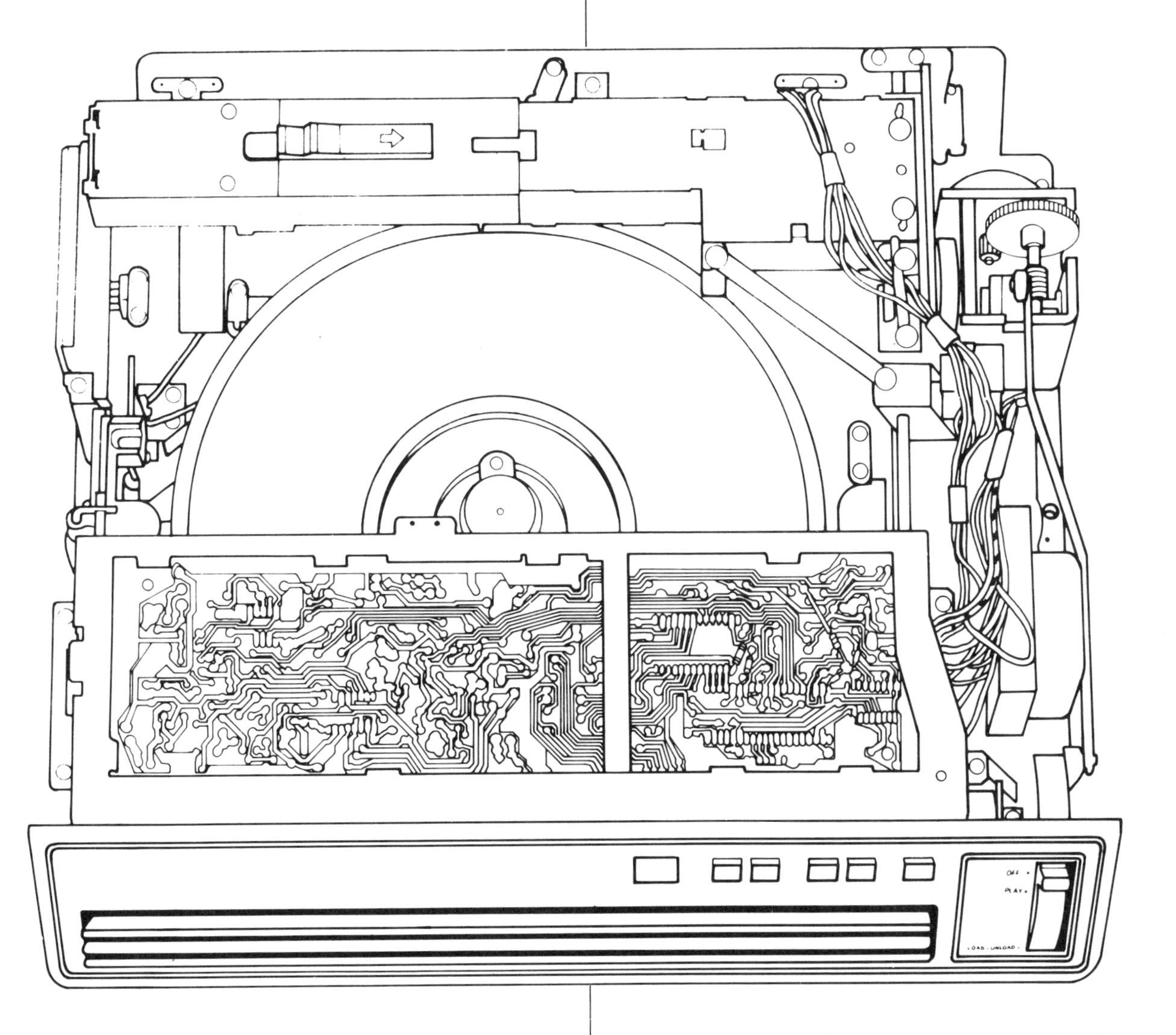

Inside the SFT-100 videodisc player—fewer features and a lower list price. One influential Wall Street investment research firm suggests that RCA's sheer marketing muscle—over 5,000 dealers, or about 9 times as many outlets as both laser systems combined—will insure that the CED format becomes the dominant consumer videodisc system of the 1980s.

©RCA

Already Hughes Aircraft has put the service manual for the Army's M-80 tank—a document that otherwise occupies 6 feet of wall space—on a single videodisc. With the ability to store up to 108,000 pages, just one videodisc may eventually be able to hold as many as 500 average-length novels. An entire college library could be replaced by a stack of videodiscs not much larger than the average home record collection.

Toshiba is currently selling a system to insurance companies, real estate firms, and government agencies that can automatically scan and store over 10,000 documents on a single disc, ready for instant access. McGraw-Hill and others are actively programming college courses onto discs for interactive video learning systems. And IBM, Xerox, Control Data, Honeywell, Exxon, Hitachi, and dozens of other companies are studying new ways of exploiting videodisc technology to store, retrieve, process, and communicate information—ways that will ultimately affect the lives, and livelihoods, of millions of people in the coming decade.

So, if the videodisc player is the wave of the future, why not have one in the house now?

Perhaps if there were just one format, so any disc would be compatible with any videodisc player, we would be tempted. But there is not. Instead, there are 3 totally incompatible formats—a fact which vastly inhibits the distribution of videodisc software since dealers simply can't afford to "triple stock" every movie or picture. The artistic quality of much of the material available in videodisc is excellent, but for the moment, as well as for the foreseeable future, the selection of programming on videodisc is limited. Especially compared to thousands of titles on VCRs.

If we were certain the videodisc players were highly dependable—or let's say, at least as reliable as VCRs—we might surely be more inclined to recommend them. But in our experience, they are not. Of the first 3 videodisc players we attempted to evaluate, only one—the RCA SelectaVision player—worked properly. Perhaps this was a fluke. Perhaps not.

Finally, if we thought that the present generation of videodisc players left little room for improved operating features, we might also be more enthusiastic about suggesting you buy one of these units. But there are major oversights on several brands. On the Magnavision 8000, for instance, there is no random-access feature—a drawback that limits the educational potential of this unit. On the RCA SelectaVision SFT 100, it is an absence of random access, direct video and audio outputs (making hookup to a stereo amplifier and speakers, or a videocassette recorder very difficult), as well as the unpardonable lack of stereo channels. Both Magnavox and RCA promise second-generation disc players that rectify some of these problems. But the CED format will never be capable of random access.

Videodisc players represent a new, and largely untried, technology. Or, more specifically, 3 new technologies, with some systems and individual models rushed into the showrooms under intense competitive pressure. It was much the same with VCRs. And with each successive generation of videocassette recorders came a marked improvement in convenience, special features, and reliability. The same is likely to happen with videodisc players. But at the same time it is also not inconceivable that one of the 3 current videodisc formats may fade from the scene.

Should you be what the textbooks call an "innovative buyer" and, bless your heart, you're still gung ho on having a videodisc player in your life, then read on...

Laser Disc Players—The VLP Format

The Video Long Play laser beam technology developed by N.V. Philips and currently employed in the Magnavision 8000 and Pioneer LaserDisc VP-1000 (as well as a second-generation Magnavox and Pioneer units scheduled to reach dealers during 1982) uses a short-wavelength laser beam to read the audio and video signals—up to 14 *billion* individual pieces of information—that have been encoded in a grooveless plastic disc. Unlike the CED and VHD formats that require physical contact with the disc, the laser optical technique relies solely on a beam of light.

There are several advantages to VLP laser format. First, there is no wear on the disc. No one knows for sure just how long a laser disc will last—perhaps centuries. Maybe even longer. That's just what the world needs. A videodisc of *Smokey and the Bandit* that will be just as preposterous in the year 4082 as it is today.

The second advantage is a bit more practical. Since the laser needs no physical contact with the disc, the actual microscopic pits that carry the audio and video signals can be sealed in a protective layer of plastic. This means that short of caressing it with medium-grade sandpaper, a laser disc can be handled with a certain degree of nonchalance.

CED and VHD discs, on the other hand, cannot be handled at all. In fact, they are sold in a cardboard and plastic "caddy" that is inserted into the disc player and then removed without the disc in it. Not only do you never touch a CED or VHD disc—you never see one.

How important is the durability of a videodisc? Not very if you play only discs from your own personal collection. Even the CED with its archaic grooved disc and diamond-tipped stylus will probably withstand a few hundred plays before any deterioration in image quality is evident due to wear. And, seriously, how many times can you really stand to watch *Smokey and the Bandit* anyhow?

When and if a viable system of disc rentals develops, there may be a slight advantage here for the laser disc with its presumed indefinite lifespan. But that is sheer speculation.

With 54,000 grooveless tracks to choose from, you might wonder how the VLP laser beam knows which exact track it is supposed to be decoding. The answer is simple. It uses another laser.

You see, not only has the laser disc been encoded with Burt Reynolds' smug smile, but there is also a second "tracking" signal that identifies each of the 54,000 individual bands. The second laser beam reads this "tracking signal" and then tells the first laser where to go. And since each track has its own 5 digit number, you can zero in on any specific segment in a few seconds by inputting an index number into the player control panel. A microprocessor and the tracking laser do the rest. Just remember the Magnavox 8000 does not have this random-access feature. Nor do any of the CED format players. Random access is, however, available on the VHD format units, although usually as an option.

How long does a laser disc play? That all depends on the speed at which it was mastered. Laser disc players, like VCRs, have multiple speeds. The Standard Play mode, which is used for virtually all instructional and educational programs, are mastered in a process known by a typically meaningless technical expression, Constant Average Velocity. What is important to remember is merely that the CAV disc speed gives you 30 minutes of playing time on each side of the disc, and that you can use all the special effects, including random access if your player has it. In the CAV mode, the turntable actually spins at a Constant Average Velocity of 1,800 rpm. And since that happens to be a good deal faster than the typical meat slicer, you can understand why the lid of the disc player locks when you engage the play button.

The second, Extended Play, speed is known as CLV, for Constant Linear Velocity. At CLV the turntable spins a bit slower (but still fast enough to slice baloney, should you ever get tired of watching movies with it). Slow enough, in fact, that you will find one hour of programming on each side, or a typical 2-hour movie on a single CLV speed disc.

Here's a rundown of the controls you will find on the VLP format, and what to expect from them:

•**Power.** This one gives your player the juice. In the Magnavox 8000 it also serves as the PLAY button. With the Pioneer VP-1000 model POWER must be on to either open or close the lid.

•**Reject/Open.** The equivalent to a STOP button on a VCR. It turns off the laser, shuts down the turntable, and unlocks the lid.

•**Play.** This, of course, is the button you're most interested in. Magnavox also offers you NORMAL PLAY REVERSE in case you like to watch the discs the same way most people read magazines, backwards.

•**Pause.** Found only on the Pioneer, and it's redundant anyway, since STILL/STEP (videodisc talk for freeze frame and frame-by-frame advance) does the same thing.

•**Still/Step.** You guessed it. Freeze frame and frame-by-frame advance. Actually these are very critical features on a videodisc player where, presumably, you or your progeny will be viewing a lot of instructional programming.

Fast X3. Triple speed. Works forward and reverse in Pioneer. Just forward on the Magnavox 8000, but don't forget the Magnavox has NORMAL PLAY REVERSE.

•**Scan** (Search forward/reverse). A much faster version of FAST X3—about 70 times normal speed. This is our favorite special effect feature in a VCR, and it's no less valuable on a videodisc player.

MAGNAVOX
Stop Action. Lets you see a museum full of art one piece at a time.
Frame-By-Frame Advance. Shows you a golf pro's swing inch by inch until you've got it down pat.
Slow Motion in Forward and Reverse. Puts you in complete control of what you're seeing.
Reverse Action. Lets you repeat a favorite scene again and again.
Laser Beam Pickup. Can't scratch or wear out video-discs.
Stereo Capability. Lets you hear concerts through your home stereo system while you're watching them on TV.
Picture Search in Forward and Reverse. Lets you quickly find precisely what you want to see.

Illustration: Magnovox

•**Slow Motion.** Variable in both directions.
•**Audio L/1, R/2.** A switch that lets you turn off one or the other of the stereo channels. Primarily intended for use with bilingual programs, also handy for interactive discs.
•**Chapter/Frame.** Index buttons that let you find the beginning of a certain "chapter" as indicated on the disc jacket or workbook.
•**Frame Search.** Available on the Pioneer VP-1000, and presumably on the new Magnavox disc player when it becomes available. This is the random-access feature that we think is so important for educational applications. With FRAME SEARCH you can locate any specific frame (merely imagine an Encyclopedia Brittanica disc with each frame representing one page) by inputting the frame index number (i.e. the page number). Maximum search time is 20 seconds. Let's see you look up "Leontocephalos" that fast.

RCA Videodisc Players—The CED Format

The Capacitance Electronic Disc techinque developed by RCA at a reported cost of $200 million has just one advantage as far as we can see—it's cheap.

In CED a diamond-tipped stylus rides across the disc in a very shallow and very narrow groove (so narrow, in fact, that almost 40 of them could fit into the typical phonograph record groove). The stylus interacts with microscopic slots that have been pressed into the electrically conductive plastic disc. The ultimate result is video image and sound reproduction that rivals anything in videocassette—at a significantly lower price.

Since the RCA disc rotates at a mere 450 rpm, 1 hour of programming can be fit on each side— a 2-hour movie on a single disc. (Which makes you wonder why the Video Long Play was selected for the laser disc system, since its 30-minute discs are actually shorter than either CED or VHD. But that's the way it goes in high-tech terminology.)

As for disadvantages of the RCA system, there are plenty. For one thing the stylus and groove technique makes random access technologically impossible. There are also serious, if not insurmountable, problems in incorporating slow motion and freeze frame (some dealers may show you how to freeze frame an RCA player by pushing the forward and reverse VISUAL SEARCH buttons simultaneously; it'll work, but we don't recommend you make a practice of it) in the CED system. There are no audio outputs on the SFT-100, which means you can't hook it up to your stereo system. And there's no stereo, even if you could.

Many of these disadvantages are merely the result of marketing decisions (to keep price as low, low, low as possible), not technical limitations. RCA has promised an upgraded model with stereo, perhaps before 1983.

Here are the features you'll find on CED videodisc systems:

•**Off/Play.** Self-explanatory.
•**Load/Unload.** This is accomplished by sliding a lever and inserting the plastic disc caddy. The disc player will automatically remove and load the disc. You remove the empty caddy. To unload, or turn the disc over to watch the other side, just repeat the process. You never handle, or see, the disc itself.
•**Visual Search.** A 17-times-normal-speed visual scan that functions in both forward and reverse. With a little experimenting you can probably make visual scan double as freeze frame—we have. But we also have no idea what this does to the mechanism of the videodisc player, or the disc itself. The problem with building a freeze frame into the CED system seems two-fold. First, by forcing the stylus to remain in a single groove we suspect it won't take long to damage the disc. Second, since each track actually contains several frames, a jerky image, or even multiple exposure, is likely to result.
•**Rapid Access.** Not to be confused with *random* access. This is merely a fast forward or reverse without picture, at about 160 times normal speed.
•**Pause.** Removes the stylus from the disc without losing your place. Great for interruptions. Does not freeze frame as does the pause button on many VCRs.

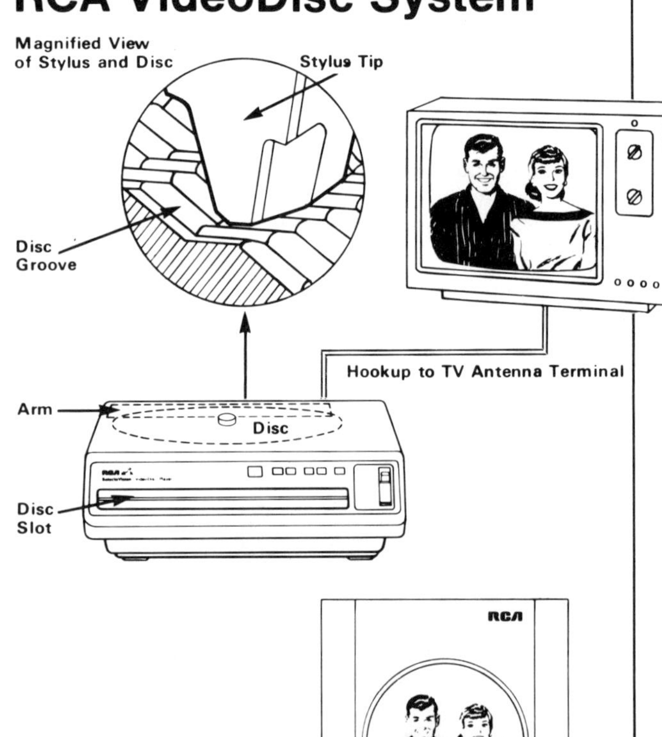

RCA's CED videodisc player—will low price and simplicity of operation capture the family market?

©RCA

Matsushita's Videodisc Player—The VHD System

The VHD videodisc player was the last of the 3 competing formats to reach the dealer's shelves. But backed by the considerable marketing clout of Matsushita's subsidiaries—Panasonic, Quasar, and JVC—as well as General Electric, which is also adopting the VHD format, the future of this newest technology looks as promising as any of the videodisc contenders.

The concept of VHD was one of inspired compromise—lower prices than laser, more features than CED. And as you might expect from the world's largest consumer electronics company, the results—although tardy—are right on target.

The catchword is "grooveless capacitance." Like the CED system there is a stylus that interacts electronically with a plastic disc. But this is no ordinary stylus, nor is it your typical disc. For one thing the stylus is extremely wide, thus subjecting the disc to far less wear and tear than CED. Also, like the laser format, there is not only a set of audio and video signals, but there is a second "tracking" signal that is encoded in the videodisc's surface. The stylus reads the primary audio and video signals, but it orients itself on the disc according to the digital index numbers of the tracking signal on each grooveless track. Hence, the second big advantage over CED—special effects and even random access.

VHD videodisc players are capable of slow motion, fast motion, freeze frame, frame-by-frame advance, speed search, and an optional microprocessor-controlled random-access feature. They also come equipped with stereo sound and audio outputs for direct connection to a component stereo system.

Here's a rundown on the key features you will find on the major brands of VHD videodisc players:

•**Start/Stop.** Obvious, but necessary.
•**Mono/Stereo.** A select switch for monaural or stereo recordings, plus a Channel A or Channel B selector for use with bilingual recordings.
•**Manual Search.** A speed search with a visible picture at 64 times normal variable down to ¼ speed slow motion. Works in forward and reverse.
•**Random Search.** This is the random-access feature. You enter the index number of the specific frame you want to see, the microprocessor guides the stylus to it within a few seconds.
•**Chapter Search.** Locates the beginning of individual program segments using the random access tracking signals.
•**Chapter Repeat.** Press this button to repeat the "chapter" you have just finished or are working on.
•**Programmable Repeat.** Program the videodisc player to repeat any segment you want at the push of a button.
•**Stop/Step-By-Step.** Freeze frame and frame-by-frame advance. Will pausing for extended periods damage the disc? Matsushita says no damage will occur if you pause for one hour or less. It's difficult to imagine a situation in which you would want to freeze frame for longer than that.

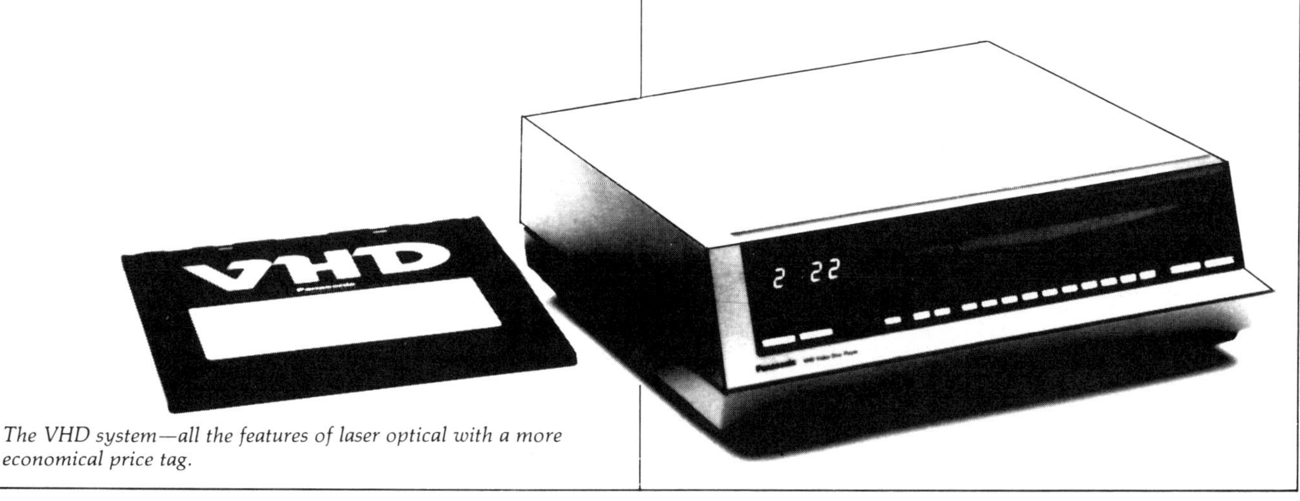

The VHD system—all the features of laser optical with a more economical price tag.

THE VIDEODISC BUYER'S GUIDE

CED FORMAT VIDEODISC PLAYERS

RCA SFT-100 Videodisc Player

List price:	$499
Format:	CED
Visual search:	Yes (forward and reverse)
Rapid search (no picture):	Yes (forward and reverse)
Slow motion:	No
Freeze frame:	No
Random access capability:	No
Time remaining indicator:	Yes (LED display)
Stereo sound capability:	No
Direct audio outputs:	No
Direct video outputs:	No
Remote control:	No
Weight:	20 pounds
Power consumption:	35 watts

Low price, simplicity of operation, and a classic-packed catalog are the main selling points of the CED system. It's disadvantages? No random access, slow motion, freeze frame, or stereo sound. Since RCA invented the CED format, we recommend going with the SFT-100, unless you find a significantly better price on one of the brands sold under license from RCA (see page 122).

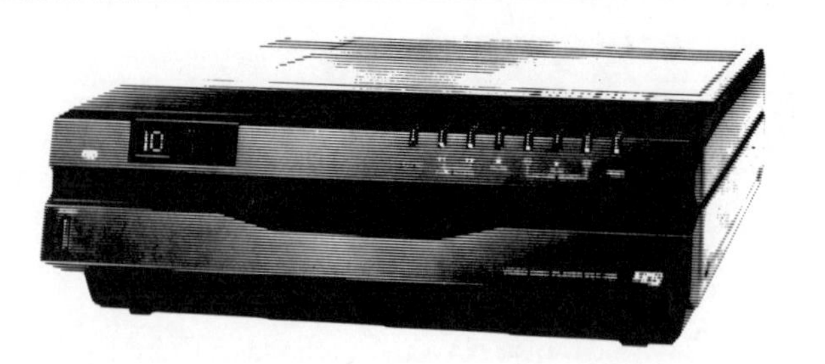

Elmo VEC-200 Videodisc Player

List price:	$500
Format:	CED
Visual search:	Yes (forward and reverse)
Rapid search (no picture):	Yes (forward and reverse)
Slow motion:	No
Freeze frame:	Yes (by simultaneously pushing "Visual forward" and "reverse" buttons)
Random access capability:	No
Time remaining indicator:	Yes (LED display)
Stereo sound capability:	Yes (built-in stereo adapter jack)
Direct audio outputs:	Yes (built-in stereo adapter jack)
Direct video outputs:	No
Remote control:	Optional
Weight:	19.8 pounds
Power consumption:	N/A

This is the first video product offered by a photographic company. The stereo-adapter jack and optional remote control are features not available on the SFT-100.

Hitachi VIP-1000 Videodisc Player

List price:	$500
Format:	CED
Visual search:	Yes (forward and reverse)
Rapid search (no picture):	Yes (forward and reverse)
Slow motion:	No
Freeze frame:	No
Random access capability:	No
Time remaining indicator:	Yes (LED display)
Stereo sound capability:	No
Direct audio outputs:	Yes
Direct video outputs:	Yes
Remote control:	Optional
Weight:	19.1 pounds
Power consumption:	25 watts

Sanyo VDR-3000 Videodisc Player

List price:	$500
Format:	CED
Visual search:	Yes (forward and reverse)
Rapid search (no picture):	Yes (forward and reverse)
Slow motion:	No
Freeze frame:	No
Random access capability:	No
Time remaining indicator:	Yes (LED display)
Stereo sound capability:	No
Direct audio outputs:	Yes
Direct video outputs:	Yes
Remote control:	Optional
Weight:	17.6 pounds
Power consumption:	35 watts

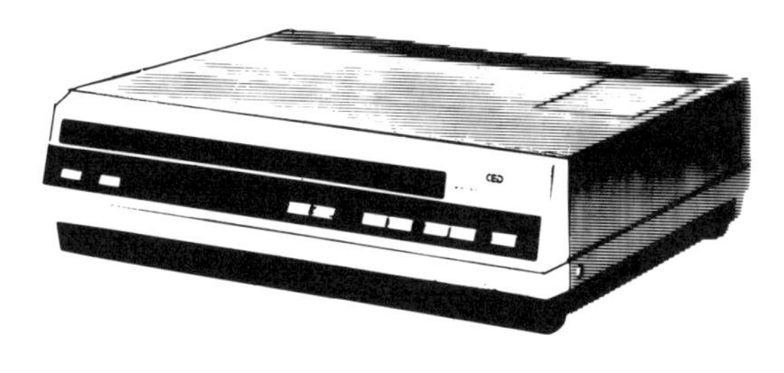

VDR-3000 Specifications

Video S/N ratio	46dB (luminance)
Audio frequency response	40-15kHz
Audio S/N ratio	50dB
Picture search	12X and 40X normal playback speed, forward & reverse
Rapid access	200X normal playback speed
Disc rotation speed	450 rpm
Pickup type	Replaceable cartridge, diamond stylus
Disc loading	Horizontal loading with CED caddy
VHF output	Channel 3 or 4, switchable
Power consumption	35W
Power source	120VAC, 60Hz
Antenna input & output	75 ohm coaxial (VHF & UHF)
Aux video & audio outputs	RCA-type pin jacks
Dimensions	17″W x 19½″L x 4½″H
Weight	Approx. 17.6 lbs.
Remote control	Wired, 8-function

Sears CED Videodisc Player

List price:	$499
Format:	CED
Visual search:	Yes (forward and reverse)
Rapid search (no picture):	Yes (forward and reverse)
Slow motion:	No
Freeze frame:	No
Random access capability:	No
Time remaining indicator:	Yes (LED display)
Stereo sound capability:	No
Direct audio outputs:	Yes
Direct video outputs:	Yes
Remote control:	Optional
Weight:	20.5 pounds
Power consumption:	22 watts

Toshiba VP-100 Videodisc Player

List price:	$500
Format:	CED
Visual search:	Yes (forward and reverse)
Rapid search (no picture):	Yes (forward and reverse)
Slow motion:	No
Freeze frame:	No
Random access capability:	No
Time remaining indicator:	Yes (LED display)
Stereo sound capability:	No
Direct audio outputs:	No
Direct video outputs:	No
Remote control:	Optional
Weight:	9.6 kg
Power consumption:	27 watts

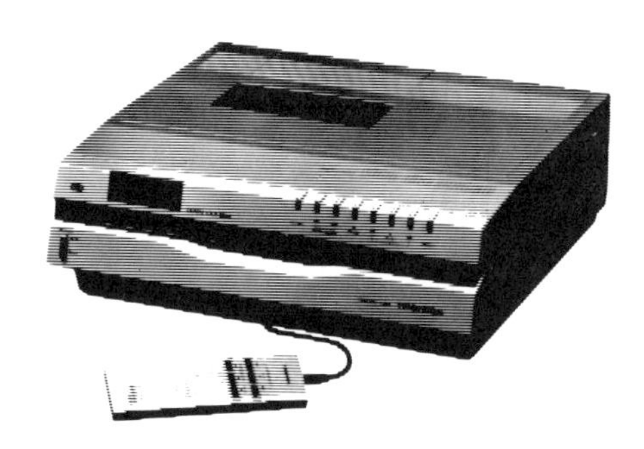

VLP FORMAT VIDEODISC PLAYERS (LASER OPTICAL)

Pioneer VP-1000 Laser Disc Player

List price: $750
Format: VLP (Laser)
Visual search: Yes (forward and reverse at 3X normal speed)
Rapid search: Yes (video playback remains viewable in high speed rapid search on the VP-1000)
Slow motion: Yes (variable from 30 frames per second to one frame every 5 seconds)
Freeze frame: Yes (STILL/STEP)
Random access capability: Yes (any of the 54,000 individual frames can be located using the FRAME DISPLAY feature. CHAPTER DISPLAY locates major subsections on certain discs)
Time remaining indicator: Yes (on screen display)
Stereo sound capability: Yes (stereo software currently available)
Direct audio outputs: Yes
Direct video outputs: Yes
Remote control: Yes
Weight: 38.6 pounds
Power consumption: 95 watts

See page 120.

Magnavox 8000 Videodisc Player

List price: $775
Format: VLP (Laser)
Visual search: Yes (3x normal speed, forward only)
Rapid search: Yes (Rapid search on Magnavox 8000 operates forward and reverse with visible video image)
Slow motion: Yes (variable forward and reverse)
Freeze frame: Yes
Random access capability: No
Time remaining indicator: Yes (display on screen)
Stereo sound capability: Yes (stereo software available)
Direct audio outputs: Yes
Direct video outputs: Yes
Remote control: No
Weight: 40 pounds
Power consumption: 80 watts

See page 120.

Magnavox 8000 Specifications

Power source	120VAC, 60Hz
Power consumption	80 watts maximum
Videodisc speed	1800 RPM
Playing time with 12 inch videodisc	Standard videodisc = 30 min/side Extended Play Videodisc = 60 min/side
RF output to TV	1200uv, VHF Channel 3 or 4 300 ohms
Video monitor output	1 Volt, 75 ohms
Audio output	220 millivolts, 1500 ohms 2 channels

VHD FORMAT VIDEODISC PLAYERS

JVC VHD Videodisc Player

List price:	N/A
Format:	VHD
Visual search:	Yes (variable from 1/4 to 64x normal speed)
Rapid search:	Yes (variable up to 64x normal speed with video playback)
Slow motion:	Yes
Freeze frame:	Yes
Random access capability:	Yes (Chapter search allows access to subsections)
Time remaining indicator:	Yes (LED diplay)
Stereo sound capability:	Yes
Direct audio outputs:	Yes
Direct video outputs:	Yes
Remote control:	N/A
Weight:	28.5 pounds
Power consumption:	40 watts

Sharp VHD Videodisc Player

List price:	N/A
Format:	VHD
Visual search:	Yes
Rapid search:	Yes
Slow motion:	Yes
Freeze frame:	Yes
Random access capability:	Yes (chapter search or individual frame access)
Time remaining indicator:	Yes (LED display)
Stereo sound capability:	Yes
Direct audio outputs:	Yes
Direct video outputs:	Yes
Remote control:	Yes

VHD Specifications

Pickup method	Electro-tracking capacitance pickup system
Playing time	Two hours (1 hour per side)
Disc material	Conductive PVC
Disc size	Diameter 260 mm (10 inches)
Track pitch	1.35 m
Stylus material	Diamond
Stylus life	More than 2,000 hours
Video S/N ratio	42 dB
Audio signal	2 channels
Audio S/N ratio	60 dB

Panasonic VHD Videodisc Player

List price:	N/A
Format:	VHD
Visual search:	Yes (forward and reverse)
Rapid search:	Yes (scans disc in 20 seconds, forward or reverse, with video playback)
Slow motion:	Yes (forward and reverse)
Freeze frame:	Yes
Random access capability:	Yes (programmable chapter search and random access to any of the 54,000 indexed images)
Time remaining indicator:	Yes (LED display)
Stereo sound capability:	Yes
Direct audio outputs:	Yes
Direct video outputs:	Yes
Remote control:	N/A
Weight:	33 pounds
Power consumptions:	N/A

VHD Specifications

Pick-up method	Electro-tracking capacitance Pick-up System without guide groove
Playing time	Two hours (one hour per side)
Disc rotation speed	900 r.p.m.
Disc diameter	10.2″
Disc material	Conductive PVC
Track pitch	1.35 micron
Stylus material	Diamond
Signal modulation	SCC (Single Carrier Composite) system
Video carrier	Pedestal 6.6 MHZ Deviation 1.8 MHz
Luminance bandwidth	3.1 MHz
Audio signal	2 channels
Power source	120 volts, 60 Hz, Ac
Player weight	Approximately 33 lbs.
Player dimensions	17 3/4″ (W) x 16 1/8″ (D) x 5 5/16″ (H)

VIDEOCASSETTE RECORDER VS. VIDEODISC PLAYER

Can't decide between a videocassette recorder and videodisc player? Then you may find a quick glance at the following chart instructive.

As you can see, there is a $200 price difference between the cheapest videodisc player (RCA's SFT-100) and the least-expensive VCR (Sanyo's VTC-9100A). The clear winner—on price alone—is the videodisc player.

Since presumably you have other things in mind than merely admiring your video equipment, we've also compared the cost of software—those prerecorded movies and programs you'll be watching. Here, too, the videodisc clearly has the price advantage. Or does it?

True, if you pick a typical feature movie, let's say *Butch Cassidy and the Sundance Kid,* you'll find the RCA video disc list for $19.95 while the videocassette costs a whopping $59.95. Disc's the winner. Right?

Well, yes. And, no.

If you buy all your movies prerecorded, you'll save a bundle owning a videodisc player. This is true even if the prices of discs go up (which is likely, since some videodiscs are being sold at a loss to help promote the sale of videodisc players) and the prices of videocassettes come down (which is also possible with newly developed mass-recording technologies).

But very few VCR owners actually buy the movies they watch. Marketing studies show only one out of every 3 VCR households own any prerecorded movies at all. What are VCR owners watching? One answer: videocassette rentals. You can *rent* a videocassette of *Butch Cassidy and The Sundance Kid* in most areas for between $7 and $13. The same goes for recently released movies.

As for videodisc rentals, we've chalked up a "NO" in that category—meaning you can't rent discs for your videodisc player. Which is not entirely true, since in a few localities disc swapping and rental clubs are popping up. We've also spoken to dealers who plan to start rental libraries once they detect a sufficient demand. But for the time being you'll have a hard time finding videodisc rentals on anything approaching the same scale and availability as in videocassettes.

The next category—"RECORDS OFF TELEVISION"—is where the videodisc really loses out. Remember, the videodisc player is like a record player for video programs. It's strictly a one-way street. You can watch shows, but you can't record them.

How about homemade programs using a video camera? Yes, if you own a VCR. No way, Jose, if you have a videodisc player. You can't make 'em, and you can't watch 'em. Not unless you want to have your Super 8 home movies transferred onto 30-minute video discs. The fee is approximately $1,500 per disc.

Well, if you can't record anything, then at least there must be a big selection of those inexpensive videodiscs to buy, right?

Wrong. And as far as we're concerned, this one's the real clincher. Today's selection of prerecorded *videocassettes* is overwhelming. More than 1,200 recent major motion picture releases, and thousands upon thousands more oldies but goodies, foreign films, serials, shorts, vintage TV, rock concerts, how-to's, and miscellaneous what-not's are available for purchase and rental on videocassettes.

Now for videodiscs. You can currently type the title of every single program available in all 3 videodisc formats on a single sheet of 8½" by 11" paper. Granted, the quality of the selections is high, especially in the RCA catalog where you'll find more megaton blockbusters per inch than on a Trident nuclear submarine. But we estimate that if the average TV household were to watch only their videodisc player, they would have viewed the entire RCA catalog within 3 months. Providing, of course, they were also willing to spend the more than $3,000 it would cost to buy all those discs.

The limited selection of videodiscs is unlikely to remain permanent. After all, it took just a little over half a decade for the selection of prerecorded videocassettes to mushroom from zero to well over 5,000 significant titles.

Will disc catalogs grow at this same pace? We don't think so. The reason has to do with disc production complexities. Anybody with a pair of VCRs and a master videotape can go into the videocassette production business (assuming, of course, there are blank videocassettes to be had, which there are). But videodisc production is a high-technology affair, requiring enormous capital investment for disc mastering and pressing facilities. RCA's disc plant is still under construction while MCA seems only to have recently ironed out all the bugs in the "DiscoVision" disc production process. The history of both the RCA and MCA disc catalogs has been one of painfully slow growth. The expectation that record pressing facilities could be converted for videodisc production has proved unrealistic. We simply don't see any sudden flood of new videodisc software on the horizon.

As for comparing features and special effect: If stereo is important to you, the videodisc offers a tangible advantage here. At the moment just one VCR has stereo capability, and very little stereo programming on videocassette. The laser disc systems and new VHD videodisc players both are equipped with stereo audio tracks, while a second generation RCA unit with stereo capability is expected to reach the market at anytime. Another videodisc advantage, perhaps the biggest of all second to price, is the random access feature found on the Pioneer VP-1000, and as an option on the Matsushita-developed VHD systems. (VHD stands for Very High Density).

These units contain a microprocessor that lets you locate any of the 54,000 frames on one side of a disc by merely touching a button. The implications of random access for educational and instructional use are genuinely staggering. Entire textbooks can be easily programmed on videodisc, complete with quizzes, exams, thousands and thousands of illustrations, and even short demonstrations and newsclips. This is the one area where the videocassette recorder is really outclassed. But take heed. Random access is not possible in the RCA system with its stylus and record-like grooved disc. Nor is it available on the Magnanox 8000. You can get random access capability on the Pioneer laser disc player and in the VHD format. But in neither case will you be getting any great bargains, since the units with random access often cost more than some videocassette recorders.

Regarding the special effects—freeze frame, slow motion, speed scan—all videodisc players have speed scan, and the higher-priced models (VHD and laser formats) are equipped with freeze frame and slow motion. In videocassette recorders you'll also find these special effects in the more expensive units. The image on your screen in these special effects modes is usually of a better quality in videodisc than in videotape. Freeze frame, in particular—which is useful for studying a printed page, illustration, or diagram—is crisp and clear in videodisc, but is often fuzzy and laden with video noise when produced on a videocassette recorder.

And how about image quality in general? We'll call this a draw, with perhaps an almost imperceptible advantage going to videodisc. The videodisc manufacturers claim the videodisc image is far superior, but frankly, we haven't been able to see much difference.

On ease of operation we'll give the advantage to the disc player. But this is mainly because it also does less. There are no recording buttons to worry about, or tuner/timers to program. If there's someone around your house who lacks the mechanical aptitude to operate an electrical can opener, then perhaps you'll find the videodisc player, especially the RCA model, a simpler device to run than a videocassette recorder. But it's been our experience that the average 6 year old requires only about 5 minutes to master the basic functions of either a disc player or cassette recorder.

Neither machine, however, is particularly brat-proof. Any juvenile, especially one armed with a couple of screwdrivers, can be lethal to either video device. Or almost anything else around the house, for that matter.

The bottom line? For the present we cast our ballot for the videocassette recorder. There's just no way you can beat the economy, the convenience, and the fun of making your own recordings. Nor can you top the availability of rentals, mail-order exchange programs, or just folks around town who are willing to trade tapes.

As for the videodisc player? Its day is coming, but it hasn't arrived yet. The only serious advantage we see for the videodisc player right now is if you have a family with small children. In which case it's a safe bet that before long you'll be finding more and more random-access kids program like the Kidisc (see page 29). Programs, we should remind you, that will only work on the higher-priced VLP and VHD players.

Videocassette Recorder Vs. Videodisc Player

	VCR	Videodisc Player
Cost (list price)	$700 & up.	$500 & up.
Cost of Prerecorded Programs	$30 to $90.	$7 to $30.
Records Off Television	Yes.	No.
Can Be Used At Home With A Video Camera	Yes.	No.
Number of Prerecorded Programs Available	At least 6,000.	Several hundred.
Features & Special Effects		
•Stereo	Very limited.	Most models.
•Random Access	No.	Some models.
•Speed Scan	Some models.	Yes.
•Slow Motion	Some models.	Most models.
•Freeze Frame	Some models.	Most models.
•Frame-By-Frame Advance	Some models.	Most models.
Image Quality	Variable according to speed. Generally quite good except in special effects.	Good to excellent. Superior to VCR in special effects.

The Supporting Cast

VIDEO GAMES

It was a tough day at work. Your secretary quit and the boss has been on your case all week. So what do you do when you get home? Mix a drink, sprawl out in front of the TV—and bomb the hell out of those Asteroids.

You feel your reflexes sharpening by the second drink as you blow that alien reconnaissance drone to smithereens. And there's enough time before dinner to intercept and destroy the Imperial battle star. Sure, saving the world like this every night can be hard on a guy. But at least the office seems light years away . . .

Successful toys are those that secretly fascinate the child in all of us. Video games, first introduced in 1972, combine skill, fantasy, competition, and the hypnotic effect of a video screen to keep a kid of 6 or 60 preoccupied for hours.

In 1977, Magnavox introduced Odyssey², the first video game unit capable of using interchangeable game cartridges. In recent years the home video game market has taken off, and there are now at least five major manufacturers of these addictive pleasure machines.

Choosing a video game is like chosing any toy. If it's done right, it will bring you and your family hours of fun. But if you choose a game unsuited to your needs, it will be in the attic in no time, next to the Hula-Hoops, the train set, and the yards of slot-car racing track.

Atari V.C.S.

Video games are not alike. Ranging in price from $130 to $300, the options and quality of the five machines vary widely. Atari has dominated the market in recent years with its Video Computer System (V.C.S.). Listed at $180 but often available for less than $150 in large department stores, the V.C.S. has the largest selection of game cartridges and the best game mechanisms.

The most attractive feature of the Atari system are the two independent "difficulty" switches, which allow users of different skill levels to compete against each other. Each game cartridge also provides several different levels of complexity for the same game. This reduces the boredom factor by assuring that the game can be made more difficult as you become more proficient.

Atari's graphics and sound effects are definitely not as sophisticated as some of the more expensive units, but we found the action of various games to be creative and fun. The system comes with two joy sticks and two Paddle Controllers, used to play some of the Pong-type games. They plug easily into the back of the unit. The Keyboard Controllers, needed to use cartridges such as

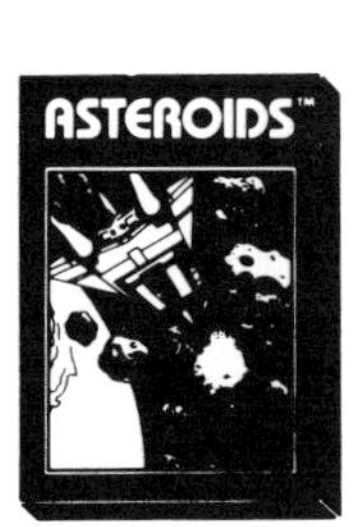

Drawings Courtesy Atari

Codebreaker, Concentration, and Basic Programming, are optional.

Atari recently introduced a remote-control video game unit that provides two cordless game controllers so you can nail those nasty aliens from your easy chair ten feet away. The new system uses all standard Atari game cartridges.

Odyssey²

Odyssey² by Magnavox competes head to head with Atari in price, but has some slightly different features. It lacks the "difficulty" switch, which means the offering of games at different skill levels is much more limited than Atari. There are also fewer cassettes available for the Odyssey².

But what we like most about the system is that it has a built-in keyboard. If you have young children, this would be a good system to consider since it's a way of introducing them to the concept of a keyboard. The unit is amazingly light, and the keyboard is a flat, touch-

SOFTWARE:
What You See Is What You Get

The quality and challenge provided by video game software ranges from awesome to icky, and there are a lot of considerations in between. Do you want games that look and sound great and call from lots of strategy? Check out Mattel or Bally. Do you want games that are easy to learn and test your reflexes? Or do you need game software that will keep your kids occupied for hours? Atari and Odyssey² are stronger here.

Cardinal rule number one is: always make sure the game cartridges will be available after you buy the system. A video game system without games ain't no fun.

Software seems to break down into several categories: space wars, earthly combat, sports, more sedentary competition like checkers and backgammon, and education. Let's look at some of the software offerings to give you a feel for what's available.

Atari has three of the sci-fi classics that have eaten up so much of your lunch money at the local arcade. Asteroids, Missile Command, and Space Invaders are all fun re-creations of their arcade ancestors, but don't expect exact replicas of the real thing. Your television set doesn't have the resolution and these game units don't have the memory necessary to create the detail found in bigger machines.

Breakout, a one-man Pong game, and Combat, an orgy of planes, tanks, and jet fighters, are two of Atari's other popular games. There are 47 different variations and skill levels on the Breakout cartridge. And Combat gives you 27 different games.

As a rule, game cassettes made by one company won't work in another brand's machine. Activision, however, is a company set up by several former Atari program designers, and it produces cartridges for the Atari V.C.S. Among those available from Activision are Fishing Derby, Bridge, Dragster, and Boxing.

Cosmic Conflict is one of over 40 games available from Magnavox for the Odyssey². Here you must line up your laser sight to blast the daylights out of battle frigates and enemy transports. But watch for those star fighters zeroing in on you from hyperspace.

Pocket Billiards is an Odyssey² game that takes care of all the racking and scoring as you chalk up the electronic pool cue in anticipation of a tricky bank shot. Alpine Skiing provides 195,000 different ski runs in a game that allows Odyssey² players to compete on slalom, giant slalom and hairy downhill courses.

Mattel's Auto Racing takes time to learn but, then again, A.J. Foyt didn't make it in a day, either. This popular Intellivision game has terrific graphics that includes cars squealing, skidding and, of course, crashing. Don't forget your seat belt.

Sea Battle is another Mattel classic. Deploy your fleet, drop mines and send your submarine against the enemy's battleship, Watch out, or one of his shells will sink you.

Among the best, and most complicated, in the sports area is Mattel's football. Complete with its own play book, the game will probably take you an hour to learn. The payoff is a challenging game of strategy where you have good control of your players. It's a great way to keep your brain in shape during the off season.

Ironically, Bally owns the Space Invaders name but gave Atari the rights to use it before they decided to go into the home video market. Bally's version of Space Invaders is, therefore, called Astro Battle. It's fast and fun but only has four skill levels, while Atari has 112 different versions of the game on its cassette.

Bally Pin is a video pinball game where the player holds two hand controls and uses them to move the flippers on either side. The colors are vivid and the action is good. Gun Fight, one of the built-in games has a couple of cowboys jumping around behind cactus trying to blow each other away. Competitors use the game control's joystick to aim their hero's revolver at the enemy. When the victim goes down for the last time, the Arcade plays a funeral dirge.

sensitive system with a plastic cover that makes it appear relatively kid-resistant.

Cassettes are easily loaded in the top front of the Odyssey². We found the Magnavox software similar in action and graphics quality to Atari, but lacking in the diversity. Alpine Skiing, for example, was a fast, bone-breaking run through an endless series of racing gates, complete with schussing sound effects. Odyssey²'s joysticks are excellent. These game controls have eight distinct positions which are easy to feel as you concentrate on the video screen.

Mattel's Intellivision

Atari and Magnavox have the two dedicated video games on the market. The other three major manufacturers differ fundamentally in that their units can be expanded into full-scale personal computers. Mattel's Intellivision is the best-known of the three. Mattel's Master Component is list priced at $300, but we found it for $268 in a department store.

Initially, it looks similar to the Odyssey² and Atari V.C.S., but cassettes are loaded in the side instead of on the top of the machine and the game controllers are permanently attached to the machine. Mattel's strength is what you see on the TV screen. The animated graphics, color, and sound effects are a knockout, and the game play is considerably more complex than the less expensive machines. Intellivision is designed to function as the main component of Mattel's home computer when it is combined with the $600 Keyboard Component. For this reason, it has more horsepower to apply to the graphics in Mattel's video games.

The unit's two hand controllers are well-built, touch-sensitive keypads. Each game cartridge comes with its own pre-printed overlays, which slide over the keypads to indicate what button should be pushed to fire a rocket or pitch a baseball. The overlays are fine for serious game players, but if young children were playing with Intellivision and constantly changing games, the overlays could easily get lost under the rug.

Mattel's graphics and game action are a kick. In basketball, the player's actions are very realistic, while in the tennis game, the ball has a shadow and the crowd applauds politely when you volley nicely.

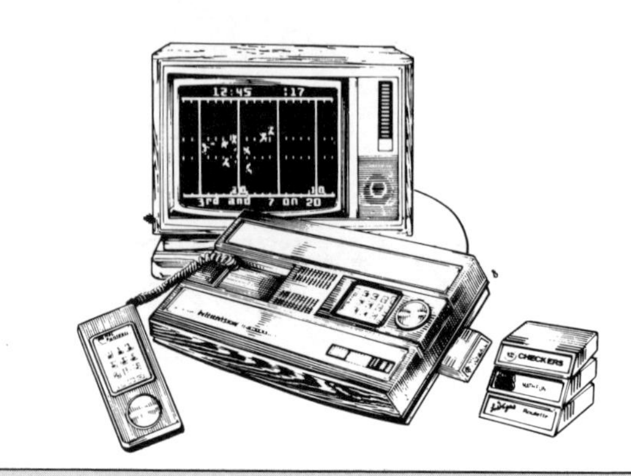

THE BUYER'S GUIDE TO VIDEO GAMES

Manufacturer	Model	List Price	Software	Special Features
APF	MP-1000	$130; cassettes $20-$30	15+cartridges in sports, strategy, arcade games. Includes Hangman, Bowling, Boxing, Roulette, Baseball.	Expandable into personal computer with addition of keyboard ($500).
Atari*	Video Computer System	$180; cassettes $19-$40	45+cartridges in sports, strategy, action, education, chance. Include Asteroids, Space Invaders, Breakout, Soccer, Concentration, Football.	Most cartridges available; many game variations per cartridge; independent difficulty switches.
Bally	Bally Arcade	$300; cassettes $20-$30	25+cartridges in sports, strategy, education, action/skill. Includes Astro Battle, Bally Pin, Football, Music Maker, Black Jack, Bally Basic.	Expandable to 32k personal computer with addition of keyboard ($600); 3 built-in games & calculator; color graphics; 4-player capacity.
Magnavox	Odyssey²	$180; cassettes $20-$30	40+cartridges in education, arcade, sports. Includes Math-A-Magic, Cosmic Conflict, Pocket Billiards, Alpine Skiing, Football, Computer Intro.	Built-in, touch-sensitive keyboard; excellent Joysticks.
Mattel	Intellivision	$300; cassettes $40	25+cartridges in strategy, sports, education. Includes Sea Battle, Auto Racing, Football, Math, Tennis, Basketball, Roulette.	Expandable to personal computer with addition of keyboard ($600); excellent color graphics and sound; most sophisticated games; software available on PlayCable.

**Sears sells the Atari V.C.S. under its own name as the "Vide Arcade" for about $160 and, of course, it uses all Atari cassettes.*

more games... Bally's line up of great video games continues to grow with the games you like best. . .right out of the game room into your living room.

more fun...

Now there's no easier way to learn about computers than with the new Bally BASIC system. This plug-in cartridge with built-in audio tape interface converts the ARCADE into a personal computer you can program yourself.The Computer Learning Lab makes it easy to learn programming while creating computer games, electronic music, and video art.

more to come!

Soon the ZGrass-32 expansion keyboard will be available to plug into the Arcade and together create one of the most powerful home video computer systems you can own.

In addition to meeting standard business and home computing requirements, the combined ZGrass-32 computer system will provide extended capabilities for creating graphs and visual displays.interactive teaching, advanced video games, business simulations, electronic music, TV commercials, video titles, and animation.

$299.95

Bally Arcade plus includes three built-in games, two hand controls, color calculator, and Bally Basic.

Drawing Courtesy Bally

The Bally Arcade

The sleeper in the home video game field is the Bally Professional Arcade. Bally has built its reputation making slot machines and pin ball machines, and when the home unit first appeared in the late Seventies it was not marketed aggresively. Astro Vision is now marketing the system, and it should give Mattel a run for its money.

The Bally Arcade is actually a programmable computer whose strongest assets are its graphics and its sound capabilities. By adding the keyboard component ($600), the system becomes a full-blown personal computer with 32K bytes of memory. Like the other video systems, the Arcade hooks easily to your television set. The difference is that when you turn it on, the Arcade is already loaded with three built-in games—Scribbling, Gun Fight, and Checkmate—as well as a built-in calculator.

For graphics buffs, Scribbling is a ball. Using the joystick, you can move the cursor around the screen, drawing and designing at will, and changing colors as often as you like. Bally claims there are 256 colors available, but it would be awfully hard to see those subtle differences on a TV screen.

Another Bally feature is the standard BASIC Programming cartridge that comes with the unit. A keypad overlay allows you to use the calculator keyboard to write programs in BASIC. A programming guide shows you how to program music, colors and even your own games into the system.

Bally offers about 25 game cassettes that, like Atari, offer a range of skill levels. Galactic Invasion, for example, has nine skill levels and is a fast-paced space shoot-out with good sound and color. One of Bally's unique features is that it can handle four players at once. (The two additional hand controls are available for $45.)

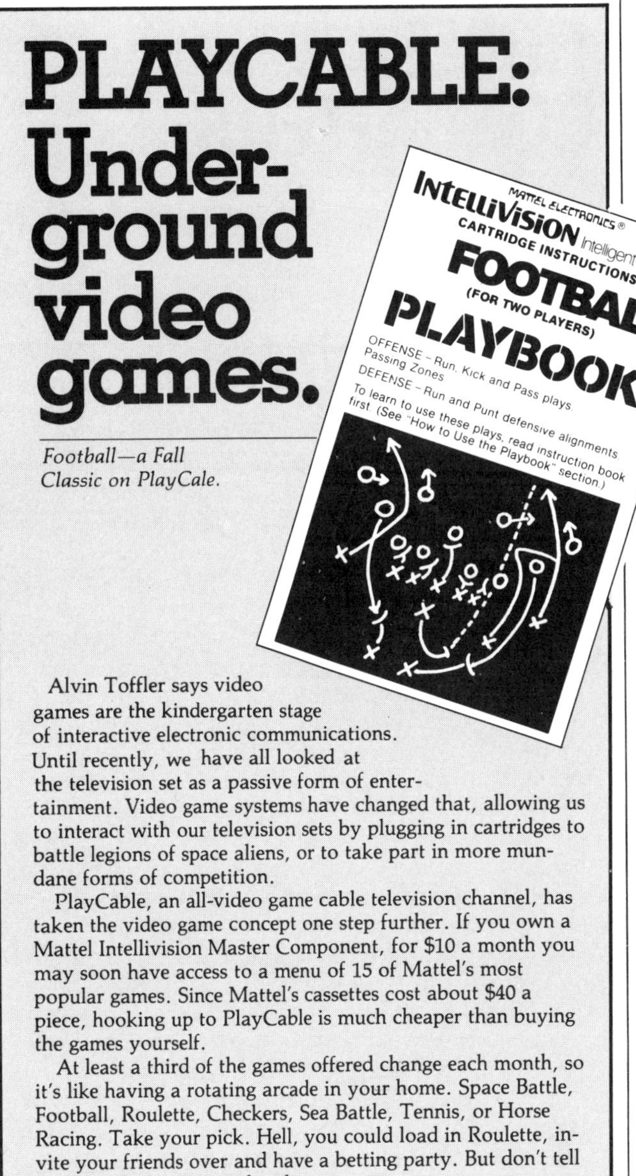

Football—a Fall Classic on PlayCale.

Alvin Toffler says video games are the kindergarten stage of interactive electronic communications. Until recently, we have all looked at the television set as a passive form of entertainment. Video game systems have changed that, allowing us to interact with our television sets by plugging in cartridges to battle legions of space aliens, or to take part in more mundane forms of competition.

PlayCable, an all-video game cable television channel, has taken the video game concept one step further. If you own a Mattel Intellivision Master Component, for $10 a month you may soon have access to a menu of 15 of Mattel's most popular games. Since Mattel's cassettes cost about $40 a piece, hooking up to PlayCable is much cheaper than buying the games yourself.

At least a third of the games offered change each month, so it's like having a rotating arcade in your home. Space Battle, Football, Roulette, Checkers, Sea Battle, Tennis, or Horse Racing. Take your pick. Hell, you could load in Roulette, invite your friends over and have a betting party. But don't tell anyone where you got the idea.

PlayCable, which is owned in part by Mattel, has recently become available in some major cities and is now trying to enter local cable markets. To find out if it's available in your area, contact you local cable operator or The PlayCable Company, 888 Seventh Ave., New York, NY 10106; (212) 265-3360.

Courtesy PlayCable

APF MP-1000

At $130, APF's MP-1000 is the cheapest programmable video game on the market. The unit is usually sold with APF's keyboard ($500), which turns the system into a personal computer. By itself, the MP-1000 comes with a pair of hand controllers that combine a joystick and a calculator keyboard into one unit.

APF has about 20 game cartridges available, but their software isn't as sophisticated as the more expensive units. The chief advantage of the MP-1000 lies in its expandability.

Now you know the basics. Go see the units in action and decide which one suits your needs and wallet the best. Just remember, sooner or later, you're going to have to let the kids play with it.

HOME COMPUTERS

Not long ago the idea of a "home" or personal computer was ludicrous. Computers filled whole rooms, required dozens of white-jacketed technicians, and cost millions of dollars. But no longer.

Today, for less than you'll pay for a decent pair of stereo speakers, you can buy a typewriter-sized home computer that will do things even the room-sized behemoths couldn't figure out. Practical things like preparing your income tax, analyzing your stock portfolio, teaching you to read music and speak French, or making the airlines reservations for your next trip. Or off-beat things, like producing your daily astrological chart, or letting you produce bizarre color "compugraphics." Or challenging things, like chess or backgammon, or even Dungeons and Dragons and Space Invaders. And if you're willing to spend a few thousand dollars for additional "memory" and "peripherals" you can easily upgrade your home computer into an incredibly powerful business tool. It might surprise you to know that this article, as well as much of the rest of THE SOURCEBOOK, was written "on" a computer.

It might also surprise you to know that *all these things can be done without your knowing how to program a computer.* True, once you own a home computer, you'll probably want to learn how to program it. Learning takes a little time, but it's fundamentally simple. And, of course, the computer can teach you.

How much money you spend depends on what you want your computer to do. So the first step in buying a home computer is to define your goals. Is it just for fun and games, or do you want to get into preparing graphics for business reports or analyzing your stocks? It can help you at home, but it can also help you at work. Word processing, accounting, and inventory control are all duck soup to the right computer. It can also help you teach your kids in ways you may never have dreamed of.

Computers are getting cheaper every year but, unfortunately, they're not free yet. Even though you may know exactly what you want to do with one, you must decide how much you can afford to spend. You can dole out $200 or $2,000 depending on whether you want a Chevy Chevette or a Porsche.

The Hard Stuff

Once you know why you want a computer and how much you can spend, you need to know a few things to make an intelligent decision. First, accept the fact that the world of computers is full of people using words you never heard before. Don't worry. You can have a good time with your computer without knowing most of the computer jargon that's thrown around.

Like the rest of the Videoworld, computerdom is divided into hardware and software. "Hardware" is the computer itself and all the wires and stuff you lug home from the store. "Software" is the computer programs that you buy or create yourself to run on the hardware. The best computer hardware in the world isn't worth squat without good software. It's like having a VCR with no cassettes. The quality of the computer programs available on a system you are considering—be they games, educational software, or programming languages—should always be of primary concern before you buy.

Now, let's go to the hardware. The brain of any personal computer is its CPU (central processing unit), otherwise known as a microprocessor. The CPU does all computations and manipulates the information it is fed. But the brain can't remember anything. Therefore, the true measure of a computer's horsepower is its memory.

There are two kinds of memory. There is memory that the computer holds permanently, the contents of which cannot be altered by the user (you). This is called ROM (read-only memory), and it's one measure of the computer's innate intelligence. ROMs are used to hold pro-

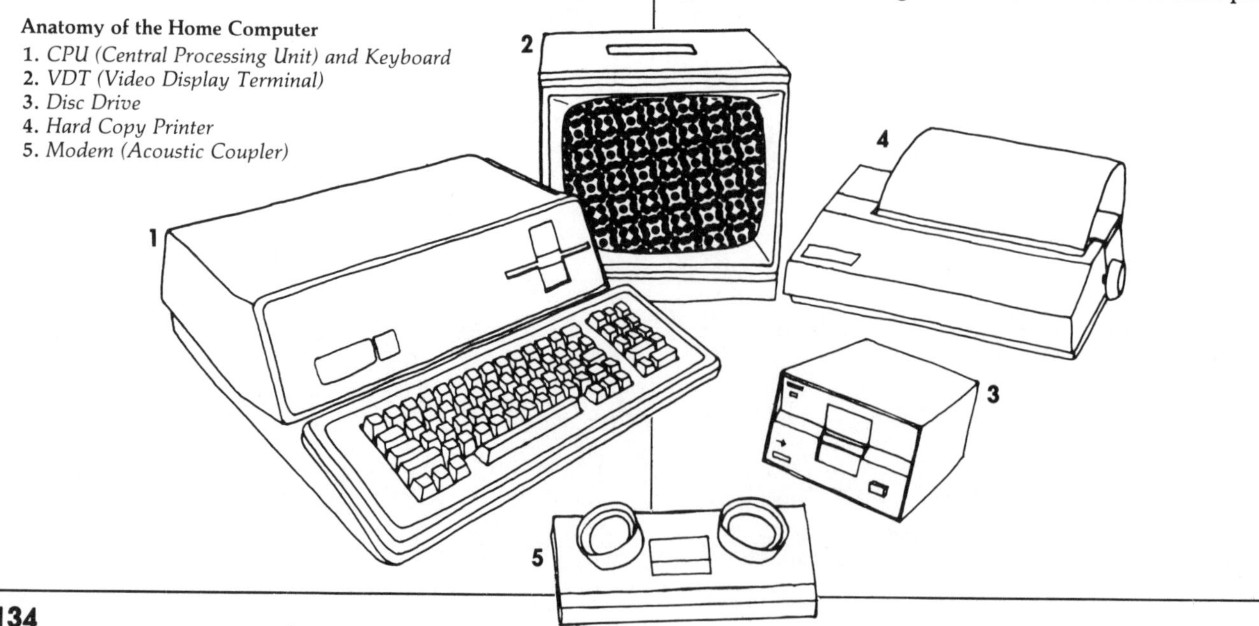

Anatomy of the Home Computer
1. *CPU (Central Processing Unit) and Keyboard*
2. *VDT (Video Display Terminal)*
3. *Disc Drive*
4. *Hard Copy Printer*
5. *Modem (Acoustic Coupler)*

grams or instructions that the computer uses frequently. For example, when you turn the machine on, the ROM has instructions to tell the computer how to load a program. Some computers have ROMs that already hold BASIC, the most common programming language. This means that as soon as you turn a machine on, you can start writing a program.

The computer's working memory is called RAM (random access memory). It stores programs and data while the computer is being used. Information in the RAM can be added to, deleted, changed, and moved around either by you or by the program you have asked the computer to run. RAM holds information only as long as the power is on. For the information to be kept permanently, it must be stored out of the computer. The size of a computer's RAM determines the amount of data and the length of the programs it can handle.

Memory is quantified in terms of kilobytes (called "K"), which represents 1,000 bytes. One letter or digit takes up one byte in the computer's memory, so 8K bytes of memory would mean enough space to store 8,000 characters. Got it?

RAM and ROM are always referred to in terms of "K." A computer with 16K bytes of RAM is big enough to play games and do uncomplicated programming. Memory is always expanded by increments of 8K. A computer with 32K RAM should be adequate for personal needs, although some more sophisticated programs now require 48K and 64K.

Now you've got this smart computer with all this memory just sitting there in a box. It doesn't do you any good unless you can communicate with it. This is where input/output (I/O) devices come in.

To get the computer to do something, you've got to give it instructions. You could handle it like you handle your kids and yell at it, but computers are hard of hearing. Like the opposite sex, they respond best to soft touches. That's where the keyboard comes in handy. All popular micros (that's cool talk for personal or microcomputers) today have built-in keyboards. The design, layout, and feel of the keyboard is an important consideration when buying a computer, particularly if you are going to be entering a lot of data or doing a lot of programming.

Communication is a two-way affair, of course, and the computer has to be able to talk back or show off, depending on how you're treating it that day. A video display terminal (VDT) is the most common means of communication. Fortunately, your television set will provide the video display you need to use your computer. It's a fairly simple procedure to hook it up.

Some of the fancier computers come with built-in video displays. If you are looking at these systems, be sure to consider the quality of the monitor. Ask yourself: Can my eyes stand looking at this screen for several hours at a time?

There is one thing you should know about before you go looking at computers. And that's "storage." Remember, computers have a working or main memory, called RAM. But when you shut that baby off, whatever program or data was loaded in there is about as accessible as last winter's snow. It's gone forever—unless you have stored it permanently outside the computer in some kind of auxiliary memory.

The two most common forms of auxiliary memory for micros are cassette tape recorders and disc drives, which hold "floppy" discs that look a lot like your old 45-rpm records. You may already have a cassette tape recorder at home that you can use to store programs from your computer. Many manufacturers, however, now make their hardware so that you must use their brand of cassette recorder.

Disc drives, which store data on plastic floppy discs which are loaded into the drive, are much more expensive than cassettes. A cassette recorder costs about $75, while disc drives are about $600 each. If the computer you are looking at is under $1,000, you are almost certainly going to be using cassette tapes for auxiliary memory.

The fundamental difference between the two kinds of storage is speed. To get to a program on the middle of a tape, you must wind the tape forward until the computer finds what it's looking for. That can take a few minutes. A disc drive, however, provides random access instantly to any program on the disc. A floppy disc is more dependable and can also store more information than a cassette tape.

We think buying a computer without having some kind of auxiliary memory is silly. Without it, you can't save anything when you turn the computer off. Yet, ask a salesperson for a price on a computer, and that figure will seldom include a cassette recorder or a disc drive. "Oh, that's extra," will be the response. You'll get used to hearing that phrase.

The Real Thing

Now you know enough to go shopping for a computer. To give you a feel for what is out there, here are seven popular micros ranging in price from $150 to about $1,400. Remember, the price is just for the computer itself. The peripheral devices and software that you will need to make the computer do what you want will cost extra and, as a rule, the more expensive the computer, the more expensive the peripherals.

All seven of these models have proven track records. If they suit your needs and your wallet, you can't go wrong. But the hardware market is changing so fast, with new models being introduced each month, that you should really use these seven systems as yardsticks. Go out and see if you can't find something better for the price. If not, buy one of these and have a great time.

The **Sinclair ZX81** is a good way to get your hands on a real computer for practically nothing. Well, $150 isn't much anymore. This tiny micro has 1K RAM, which is expandable to 16K with an add-on memory module for another $100. It also comes with an 8K BASIC ROM, which gives it excellent programming capabilities for its size. The computer's touch sensitive keyboard includes 22 graphics keys which allow you to create animated graphics on your television set.

The ZX81 is manufactured by Sinclair Research Ltd. of Cambridge, England, and is available by mail order. For ordering the catalogue information write: Sinclair Research Ltd., One Sinclair Plaza, Nashua, NH 03061.

Byte for byte, the **Commodore VIC 20** is one of the best computer buys on the market. At $299, the VIC has a full touch-type keyboard which makes it unique in this price range. The computer comes with full BASIC and access to over 60 graphics characters. The user also has full control over the combination and placement of the VIC's eight colors on the video display, which would most likely by your color television set. This little Commodore comes with 5K of programmable memory and it is expandable up to 32K. You will need a Commodore cassette recorder ($75) for auxiliary storage.

The biggest drawback of the VIC is its 22-character lines—most competitive computers are 32 characters—which can be painfully short when programming or printing out data.

The **Radio Shack Color Computer** is $399 and offers many of the same features as the Commodore machine. It has 4K bytes of RAM and can be expanded to 16K for another $119. Extended BASIC for more sophisticated music and graphics can be added for $99. The quality of the computer's characters on your television set will appear slightly better than those of the VIC, but the keyboard's square keys make it feel more like a calculator than a typewriter.

One of Radio Shack's strengths is vast array of software that is constantly being developed for its products. Since the company has a network of stores, it shouldn't be hard to find one nearby if repairs are necessary.

If you have trouble plugging in your toaster, and the just idea of hooking up a computer at home throws you into apopletic shock, then the **Atari 400** is for you. Atari has done its best to build computers that are people-proof, or "user friendly," as they like to say. Everything is of modular design, which allows you to hook up the system's components more easily than your stereo system.

The 400 is $399. It comes with 16K bytes RAM and is not expandable. However, hooked up to your color television, Atari has the best graphics and music available for the price. Games like Star Raiders will keep you going for hours. The biggest drawback of the 400 is its monopanel keyboard, which doesn't allow for touch typing. If you are doing a lot of programming this would be a drawback.

Moving beyond the games, education and simple programming spectrum, the **Radio Shack Model III** is a good buy if you want to get into serious programming at a reasonable cost. The Model III is $999 and it includes 14k bytes ROM (with a good BASIC) and 16k bytes RAM, expandable to 32K. This system has a built-in black & white monitor. As with most systems, you will need a cassette recorder for storage, or you might want to consider installing a disc drive for $839. Just an idea.

The **Atari 800** is the 400's big brother. It lists for $1,080 and comes with 16K bytes RAM but, unlike the 400, is easily expandable up to 48K by popping in RAM "Memory Modules." These cost $100 per 16K. The Atari 800 has a touch-type keyboard and its expanded memory allows it to run more business-oriented accounting and word-processing programs. And Atari is still best for color graphics, music, and games.

Apple fever hit the U.S. last year and turned the **Apple II Plus** into the hottest thing since Bo Derek. People were breaking down the doors of their neighborhood computer shop to get one. At $1330, the II Plus is a top-shelf personal computer and, if money is no object, go for it. This Apple model is expandable to 48K. As usual, you will need a cassette recorder or disc drive to load and save programs. You will also need an RF modulator ($40) to hook the Apple up to your television set.

There is probably more fancy software available for the Apple than any other personal computer. Everything from the Dow Jones Portfolio Evaluator to Apple Music Theory make this one of the sexiest and most expensive personal computers you can buy.

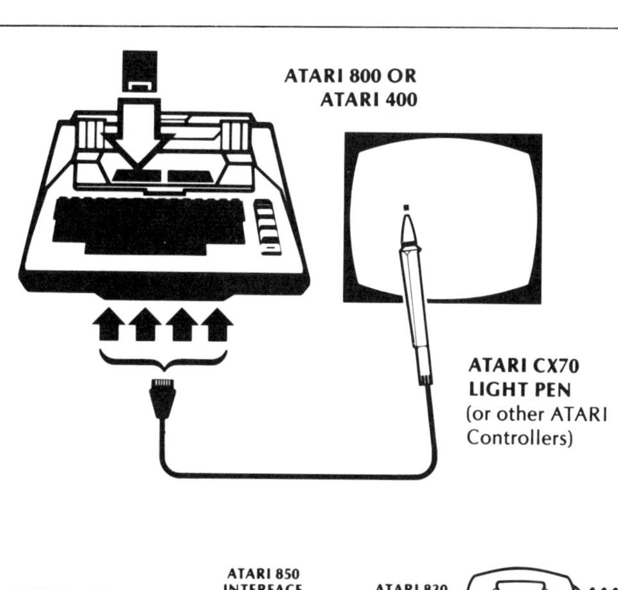

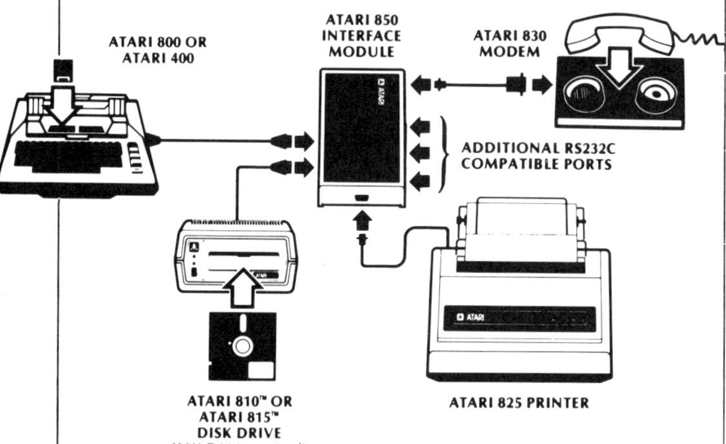

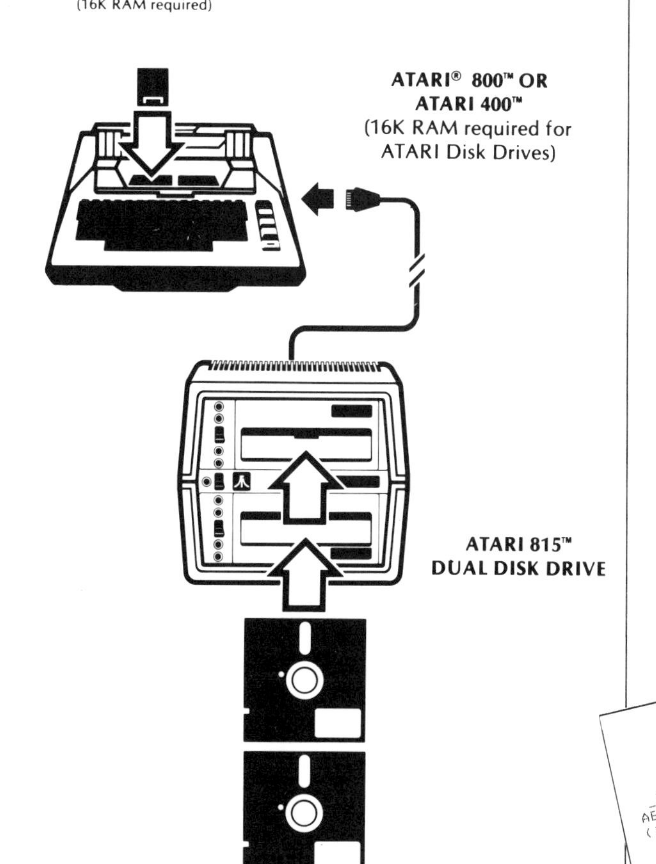

What's a Peripheral? or, "Oh, That'll Cost You Extra"

"Peripherals" is a euphemism for the hidden costs of computers. It's not too expensive to get yourself a computer, and it's also cheaper to buy a car without tires. The simple fact its: If you want to make the most of your computer's potential, you are going to spend at least as much on peripherals as you did on the computer itself and probably twice as much. Be prepared for that.

The primary peripheral or "major option," as computer manufacturers like to call them, is auxiliary memory. The alternatives here are cassette tape recorders and disc drives, and since we don't really consider this an optional purchase we dealt with it in the main text.

Never assume that the monitor sitting on top of the store's computer is included in price of the system. If the computer doesn't have a built-in video display, check to see what's necessary to hook it up to your television set at home. To use your home set, you will need an RF modulator, a small box which is clipped onto the television's antenna and wired to the computer. The RF modulator is built into the Atari but costs extra with the Apple ($40). Always ask about it.

People who get serious about their computers will want a printer. Basically, they break down into dot matrix and fully formed impact printers. Unless you must have typewriter-quality copy for business letters, stick with dot matrix. You must also decide whether you want the printer to be able to handle graphics and whether it should print 40 or 80 columns. Atari's printers range from $300 (40 column) to $1,000 (80 column), while Radio Shack's line goes from $219 to $1,960. Sinclair is introducing its tiny ZX Printer this year for $100. No matter what printer you buy, make sure it is compatible with your computer. The interface necessary to connect the two usually costs extra.

Linking computers to the large data banks now available to the home user requires an acoustic coupler or modem. This device holds your telephone receiver and converts electrical signals so that your computer can talk with another computer. Modems range from about $150 to $300.

These are just the basics. There are new gadgets coming out every day that you can attach to your computer.

SAMPLE PRINT MODES FOR ATARI 825™ PRINTER

PRIMARY CHARACTER SET [ESC,CTRL S]

ELONGATED PRIMARY [START=ESC,CTRL N STOP=ESC,CTRL O]

CONDENSED CHARACTER SET [ESC,CTRL T]

ELONGATED CONDENSED [START CONDENSED THEN START ELONGATED]

Eight Tips on Shopping for a Computer

● Do your homework. One of the best sources of information, is *The Secret Guide to Computers*, published by Russ Walter, 92 Saint Botolph St., Boston, MA 02116; phone: 617-266-8128. Walter has published eight volumes at $3.70 each, and buying the first four Volumes is probably the best investment you can make before buying a computer. Each volume is highly entertaining and is jammed with unbiased information on why this computer or program language is better than that one. Once you buy one of his books, Walter will answer any computer questions you have, 24 hours a day, free of charge.

● Know what your needs are and what you want to do with a computer before you go looking. Determine the capabilities you must have and those you are willing to trade off.

● Make an appointment before going to a computer store. You will get much better attention if they know you are coming to look at a particular computer.

● Understand that many computer salespeople don't know much more than you do, and they try to cover up their ignorance by snowing you with jargon. Good salespeople will patiently explain things and will urge you to take your time and compare systems. If you find a good salesperson, take him or her to lunch, and maybe introduce him or her to your best friend. Good help is hard to find.

● Know that most computer store-salespeople are trained in intimidation. They are great at making you feel like a jerk for even considering a computer sold by the competition. If you don't believe us, go tell a guy selling Apples that you really liked what you saw at the Radio Shack store—and vice versa.

● Always ask about service and support for both hardware and software. Will the store service it when it breaks down? Who will answer your questions about software? After-sale support is one of the things you are paying for at a good store. It's worth every penny.

● Consider discount houses. You can save a bundle if you buy hardware from discount dealers, usually by direct mail. But remember, the money you save would have gone to after-sale support. Be sure you know what you are doing with the equipment, or that you have a neighbor who is a computer expert.

● Take your time. Buying a computer is a process, and shopping for one can be great fun.

THE BUYER'S GUIDE TO HOME

Computer/Company	List Price	RAM included	Maximum RAM possible	ROM included	Keyboard	Video screen size columns/lines
Apple II Plus Apple Computer, Inc. 10260 Brandley Dr. Cupertino, CA 95014 408-996-1010	$1,330	16K	48K	12K	full-size keyboard	40x24
Atari 800 Atari Computer Division P.O. Box 427 Sunnyvale, CA 94086 800-538-8547	$1,080	16K	48K	10K	full-size keyboard	40x24
Radio Shack Model III	$999	16K	48K	14K	full-size keyboard	64x16
Atari 400	$399	16K	16K	10K	monopanel; slightly smaller than normal	40x24
Radio Shack Color Computer	$399	4K	16K	8K	full-size square keys like calculator	32x16
VIC 20 Commodore Business Machines 950 Rittenhouse Rd. Norristown, PA 19403	$299	5K	32K	N.A.	full-size keyboard	22x23
Sinclair ZX81 Sinclair Research Ltd. One Sinclair Plaza Nashua, NH 800-543-3000	$150	1K	16K	8K	touch-sensitive; flat panel, smaller than normal	32x24

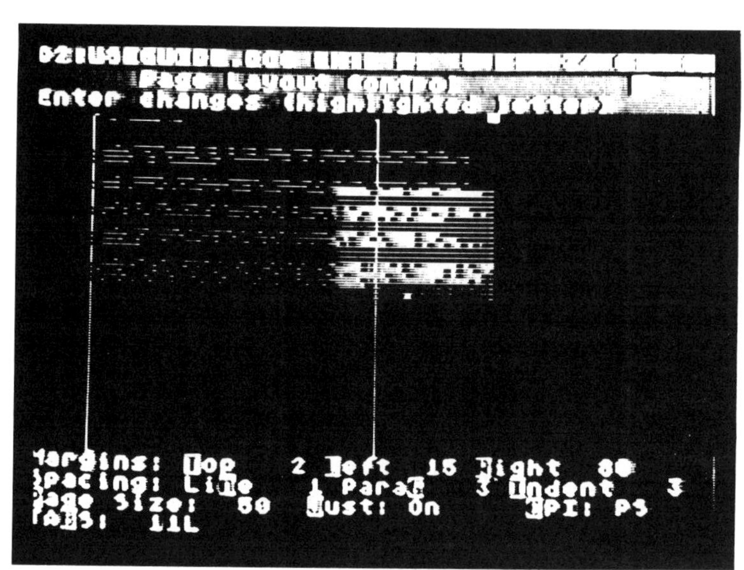

With the addition of a word processing package a modest-priced home computer like the Apple II or Atari 800 can be upgraded into a powerful text editor.

COMPUTERS

Color available	Graphic characters available	High-resolution graphics available	Music available	Microprocessor used
Yes	No	Yes, 280x192	Yes	6502
Yes	Yes	Yes, 320x192	Yes	6502
No	Yes	No	No	Z80
Yes	Yes	Yes, included 320x192	Yes	6502
es	No	Yes, extra cost 256x192	Yes	6809E
es	Yes	Yes, extra cost 176x176	Yes	6502A
o	Yes	No	No	Z80A

A WORD ABOUT SOFTWARE

Software is the computer programs that you must buy to make your computer more than a piece of furniture. Without software, computer hardware is useless. If you are going to be doing any programming, the most important software you will buy will be BASIC, the programming language universally used by beginners. There are different dialects of BASIC in different computers. If you think you are serious about programming, seek some advice on the pros and cons of the various versions of BASIC available.

The software market has grown up largely independent of the hardware market. While the differences in the quality of hardware systems have been decreasing, the quality of software has grown more diverse. A clear advantage of buying a computer from a major manufacturer like Radio Shack or Apple is the incredible array of software that is available for these machines. However, unless the particular program is being marketed by the manufacturer, it's up to you to determine the quality of the program you are buying.

Software comes in either cassette tapes or cartridges or on discs. Obviously, you must check to make sure a program is compatible with your hardware. "Documentation" is the key word in finding good software. No matter how good a program is, if you can't understand the user's manual that comes with it, the program is useless. Always try to find out how good the documentation is before buying a piece of software.

The other important consideration is "support." Will the developer support the software once you have it? Will someone be available to answer your questions (usually by phone) when you have problems and can't find the answer in the user's manual? This has been a critical problem in the early stages of the personal-computer software industry. With rare exceptions, your local computer store doesn't have the time or expertise to help you get each new program running. You must depend of the software developer for help. Always ask before you buy if there is a phone number where you can get help.

PROJECTION TV

Face it. Once you're on a cable hookup, or own a VCR or videodisc player, it's just a question of time until you grow dissatisfied with your present TV set. Somehow *Being There* isn't like being there at all—when Peter Sellers and Shirley MacLaine are only 11 inches tall.

You are now ready to attain the third level of video-consciousness. You are awakening to the joy of projection TV.

All big-screen TVs do the same basic thing. They project the video image through an optical lens into a screen—a screen many times larger than any conventional TV. But not all big-screen TVs produce the same results.

The simplest projection TV takes a standard picture tube—usually from a 13 or 19-inch portable—and magnifies the image onto a reflective screen using a single large lens. This method has one big advantage—it's cheap. Some "conversion" kits sell for under $500.

Unfortunately, the single-lens system has one drawback. It doesn't work very well.

For one thing, the laws of optics guarantee that the projected image will only be about 1/10th as bright as the tube you are using. How bright is that? Try watching your set through 3 pairs of sunglasses and you'll get the idea. Just about any light source brighter than a penlight in the same room with a single-lens system will leave the image looking faint and washed-out. Which is no problem, if you happen to live in a cave, or do all your TV viewing after sunset.

But that's only half the problem. Conventional TV tubes have a black "shadow mask," or grille, which is not noticeable during ordinary viewing—but enlarge that mask 3 or 4 times and not only is the image faint, it is also fuzzy and ill-defined.

In all fairness, there are thousands of people using single-lens projection TVs right now who are perfectly content with the results they are getting. Maybe you will be, too. But if you have your heart set on converting your TV room into an honest-to-goodness video theater, we heartily uggest you consider a .3-lens 3-tube projection TV.

While still not comparable to the picture you'll see on the silver screen at the Downtown Triplex, a 3-lens projection TV has even the best 25-inch console beat every way but sideways.

And why not sideways too? It seems that all projection TVs, even the $5,000 top-of-the-line models, suffer from image "fall off" and "color shift" if viewed from too far to either side. The optimum viewing position for big-

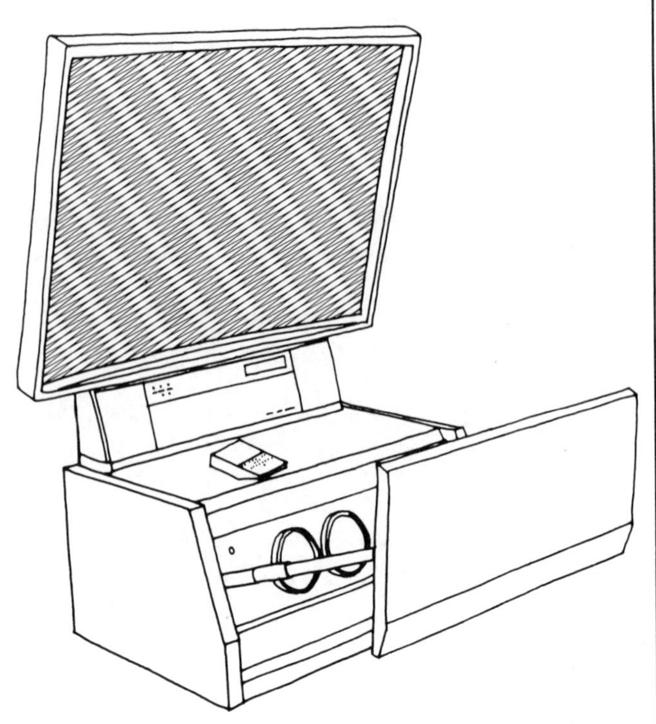

The one-piece front-projection model leaves you more room for living.

screen TV, in other words, is right on the 50-yard line. As you move toward the end zone, the corners of the screen grow dim and the colors begin to blur. This is true whether you are watching the Cowboys drubbing the Saints, or Debbie doing Dallas.

The triple-lens system uses 3 picture tubes—one for each of the primary colors: red, green, and blue. This method, invented by Henry Kloss in the mid-1960s, has the advantage that each tube is specifically designed for projection. There is no mask to blur the video image, and 3 tubes are obviously going to be a whole lot brighter than one.

Generally, the highest level of picture quality is achieved with a "2-piece," front-projection system—although these cumbersome 2-part units with their separate projection console and huge viewing screens are also the most difficult to live with since they tend to dominate any room smaller than the Astrodome. Advent, Kloss Video, Mitsubishi, and Panasonic all market well-respected 2-piece system. List prices range from $2,500 to $5,000. The minimum distance between the projection console and the screen varies from 8 feet for some models to as much as 14 feet for others. Since you'll want to sit *behind* the projector, you're going to have to dedicate between 80 and 200 square feet of living space to your 2-piece video theater. Considering the average new *house* these days has only about 1,200 square feet, this is a serious commitment.

In addition, most 2-piece systems have a free-standing reflective screen that is at least 72 inches (measured diagonally, like small screens), or about the size of the average 6-place dinner table. Unless you take great pains to conceal it—such as placing the screen behind full-length curtain—the screen will inevitably become a permanent, and visually dominant, part of your living environment.

Fortunately, the projection console part of the 2-piece system is far less obtrusive. Projection consoles on models such as the Sony KP-5020 and the Kloss Novabeam Model One are functionally disguised as contemporary coffee tables.

Since they use an optical lens system, all projection TVs must be carefully focused—and with 3 lenses to coordinate, this can be a demanding task. Two-piece systems, particularly, require meticulous aiming and color tuning. Should you accidentally nudge either the screen or the projection console, you'll have to do it all over again—or else suffer with a picture that looks more like the light show at The Electric Circus than "Good Morning America."

If you don't want to sacrifice picture quality but are willing to settle for a slightly smaller screen (in the 48 to 60-inch vicinity) you should investigate the one-piece front-projection TVs marketed by Advent, GE, Mitsubishi, Quasar, RCA, Sanyo, Sears, Sony, and others. The one-piece front-projection systems use a mirror assembly to allow the projection tubes *and* screen to be combined in one cabinet. They occupy about as much floor space as a small sofa or love seat and can be comfortably viewed at distances within 15 feet. With some one-piece consoles—the Sears Projection TV is an example—the screen can be removed. Another one-piece front-projection TV, the "hutch" designed Mitsubishi VS510UD, can be completely concealed within a handsome hardwood buffet.

If limited space is a serious problem, your best bet in big-screen TV may be a *rear-projection* model. GE, Sylvania, Quasar, Panasonic, and Magnavox all offer units that won't steal much more floor space than a conventional console TV. Rear-projection TVs also have what some may consider a slight aesthetic advantage—they actually look like a TV set, only bigger. Many of the front-projection systems, on the other hand—with their 5 to 6-foot concave reflective screens—look more like something off the flight bridge of the starship *Enterprise.*

As with the one-piece front-projection models, the rear-projection TVs also use a mirror assembly to combine the projection tubes and screen into one cabinet—only now the image is projected *through* a translucent plastic screen. As a rule, rear-projection TVs have the smallest screens (45 to 50 inches), and the image may not be as bright as on some front-projection models. This is because both the mirror and the plastic screen on rear-projection models tend to reduce light transmission. Several deluxe rear-projection TVs, such as GE's Widescreen 4000 and Magnavox's 8505, are offered with optional furniture "coordinates" that will house—and display—stereo systems, VCRs, videodisc players, home computers, and video games.

All 3 types of multi-lens projection systems are, of course, compatible with almost any home video device or accessory. Also, in anticipation of the commencement of stereo TV broadcasts by the mid-1980s, several big-screen TVs—including models sold by Sears and GE—are being marketed with built-in stereo amplifiers. Until the real thing is available, the Sears model will even provide you with "simulated" stereo by feeding the monaural signal from you VCR or broadcast TV program through a special "phase/delay" circuit. If you own a videodisc player with stereo capability, a big-screen TV with stereo will let you take full advantage of LaserDisc programs like *Saturday Night Fever.*

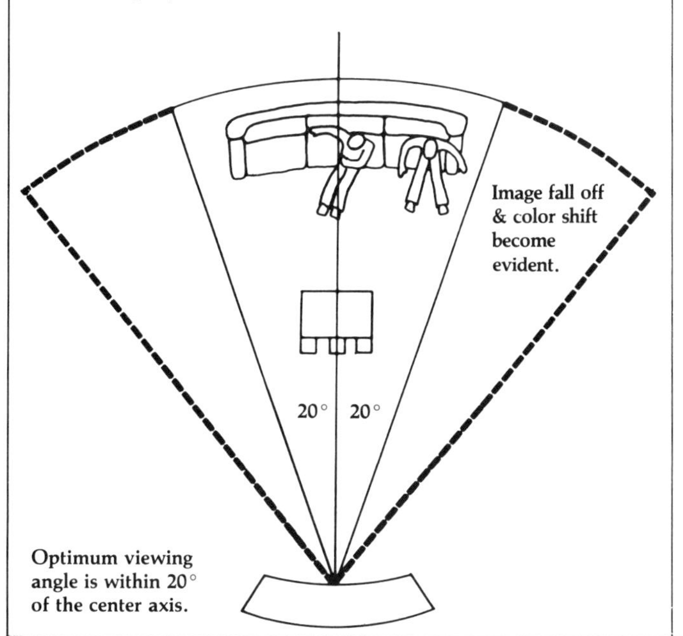

Optimum viewing angle is within 20° of the center axis.

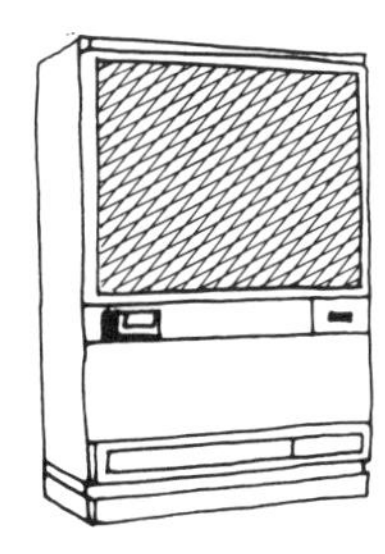

Rear-projection TV saves space, but offers a smaller screen.

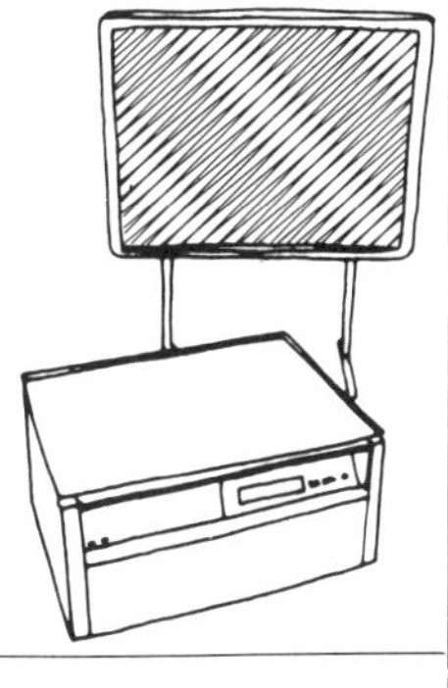

The 2-piece front-projection model offers the ultimate in big screen viewing—if you have a room for it.

Shopping For A Projection TV

Don't judge today's generation of projection TVs by those faint and fuzzy-screened space-hogs you may have seen around sports-oriented bars in years past.

While it's true there have been no big breakthroughs in projection TV since the 1960s, there have been improvements. Sony uses liquid coolant in its deluxe units to keep the tubes cool and the picture bright. Henry Kloss has refined his original projection tube, producing a significantly brighter—and sharper—big-screen video image. A new electronic circuit, called the comb filter, is also being used to upgrade big-screen resolution. Front-projection screens are now being made of a washable, ultra-reflective coated aluminum, while rear-projection screens are beginning to employ a recently engineered Fresnel field lens to help reduce fall off. All of these factors help to make projection TVs brighter, sharper, easier to live with, and easier to afford than ever before. But, as we mentioned earlier, don't expect big-screen video to compete with 70mm Technicolor. It can't. Not yet, anyway.

Shopping for a projection TV won't be easy. With total annual sales in the vicinity of 100,000 units, the big screen TV isn't exactly a mass market item. To assist you in your quest for the perfect big screen unit for your circumstances THE SOURCEBOOK has compiled the 'Ten Commandments" of projection-TV shopping.

I. *Thou shalt read the specifications.* As with VCRs, you can learn a lot by reading the fine print. Pay special attention to "Screen Size," "Dimensions," "Weight," "Circuitry," "Projection System," and "Picture Brightness." This last spec—picture brightness—is the most important of all. Our favorite model, the Kloss Novabeam, claims a brightness of more than 80 foot-lamberts. This is a good basis for comparison. The lower the number of foot-lamberts, the dimmer your screen will look.

II. *Thou shalt know thy living space.* Don't wait until a 2-piece unit has been delivered to decide you just don't have room for it. Decide where your projection TV is going, then find a unit that fits your available space.

III. *Thou shalt know thy warranty.* Some big-screen manufacturers stand behind their product with full parts-and-labor warranties for one year. Some don't. If you buy from a video dealer with a large service department, he may offer his own warranty.

IV. *Know they dealer's service capability.* Check out your dealer's service department. It is orderly and well equipped, or is the lab bench propped up with a stack of unread service manuals? If you buy from a discounter or a department store, make certain there is a factory-approved repairman in your vicinity. Some rear-projection units weight over 400 pounds, so shipping your big-screen TV back to Peoria for service is no joke.

V. *Thou shalt not buy an Ektalite screen.* You may run into a model that is being sold with a Kodak Ektalite screen. Ektalite is an excellent reflective material, with one disadvantage—touch it and the imprint is there forever. We think the new coated aluminum screens are just as reflective but far more durable. Especially if you have curious children in the house.

VI. *Know thy viewing angles.* When deciding where your projection TV is going to be installed, keep in mind that the best viewing is at the center of the screen. Plan your seating area accordingly.

VII. *Know thy reception quality.* A big-screen image is only as good as the signal you put into it. In fact, if you've got little problems with you TV reception now, you'll have big problems with your projection TV. Those tiny flurries of video snow you see on your 19-inch portables will become a full-fledged Klondike blizzard on the 72-inch screen. Likewise, you may be able to get away with a so-so quality videocassette on your portable, but not on the big screen. Unless you've got top quality reception you're better off forgetting about the big screen, unless the only thing you plan to watch are prerecorded tapes or discs.

VIII. *Thou shalt consult thy housemate.* As you will quickly discover, there are many unseen advantages, and disadvantages, to owning a projection TV. Your uninvited-guest frequency is likely to skyrocket, especially on big movie nights and football weekends, and your domestic harmony-to-noise ratio is likely to plummet—unless the other members of the household share your enthusiasm for the big-screen lifestyle.

IX. *Thou shalt beware of they dealer's showroom viewing conditions.* Sometimes the conditions under which a big screen TV is demonstrated are ludicrous. Our local big-name department store has one Sony big screen directly under a bank of flourescent lights. Obviously, the picture looks terrible, although the unit in question happens to be an excellent projection TV. Then there's the opposite extreme. The swanky viewing room with chocolate walls, a midnight blue rug, and recessed lighting on a dimmer switch. Chances are that slightly overweight salesgirl will look terrific in there—and so will that slightly under-powered projection TV.

X. *Thou shalt inveigle thy dealer for a trial run.* It wasn't that long ago that you could buy a mid-priced car for what you'll be paying for a projection TV. If you do decide to buy from a video dealer (as opposed to a department store or discount outlet) there's a good possibility the owner will let you try out a model on approval. Do it.

©1982 THE HOME VIDEO SOURCEBOOK

The Kloss Novabeam
Model One
Projection Color Television
List price: $2,995
Kloss Video Corporation
145 Sidney Street
Cambridge, MA 02139
617-547-6363

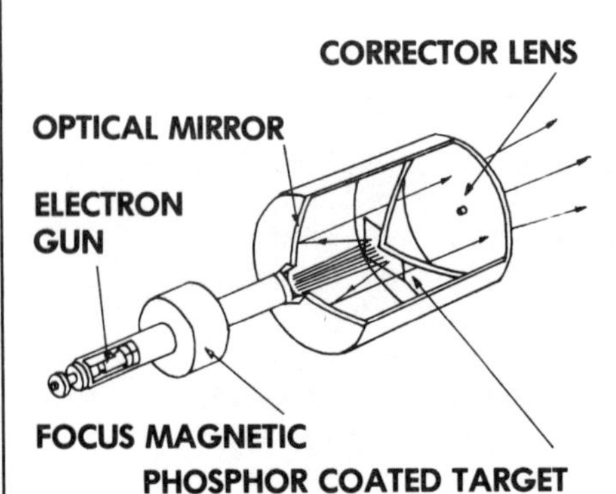

Before you invest anywhere from $2,500 to $5,000 in a projection TV we recommend you look at the Kloss Novabeam.

True, the name Kloss isn't exactly a household word. But it was Henry Kloss who pioneered the 3-tube projection television at Advent in the 1960s. Today Kloss is on his own, and we think the 2-piece projection system that bears his name merits a close inspection.

The key feature of the Novabeam is a new projection tube engineered by Henry Kloss and his staff. It differs from most other projection-TV tubes in two ways. First, the "Novatron" tube uses magnetic—instead of the conventional electrostatic—focusing. Kloss claims this system produces finer picture detail. After seeing the Novabeam, we agree. In addition, Kloss has managed to eliminate most of the glass-enlarging lenses (which are expensive and cut down on picture brightness) from his projection system. In their place Kloss has designed a mirror reflex system—similar to the one used in high-speed telephoto surveillance lenses—which is actually built into the projection tube. The result is a simpler and more economical tube that produces an image that is as sharp and brilliant as anything we've seen in projection TV. Unless there are direct rays hitting the screen, you'll find the Novabeam bright enough for viewing even in a sunlit room.

As for other features, the Novabeam uses a Magnavox chassis in its projection console—always a good sign since the Magnavox is widely considered to have some of the most advanced color circuitry around. There is also a "Detail" switch that helps improve a poor picture by filtering out some of the high-frequency video noise.

As with any 2-piece projection TV, the Novabeam's biggest advantage is also its biggest drawback. Separating the projection console and the screen gives the brightest possible picture. It also takes up a lot of living space and creates the need for frequent focusing and color tuning.

Kloss deals with these problems in two ways. First, the projection equipment is housed in a relatively small cabinet (27½ inches wide by 22 inches deep and 18 inches high) which can do double-duty as a small coffee table—although we have our reservations about actually placing anything here which might spill into the control panel. The distance required between the back of the projection console and the back of the free-standing screen is slightly over 9½ feet—which means the Novabeam could be located in the average den or TV room. As for focusing and color tuning, Kloss has managed to make this process as painless as possible. The Novabeam even projects its own cross-shaped test pattern to assist in making the lens "convergence" adjustments.

With the Novabeam what looks good also sounds good. Those familiar with audio components may recognize the Cambridge, Massachusetts, firm of KLH. You guessed it. The "K" stand for Kloss. Let it suffice to say that the sound you'll get from the Novabeam will be a quantum improvement over the speaker in your present TV. Kloss actually uses the video screen to reflect both the image, and the sound, back to the audience. This way Peter Sellers and his voice seem to be coming from the same place, not separate corners of the room.

Other features worthy of note include the comb filter circuitry for high resolution and a wireless remote controller with electronic random access to all VHF and UHF stations.

BUYER'S GUIDE TO MULTI-LENS PROJECTION TV'S

Brand/ Model	List Price	Screen Size	Design	Features
Advent				
VBT-100	$2,500	50 inches	Front-projection console (1-piece).	*3-Tube, 3-lens, "table-top" styling.*
VB 225	$3,330	72 inches	Front-projection 2-piece.	*Video/RF inputs, remote control, mfr. claims 70-foot-lamberts.*
VB 125	$4,000	60 inches	Front-projection console	*Video/RF inputs, electronic tuning, remote control.*
Curtis Mathes				
F517R	$4,000	60 inches	Front-projection console.	*Simulated walnut finish, 105-channel remote control tuning, comb filter, claims 120 foot-lamberts.*
General Electric				
3000	$3,500	50 inches	Rear-projection console.	*Remote control, 4-speaker sound system.*
4000	N/A	45 inches	Rear-projection console.	*Stereo sound, remote electronic tuning, comb filters, average brightness of 45 foot-lamberts.*
Kloss Video				
Novabeam One	$3,000	78 inches	Front-projection 2-piece.	*See sidebar.*
Magnavox				
8505	$3,500	50 inches	Rear-projection console.	*105-channel electronic tuning with remote, rapid scanning, stereo, comb filter.*
Mitsubishi/MGA				
VS-510	$4,000	50 inches	Front-projection console.	*Remote control, time channel on-screen display, "hutch" designed hardwood cabinet with optional hinged screen door.*
VS-707	$3,500	72 inches	Front-projection 2-piece.	*Remote control, 80 foot-lamberts.*
Panasonic				
CT-4500	$3,300	50 inches	Rear-projection console.	*Remote control, direct audio/video inputs.*
CT-6000A	$4,000	60 inches	Front-projection console.	*Remote control, electronic tuning, 50 foot-lamberts.*
Quasar				
PR4800	$3,750	50 inches	Rear-projection console.	*Remote control, CATV tuner.*
CT-6000A	N/A	60 inches	Front-projection console.	*On-screen channel display, electronic tuning, remote control*
RCA				
PFR-100R	$3,200	50 inches	Front-projection console.	*Wireless electronic remote, 12-channel scanning, stereo.*
Sanyo				
PV-5080R	N/A	50 inches	Front-projection console.	*Stereo, public address system, wireless remote control, comb filter.*
Sears				
5450	$3,000	50 inches	Front-projection console.	*Electronic remote, stereo sound, comb filter.*
Sony				
KP-5040	N/A	50 inches	Front-projection console.	*3-Tube system, CATV switch, stereo sound, 45° viewing area, remote.*
KP 5000	$3,700	50 inches	Front-projection console.	*Remote control, VIR tuning.*
KP 7220	$3,000	72 inches	Front-projection 2-piece.	*Remote control, VIR tuning, smoked glass top on projection assembly. Model 5220 available with 50 inch screen.*
Sylvania				
Super-Screen	$3,500	50 inches	Rear-projection console.	*Remote control, 4 speakers.*
Video Concepts				
5060/3R	$4,000	72 inches	Front-projection 2-piece.	*Projection assembly in oak coffee table, on-screen read-outs.*
6050/3R	$3,800	60 inches	Front-projection console.	*Oak cabinet, remote control, on-screen read-outs.*

VIDEO CAMERAS

$750 to $1,300 for a color camera? Yes, video cameras *are* expensive. But before you write off the idea of shooting your own video tapes, consider this: Film and processing for 4 hours of Super 8 will cost you well over $1,000. How much would you pay for 4 hours of video tape? Under $30.

You can also look at it this way: A few years ago all color cameras had to be custom ordered and were hand assembled, one at a time. The price? Often in excess of $5,000. By comparison, today's deluxe $1,300 color video camera is a big bargain.

Will video cameras get any cheaper? Yes, but probably not by much. The next step will be to a camera that not only translates visual images into video signals, but actually does the recording itself—eliminating the need for a portable VCR. Sony has demonstrated a prototype but says mass production is still 4 or 5 years away.

And what if you don't have $750 to $1,300 to spend on a color video camera, but want to do some taping now? One way is to rent a camera. Check in your Yellow Pages under "Video." Prices can range from $35 to $75 a day. This isn't a bad way to do it if you want to get some occasional footage of your kids while they're still kids but you don't really want to lay out the cash for a camera you'll hardly ever use. If you're willing to settle for black and white, you can pick up a camera for not much more than the cost of a couple of prerecorded movies. JVC, Sanyo, and Sony each offer one black-and-white model. The cheapest of the bunch, Sanyo's VC-1400, doesn't have many features, but the price is right—$199 retail.

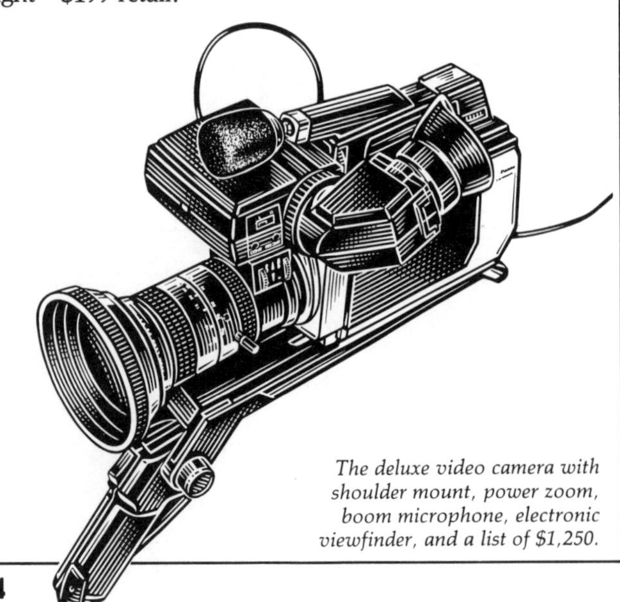

The deluxe video camera with shoulder mount, power zoom, boom microphone, electronic viewfinder, and a list of $1,250.

Contrary to what the manufacturer's sales literature may lead you to believe, any color video camera can be used with any VCR—portable or console. You can even use a Sony camera (Sony—read Beta format, right?) with a VHS-format recorder, or a JVC camera with a Beta-format VCR. The only hangup—there's no standardization of cable pins, so you may need an adapter and possibly some minor rewiring.

You may also find that a $750 color camera will serve your needs just as well as a $1,300 model. Here's a primer on how the color video camera works—and why you may be able to get along fine without some of the more expensive features.

Inside A Color Video Camera

In some ways the video camera is the mirror image of your TV receiver. While the TV converts electronic signals into recognizable patterns of light (video images), the video camera turns light into electronic signals. It's that simple.

The basic features of all video camera are essentially the same. At the front is the lens. At the back is a viewfinder that enables you to aim and focus. Within the camera body you will find the so-called "vidicon" tube which actually creates the video signal. In addition, there is a maze of amplifiers, limiters, and circuits to shape—and color—the video signals, as well as a microphone to capture the sound and a button to start and stop your recording. Since video cameras don't, as yet, paint the actual electronic signals onto the video tape, you'll also find an "umbilical" cable that brings operating power to the camera, and a video signal to your VCR.

Lenses and Lens Mounts

Many of today's video cameras use what is known as a C-mount to attach the lens to the camera body. You may still encounter some models that have a lens permanently attached, and no C-mount. Our advice: Forget it. Find yourself a camera with a C-mount that allows you the flexibility of changing lenses to suit the lighting conditions and the scope of the scene you are filming.

There are two basic styles of lenses. The so-called fixed-focal-length lenses (extra-wide angle, wide angle, normal, telephoto, etc.) and the zoom lenses. In video cameras a "normal" lens—one which gives you a picture fairly much the same as you perceive it—is about 17mm. A wide-angle lens is 12mm, while a telephoto lens is 25mm or more. You may find that you can buy a C-mount adapter that will let you use the lenses from your 35mm still camera. It's worth a try, but the results are usually rather poor.

The alternative to owning a collection of fixed-focal length lenses is to buy one zoom lens. For most home video applications, we think this is the best bet. A typical 6 to 1 ratio video zoom lens will cover focal lengths from about 12 to 75mm—or everything from a wide-angle shot that will take in most of the football stadium to a telephoto that will give you a tight close-up of your favorite cheerleader. Some zoom lenses are also available with "micro" capability which will let you focus on an area not much larger than a postage stamp.

Evaluate zoom lenses in terms of their smoothness of operation and ability to keep a subject in focus while zooming. Power-zoom with a manual override is a

TOSHIBA
IK-1850

Portable Color Video Camera.

- F 1.4 6X zoom lens 11 to 70 mm
- Electronic viewfinder
- Low light operation 50 Lux
- Remote trigger pause control
- Built-in microphone
- Auto shut off vidicon burnout protection

popular option. Auto zoom, however, can cause distinct hum which may turn up later on your audio track.

Video lenses have an iris that allows you to adjust the amount of light that enters the camera. On most lenses the iris, or aperture, settings are calibrated in *f-stops.* With f-stops, the *lower* the *number* the "faster" the lens is said to be. And the faster the lens, the less light you need to tape. Lenses with a speed in the vicinity of f 1.8 to f 2.1 are common on most basic cameras. On more expensive models you will see f-stops as fast as f 1.6 or f 1.4. Obviously, the faster lenses will let you shoot in darker conditions, but for general purposes we don't think the added speed warrants the added expense. Unless you've got money to burn, we suggest you stick with the slower lenses.

One reason we don't think you need to spend the extra money on a faster lens is that video cameras are far more sensitive to light than color film cameras. In the "Buyer's Guide to Video Cameras" you will notice a column that reads "Minimum Light Sensitivity." The figures in that column are given in a unit of measurement known as footcandles (f.c.). Since the typical 75-watt lightbulb puts out over 500 footcandles—but most video cameras need as little as 10 footcandles of light—it should be obvious that you'll be able to use your camera almost anyplace that there's enough light for you to see to focus.

Your Vidicon Tube

The secret to this remarkable sensitivity is an ingenious device known as the Vidicon tube. The vidicon is a glass vacuum tube with a light-sensitive surface at one end. Just as a film camera focuses an image on the "film plane," the video camera focuses an image on the light-sensitive area of the vidicon tube, where light values are electronically scanned and processed into an electronic signal.

Since the vidicon tube is so ultra-sensitive to light, you must be very careful never to point your camera at the sun, or any other source of brilliant illumination (most of the romantic cinematography you see using direct pictures of the sun is done on film, not videotape). Ignoring this warning will result in a "burn," which remains on your vidicon tube—and hence on your video tape if you continue to record—for as long as an hour after the tube has been exposed to the sun's direct rays. In the case of prolonged exposure, the sensitivity of the vidicon tube may be permanently impaired. Replacing the vidicon can cost several hundred dollars.

When panning with a video camera, the vidicon tube will also often "burn out" the highlight areas of your picture, leaving what is known as "lag" or "ghosting effect." Some cameras are more prone to ghosting than others, which is one of the reasons we strongly recommend you try out several different cameras in the store before making up your mind.

Viewfinders

Under "Lenses" we suggested you avoid any camera that does not have interchangeable lenses. Such cameras also often lack a proper viewfinder—which is a second good reason to pass them by.

What we mean by "proper" viewfinder is a system which allows you to SEE whether your lens is in focus or not. Both the optical *through-the-lens* and the *electronic* viewfinders are fine. What isn't fine is a plastic box on the top of your camera that lets you aim it in the general direction of your subject, but nothing more. Unless you're living on dog food while you save up for your camera, avoid cheapo viewfinders.

If most of your taping is going to be done in the living room using your console VCR, you might as well save some bucks and buy a model with an optical TTL (through-the-lens) viewfinder. You won't be able to play back what you've just recorded through the viewfinder, but you won't need to—you'll have your VCR and television set right there.

On the other hand, if you're serious enough to have invested the extra money in a portable VCR, we think you might as well spend a few hundred more and purchase a camera with an electronic viewfinder. This is actually a tiny 1½-inch-square b&w TV screen that is mounted inside the video camera. You see the image you are recording in just the same way as the vidicon sees it, only your version is in black and white. The big advantage of the electronic viewfinder is if you want to see what you've recorded before you get back to your TV set. Just hit the play button and your viewfinder becomes a mini monitor. Unlike the TV producer on the soundstages of The Burbank Studios, you won't even have to wait for the "rushes" to find out if what you shot is what you wanted.

Auto Iris

This is akin to automatic exposure in a still camera. A light meter is electronically coupled to the iris—whenever the light conditions change, your aperture automatically adjusts. It's a great convenience (just one less thing to worry about) as long as there is a manual override so you can over or underexpose should you decide to try some creative effects like fading to black. If you can afford it, wonderful. If not, you'll survive fine without automatic iris control.

Color Temperature

Just like you and me and all the creatures great and small,even daylight has a "temperature." We call it "color" temperature.

To get a handle on color temperature, you merely have to think of mid-day sunlight as being made up of a full spectrum of colors. Other light sources, however, such as fluorescent and incandescent lights or sunsets and sunrises, lack certain wavelengths, and hence often appear warmer or cooler than mid-day sunlight. Our eyes rapidly adjust to differences in color temperature (unless the differences are extreme, like under mercury vapor lamps where everything looks an eerie green), but color film—and color video tape—does not. To make the color temperature of a scene appear "natural" on screen, it is often necessary to filter the color of the light to make it as close to the color temperature of daylight as possible. Originally this was done on video cameras, as on still or motion-picture cameras, with color filters placed over the lens. Now it is done with a flick of a switch that automatically adjusts the camera's color circuitry. Usually your color temperature corrections can be made by setting a selector switch at "outdoor bright," "outdoor cloudy," or "indoor."

You may also find a separate "White Balance" switch. If you are not getting proper color—and you're certain the problems is not with your TV set—you can follow the instructions set forth in your owner's manual (ah, yes, you should always save the owner's manual because you never know when it's going to come in handy) to correct the white balance of your camera.

Sound and the Video Cameraman

All video cameras have built-in microphones. Some are omni-directional and will pick up any comments you may make as well as anything your subject is saying. On the more expensive models you are likely to find the microphone extended on a short boom. This helps cut down on unwanted camera noises and staging directions. You may, however, want to replace your omni-direction mike with a one-direction model, particularly if you are doing a great deal of telephoto shooting where your subject is at a considerable distance.

Anatomy of Color Video Camera

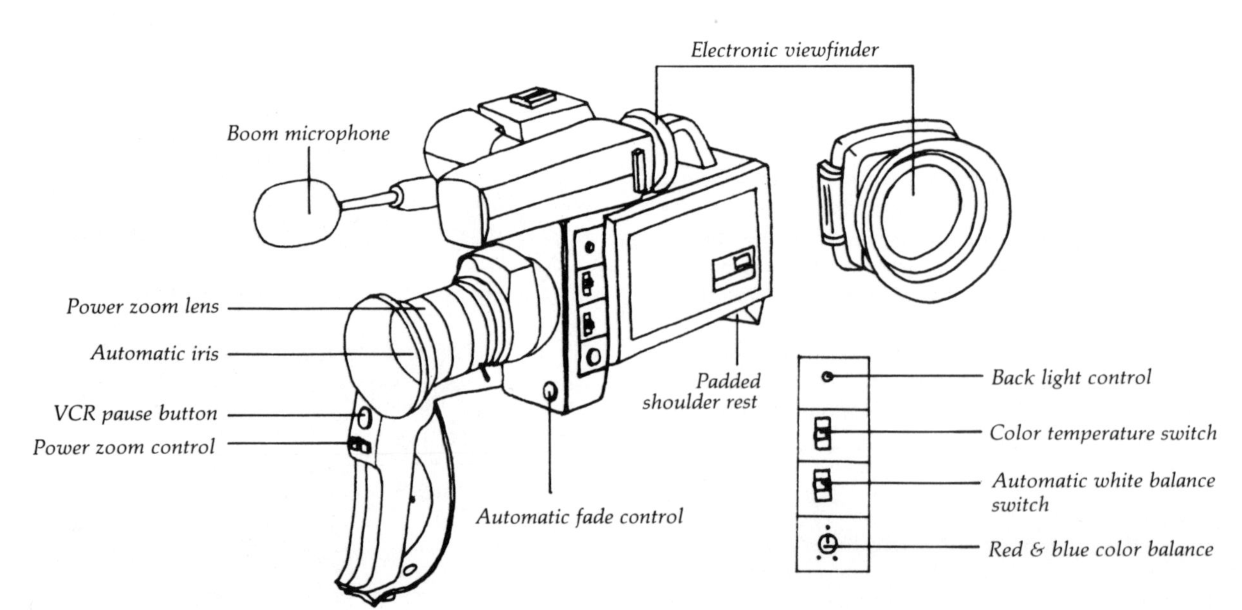

Black & White Video Cameras

Brand/ Model	MFR/ Weight	List Price	Lens/ Mount	Controls	Sig-Noise Ratio/ Horizontal Res/ Power Consumption	Viewfinder/ Minimum Light Sensitivity
JVC GS-1000AU	JVC 3.31 lbs	$375	f2.5, 16-32mm Zoom.	Auto/Manual Shutter, Iris.	40dB/450/8W	Optical/4.7f.c.
Sanyo VC-1400	Sanyo	$199	f1.6, 16mm, C-mt.	Manual Iris	42dB/500/NA	Optical/1f.c.
Sony HVM-100	Sony 2.75 lbs	$300	f2.5, 16-32mm, C-mt.	Manual Iris	42dB/450/4.8	Optical/5f.c.

Other Features

There are a few miscellaneous items you are likely to encounter on some of the newest cameras. These include a "back-light" (BCL) switch which overrides the auto iris for better exposure in back-lit situations. A "fade control" will automatically let you fade to black when finishing a scene or sequence. And a "zoom speed" selector that lets you vary the speed of your power zoom. All nice, but nonessential.

Reading the Specs

Once again, the manufacturer's specifications can tell you more about the camera you are considering than many people realize. Here are the key items to look for:

Horizontal Resolution. The higher the better. 230 lines is adequate, but some models go as high as 350 or more.

Signal-To-Noise Ratio. Just as important here as in VCRs. If the manufacturer doesn't give this spec, you have to wonder why. The higher the number of dBs the better—40 dB is about the minimum we would accept.

Power Consumption. 7.5 watts at 12 volts DC is typical. Higher power consumption rates will run your battery down faster, rates lower than 7.5 watts will let you shoot longer without changing or recharging batteries.

Weight. Modern video cameras are light. In fact most are *too* light. Christopher Reeve might have the nerves of steel needed to keep a video camera rock steady, you probably don't. Unless you—and your viewers—don't object to scenes that pitch and yaw amateurishly across the screen, we strongly urge the use of a tripod or shoulder mount.

Minimum Illumination. This spec gives the absolute minimum amount of light in which you can hope to record a recognizable video image. For most cameras this is in the vicinity of 10 footcandles, which is very little light indeed. True, with video tape you will be able to shoot under lighting conditions that would have a Hollywood cinematographer screaming for klieg lamps. But don't count on great quality. If you want the best results—good contrast and pure colors—you too will have to shoot in daylight, or else use a couple of well-placed flood lamps. A few cameras have a "sensitivity switch" which activates a set of low-light circuits. But even this feature is no substitute for bright, even light.

Before You Buy

If you are already familiar with still photography or traditional filmmaking, selecting the right video camera shouldn't be hard. If not, we suggest you rent a camera for an afternoon. You'll probably be surprised at how easy it is. For one thing, you won't have to wait a week for the film to come back before you can see what kind of mistakes you are making. And the chances are good the dealer will credit the rental against any eventual purchase.

Use "The Buyer's Guide to Video Cameras" to get a feeling for the video camera marketplace. But we urge you not to buy until you've seen—and are satisfied with—the manufacturer's specifications. In video cameras, as with virtually all consumer electronics products, new models are being constantly added, and old models continuously changed to keep pace with the competition.

You may feel more confortable buying a camera of the same brand as your VCR. If you like someone else's camera better, that's fine too. Just remember what we mentioned earlier—there is no standardization of cable pins between brands. Some will fit your VCR, some won't. And some that fit may not be wired the same as your VCR. If you mix brands (and there's no other good reason why you can't), just make sure the camera plug and your VCR receptacle match *and* that all the camera features work the way they are supposed to. If the plug and receptacle don't match, a simple adapter may do the trick—or it may not. Once again, make sure *all* the lights, buttons, and gizmos work. If they don't you are either going to have to rewire the camera plug or find yourself another camera.

One final word. Remember George Lucas, Steven Spielberg, and Francis Coppola have the luxury of being able to "edit" their footage. But unless you have access to a second VCR to use for editing, you are going to have to be satisfied with what you shoot, exactly the way you shoot it—glitches, goofs, and all. You may quickly discover that the secret to videotaping—like family planning—has more to do with preparation than inspiration.

Color Cameras

Brand/ Model	MFR/ Weight	List Price	Lens/ Mount	Controls	Sig-Noise Ratio/ Horizontal Res/ Power Consumption	Viewfinder/ Mimimum Light Sensitivity
Akai						
VC-30	Akai 3 lbs	$800	f1.9, 15-45mm Zoom, C-mt	C/W Bal, Auto/Man Iris.	45dB/240/7W	Optical/10f.c.
VC-65	Hitachi 4.8 lbs	$999	f1.8, 13.5-81mm Zoom, C-mt	Same as VC-30.	45dB/240/7.5W	Electronic/10f.c.
VC-X1	Akai 5 lbs	$1,300	f1.4, 11-70mm Macro Power Zoom	Auto/Man C/W Bal, Auto/Man Iris, Auto Focus & Fade.	46dB/270/8.4W	Electronic/5f.c.
Hitachi						
VKC-750	Hitachi 3.97 lbs	$750	f2.1, 13.5-37.8mm Zoom	C/W Bal, Iris, 2-Position Contrast.	45dB/250/5.8W	Optical/10f.c.
VKC-770	Hitachi 5.1 lbs	$950	f1.6, 14-84mm Zoom, C-mt	Same as VKC-750.	45dB/250/8W	Electronic/10f.c.
VKC-800	Hitachi 6.83 lbs	$1,395	f1.4, 12.5-84mm Power Macro Zoom C-mt	2-Positn C; W Bal; Auto/Man Focus; Auto Iris & Fade	45dB/250/9.5W	Electronic/5f.c.
JVC						
GX-33U	JVC 3.1 lbs	$750	f1.9, 15-45mm Zoom	C, Tint, Auto/Man Iris, Sens.	45dB/250/7.5W	Optical/9f.c.
GX-66U	JVC 3.3 lbs	$850	f1.9, 12.5-75mm Macro Zoom	Same as GX-33U.	45dB/250/7.5W	Optical/9f.c.
GX-88U	JVC 4.3 lbs	$1,000	f1.4, 12-72mm Power Macro Zoom	C, W Bal, Auto/Man Iris, Filter.	40dB/270/6.5W	Electronic/9.3f.c.
Magnavox						
8244	Matsushita 4 lbs	$950	f1.4, 33-65mm Power Macro Zoom	C/W Bal, Auto/Man Iris, Filter.	40dB/270/4.5W	Optical/9.3f.c.
8245	Matsushita 5 lbs	$1,150	f1.4, 12-72mm Power Macro Zoom	Same as 8244	40dB/270/6.5W	Electronic/9.3f.c.
Panasonic						
PK-530	Matsushita 3.8 lbs	$750	f1.8, 11-33mm Zoom	3-Positn C, Tint, Auto/Man Iris.	43dB/240/5.4W	Optical/10f.c.
PK-700	Matsushita 4.5 lbs	$996	f1.8, 12.5-75mm Power Zoom	Same as PK-530.	43dB/240/7.2W	Electronic/10f.c.
PK-800	Matsushita 5.3 lbs	$1,250	f1.4, 12.5-75mm Power Zoom	Same as PK-530.	45dB/240/7.5W	Electronic/7f.c.
RCA						
CC-005	Matsushita 3.3 lbs	$750	f1.8, 14-42mm Zoom	2-Positn C, W Bal, Tint, Auto Iris.	45dB/270/7W	Electronic/10f.c.
CC-007	Matsushita 5.5 lbs	$995	f1.4, 12.5-71mm Power Zoom	Same as CC-005 + BLC, C Bal, Fade.	45dB/270/7.6W	Electronic/7f.c.
CC-010	Matsushita 5.8 lbs	$1,050	f1.4, 12.5-75mm Power Macro Zoom, C-mt.	Same as CC-007.	45dB/270/7.5W	Electronic/7f.c.
Sanyo						
VCC-542P	Sanyo 5.7 lbs	$950	f2.0, 25mm, C-mt.	4-Positn C, W Bal, Auto Iris.	43dB/250/NA	Optical (Not TTL)/10f.c.
VSC-450	Sanyo 4.4 lbs		f1.4, 12.-5-75mm Power Zoom, C-mt.	Remote Transport, 4-Positn C, Tint, Auto Iris, Sens.	45dB/250/NA	Electronic/10f.c.
VSC-545P	Sanyo	$1,340	f1.8, 17.5-105mm, Power Zoom, C-mt.	4-Positn, C, W. Bal, Auto Iris.	43dB/250/NA	Electronic/10f.c.
Sears						
5308	Hitachi 3.96 lbs	$900*	f1.8, 14.5-52mm Macro Zoom	C/W Bal, Auto Iris, Sens, Sun Shutter.	45dB/250/7W	Electronic/7.5f.c.
53812	Hitachi	$1,000*	f1.6, 14-84mm Macro Zoom	same as 5308.	45dB/250/7W	Electronic/7.5f.c.
Sony						
HVC-2010	Sony. 4.4 lbs	$800	f1.8, 2-Positn, normal, tele.	C/W Bal, Auto/Man Iris.	45dB/300/6.5W	Optical/7.5f.c.
HVC-2000	Sony 5.3 lbs	$1,250	f1.8, 12.5-75mm, Power Macro Zoom.	Same as HVC-2010+ Auto Fade, Sens.	45dB/300/8.3W	Electronic/7.5f.c.
HVC-1000	Sony 4.78 lbs	$1,400	f1.8, 14-42mm Zoom.	C/W Bal, Auto/Man Iris, Filter.	45dB/300/7.7W	Optical/10f.c.
HVC-2200	Sony 6.38 lbs	$1,300	f1.4, 11-70mm Power Macro Zoom.	Same as HVC-2000.	45dB/300/8.3W	Electronic/4f.c.
Toshiba						
IK-1850	Toshiba 4.6 lbs	$1,050	f1.4, 11-70mm Macro Zoom, C-mt.	C/W Bal, Auto/Man Iris.	48dB/250/9.4W	Electronic/10f.c.
IK-1850AF	Toshiba	$1,400	Same as IK-1850	Same as IK-1850+ Auto Focus.	48dB/250/11W	Electronic/10f.c.

Selected Video Cameras, with manufacturer supplied features and specifications.
**Approximate List Price*

Abbreviations: Bal=Balance, BLC=Backlight Control, C=Color, Sens=Sensitivity Switch, W=White.

VIDEO SOFTWARE & ACCESSORIES

Before you buy—check spindles for adequate "float" and tape pack for signs of "cinching."

In the mid-1970s Universal and Disney went to court to halt Sony from selling the home videocassette recorder. Their reasoning: People would use these machines to record movies and other programs off TV without paying royalties. This, the studios felt, would be an infringement of their copyrights.

The district court found in favor of Sony. Home videotaping of TV programs, the court held, was a "fair use" of copyrighted material, and therefore legal.

The U.S. Court of Appeals did not agree. In October 1981 the appeals court overturned the Sony decision. "Absent a clear [signal] from Congress," the appeals court decided, home videotaping of TV programs DOES NOT constitute a "fair use." There is no "blanket exemption for home video recording," the court held, "even when the recording is not for a commercial purpose." The appeals court decision leaves both manufacturers and VCR owners liable for copyright damages for recording off television.

What next? The final decision will probably rest with the Supreme Court. An Act of Congress could also legalize taping off TV. One Congressman, Stan Parris of Virginia, has already introduced such a bill.

As for purchasing a bootleg tape, you're probably safe from prosecution here—although you might end up with your tape confiscated. Tape piracy, both audio and video, has become very big business, with the FBI making a considerable effort recently to nab bootleggers who, in some cases, are producing both tapes and packaging that are almost indistinguishable from the legal version. Unfortunately, many great films—from W.C. Fields' *Bank Dick* to *Gone With the Wind* and *The Empire Strikes Back*—haven't been legally available on videocassette. That only increases the temptation for bootleggers—and collectors.

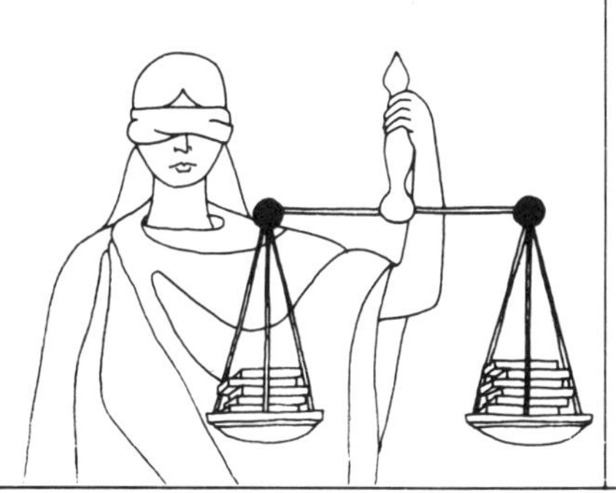

Video Tape & Videocassettes–The Basics

Just like an open-faced grilled-cheese and tuna, video tape is a sandwich. Except this entire sandwich happens to be thinner than a human hair—just slightly under 1/1,000th of an inch.

The main ingredient is a layer of metallic oxide (a binary compound of oxygen with an element like iron) which is flexible, durable, and highly responsive to magnetic energy. The fewer impurities in this oxide layer and the greater the density of the metallic molecules, the better your ultimate video image will look.

In home videocassettes, this oxide layer is bonded to a ½-inch polyester ribbon, which is then wound around a plastic spool and inserted in a precision-machined cassette housing. The entire manufacturing process requires accuracy of an almost incredible magnitude, and must be done in a highly controlled environment. Even a virtually invisible amount of airborne dust can contaminate an entire "batch" of metallic oxide, affecting the quality of hundreds of videocassettes.

Unfortunately, you can't tell if the tape in your cassette is from a bad batch until you get it home and discover problems such as "dropouts" and "video snow" in your recordings. But you can often spot a mechanically faulty cassette before you leave the store. First, check the "float" in the plastic spindless, or spools. Unless they have about 1/8th inch of play when you press them down, return the cassette. Insufficient play will cause excessive strain when you rewind or fastforward your tape. Next, examine the "tape pack" through the clear plastic window. Is the surface smooth and free of "ruffles" or bumps? If the tape pack is ruffled or "cinched," return it. Eventually you'll have problems with video snow because the tape cannot pass evenly over the video heads. You may also get tape stretch or even horizontal roll from a poorly loaded tape pack.

Tape Designations And What They Mean

When Sony introduced the first home videocassette in 1975 the designation L-500 was chosen. "500" referred to the length (hence the "L") of the video tape inside—approximately 500 feet. When JVC jumped into the video ring with its VHS format the following year, the designation T-120 was selected, not illogically, since the cassette had a playing time (hence the "T") of 2 hours or *120* minutes.

Sony and the other Beta format video tape suppliers have continued to use the L-designations, with videocassettes now available in 6 lengths ranging from 125 feet to 830 feet and corresponding playing times at the Beta III speed of 45 to 300 minutes.

©1982 THE HOME VIDEO SOURCEBOOK

Most, but not all, of the VHS tape suppliers have stuck with the T-designation system—although the addition of LP and SLP speeds have rendered the idea of playing time somewhat obscure, since a T-120 videocassette can now actually play for up to *360* minutes at the SLP mode. RCA, obviously listening to the beat of a different drummer, calls its "T-120" videocassette the VK-250, apparently referring to the fact the cassette holds about 250 meters of video tape. Currently, there are 4 commonly available VHS cassette lengths ranging from the T-30 to the T-150

Which designation is best for you? Here's our advice: In the Beta format you'll get the most playing time for your money with either the L-750 or L-830 videocassettes. If you plan to erase and record a great deal, stick with the L-750—theoretically it's a stronger tape than the L-830. If you want archival storage—that is, a tape for collecting shows you'll only watch occasionally, the L-830 is a good bet.

In VHS your best buys are the T-120 and T-150 videocassettes. Here, too, we like the slightly shorter tape—the T-120—as an everyday workhorse, with the T-150 reserved for applications that entail less wear and tear.

What have we got against the shorter videocassettes, such as the Beta L-500 or VHS T-60? It's purely a matter of economics. The precision housing for today's videocassettes is extremely expensive, so the more tape you can put in the housing, the lower your cost per hour.

To compute the total recording time you can get from the various videocassettes at different recording speeds, simply use the following "Videocassette Comparison Chart."

Videocassette Comparison Chart

Cassette Model	Playting Times/Playing Speeds		
Beta Format Tapes	**Beta I***	**Beta II**	**Beta III**
L-125	15 min	30 min	45 min
L-250	30 min	60 min	90 min
L-370	45 min	90 min	135 min
L-500	60 min	120 min	180 min
L-750	90 min	180 min	270 min
L-830	100 min	200 min	300 min
VHS Format Tapes	**SP**	**LP**	**SLP**
T-30	30 min	60 min	90 min
T-60	60 min	120 min	180 min
T-120	120 min	240 min	360 min
T-150**	150 min	300 min	450 min
T-180***	180 min	360 min	540 min

*Beta I is available as a *playback* speed only.
**Approximate playing times for the T-150 videocassette.
***At publication the T-180 was not available in the United States.

Buying Blank Videocassettes

As you well know if you've done much audio recording, not all tapes are the same. And what's true for audio tape also goes for video tape. In fact, since video tape has to store up to 200 times more information, the slightest imperfections can have a profound impact on the image on your TV screen. Profoundly negative.

The main culprits are dropouts and video noise. Dropouts are actual physical imperfections in the magnetic coating—microscopic dust particles, paper fibers from packaging materials, and tiny holes caused by shedding of the magnetic oxide itself. Dropouts can cause your picture to streak, breakup, and—in extreme situations— roll uncontrollably. The second problem common to inferior video tape is video noise, which is much like audio tape hiss, except that it shows up as a gentle snow flurry or a full-blown blizzard, depending on whether the quality of your tape is merely marginal or really rotten.

Unfortunately, it is hard to make definitive comparisions between the many brands of video tape because the quality often varies dramatically from one batch to the next (the "bad batch" syndrome). With that said, here are a few generalizations to guide you down the blank tape aisle.

- **TDK videocassettes.** Most video buffs agree that it's hard to go wrong with TDK. While certainly not the only brand with an above-average product, TDK seems to give consistently good results. List prices are somewhat higher than most other video tapes, but the discounts are often generous. List price on TDK's T-120 VHS tape, for instance, is $26, yet we found it widely sold for under $15. We especially recommend TDK's T-120 and L-750 videocassettes if you intend to record at the SLP or Beta II speeds where dropouts and video noise are the most obnoxious.
- **HG (High Grade) tapes.** TDK, Maxell, and RCA all market a top-of-the line videotape with the designation HG. These tapes generally have superior signal-to-noise ratios (meaning a minimum of tape-induced video noise) and lower dropout rates (meaning fewer dropouts). They also cost considerably more. We found Maxell's "standard" T-120 videocassette selling at discount for $14.95, with the HG version costing $17.95. In our opinion, unless you are a true video connoisseur with the finest of equipment, the slight increase in image quality doesn't justify the big jump in price. This is doubly true if for some reason you can't buy your tape at a discount and must pay near list price.
- **Beta videocassettes.** In the Beta format we are enthusiastic about two brands in addition to TDK. These are Sony and Maxell. Others that should give at least adequate results are Ampex, Fuji, Toshiba, and Zenith.
- **VHS videocassettes.** Just as there are more VCR brands available in VHS, so too are there more tape brands. Fuji, Hitachi, JVC, Maxell, Panasonic, Quasar, RCA, and Sylvania are tapes from which we think you can count on for average to above average results.
- **Off-brands.** Perhaps there are some off-brands available that can be relied upon, but we have yet to find one. Our advice on no-name videocassettes is simple: Don't. Many are manufactured in Taiwan or Hong Kong with woefully inadequate equipment and nonexistent quality control. Not only can some of them be unspeakably bad—if they work at all—but they are even sold in packaging that is deceptively similar to the well-respected brands. If your dealer is stocking one of these, he should know better.

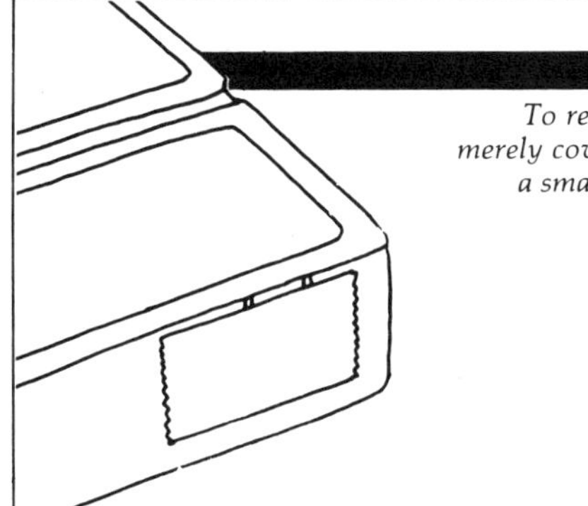

To recycle prerecorded cassettes merely cover the safety tab hole with a small piece of cellophane tape.

Using And Storing Your Videocassettes

● **Safety tabs.** As with audio cassettes, videocassettes are equipped with a safety tab to prevent accidental erasure. Once this tab is removed (as it is on most prerecorded tapes you buy) you will NOT be able to engage the record button on your VCR.

● **Making a tape erasure-proof.** This requires the safety tab off the cassette with a screwdriver. On Beta cassettes you will find the tab located near the edge of the spindle on the bottom of the casing. With VHS tapes the tab is situated on the spine of the cassette.

● **Suppose you change your mind.** If you later decide you *do* want to record on a tape from which the safety tab has been removed, merely cover the hole with a small piece of cellophane tape. The same goes for any prerecorded movies you're ready to recycle as blank tape.

● **One direction only.** This may seem a little elementary if you're used to using a VCR already. But unlike audio tapes, videocassettes play and record in one direction only. When your tape has reached the end, that's all there is. You can't turn the tape over to play the other side. There is no other side. The tape runs from left to right, and stops when the right hand "take up" spindle is fully wound. You must rewind the tape before it can be used again.

● **Partially wound tapes.** It's always tempting, and a lot easier, to store a partially wound tape. This way if you want to use the remaining unrecorded portion all you have to do is drop the cassette in your recorded and bingo, you're ready to roll. Our advice: Avoid this temptation and NEVER STORE A PARTIALLY WOUND TAPE. The reason? The clear plastic leader at the beginning of a tape is tough stuff designed to stand up to a certain amount of abuse, as well as extremes in humidity and temperature. This is not true of the magnetic tape itself. Thus, storing a tape with a section of magnetic tape exposed is just asking for trouble. At the least you can expect to pick up dust-caused dropouts on the exposed section of tape, at the worst you might accidentally damage the tape, ruining your cassette and everything on it.

● **Never touch the magnetic tape.** Why not? For the same reason you don't touch phonograph records. Your skin leaves a deposit of oils, and oils eventually attract dust. On your phonograph record you'll recognize your fingerprints as a "crackling" sound. On videotape you'll see them as video snow and even picture breakup.

● **Never attempt to splice a damaged tape.** Like audio cassettes, videocassettes do occasionally jam, and the result is usually a damaged or torn tape. Unless you're a real pro (and willing to back up your expertise with $300 to replace your video heads in case you aren't as good as you think) we suggest you take any damaged tapes to your video dealer for repair. Chances are that even then you should only play the spliced tape back just one more time—long enough to make a dupe.

● **Loading.** Don't repeatedly load and unload a cassette without playing the tape. This may cause excessive tape slack, and ultimately result in damage to the tape.

● **Temperature and humidity.** Some folks tell us videotape is a lot tougher than we give it credit for being. Maybe so. But if you've invested $60 or $70 in a prerecorded film, or 5 or 6 hours of your time making commercial-free recordings of some of your favorite TV flicks, why take chances? The rules are simple. Never store videotape anywhere that the temperature can go above 130°F—such as the inside or trunk of your car. Never attempt to play a tape which is "hot" to the touch, say one that has been left in the sun or near a heater or radiator. Never play a tape in excessively humid conditions, unless you have a DEW indicator, and know that it's working. Never play a tape that has been in a cold place, such as an unheated room or your car on a winter day, until it has reached room temperature. (Here, too, the problem is moisture. This time from "sweating" or condensation as the tape warms up.)

● **The limits of tape.** At 130°F the binder that holds the metallic oxide to the polyester ribbon begins to deteriorate. At 140° the polyester itself begins to shrink—permanently. As for cold, temperature conditions found in the continental U.S. shouldn't harm videocassettes, as long as they are allowed to reach room temperature before they're played.

● **Storage conditions.** Tapes like about the same temperature and humidity range as people. 50 to 90°F and 40 to 60% humidity. And DUST-FREE. Always keep your tape collection AWAY from magnetic fields, such as those created by TRANSFORMERS (in stereo amplifiers, for instance), ELECTRIC MOTORS in VCRs and stereo turntables, or MAGNETS (sometimes found on the latches on cabinet doors). Always store cassettes in sealed boxes.

● **Vertical storage.** Videocassettes should always be stored vertically, like books on a shelf. Not only will your tapes be easier to find, but the spindles won't be prone to warp—which can happen to tapes that have been stored horizontally for extended periods of time.

● **Mailing.** Never use a Jiffy Mailer or other fiber-padded mailers for shipping videocassettes. The tiny fiber particles will almost certainly find their way onto the tape, causing horrendous dropout problems.

You can make your tapes erasure-proof by breaking off the safety tab with a screwdriver.

©1982 THE HOME VIDEO SOURCEBOOK

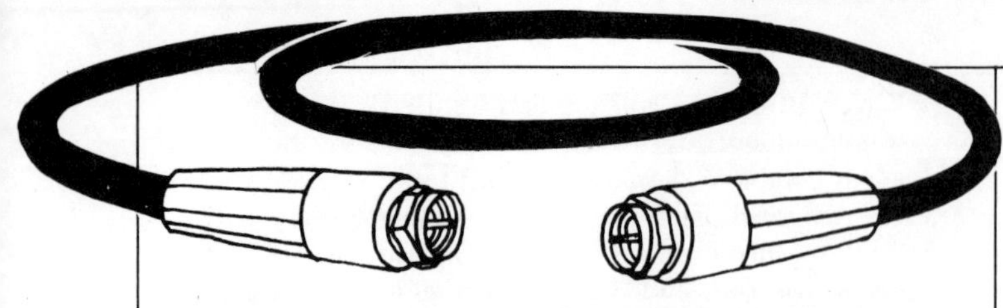

Cables, Connectors, And Black Boxes

Your new VCR may—or may not—come packed with all the necessary cables, connectors, and accessories to hookup with your TV set, external antenna, cable TV tuner, and/or subscription TV decoder.

If it's just a simple installation—say a TV set with rabbit ear antenna—a few minutes spent with the owner's manual and you should be home free. But if you have an exterior antenna with separate VHF and UHF leads (Very High Frequency and Ultra High Frequency), or if you are on a cable network, or possibly on a cable network with subscription TV as well, or have an external antenna *and* a subscription TV decoder box, your VCR installation may become dicey indeed. Here's where you're better off buying from a dealer who specializes in video equipment, particularly one who offers on-the-premise repairs and a reasonably priced installation service.

Should you decide to install your VCR yourself anyway, more power to you. Among the additional accessories you may have to purchase are a couple of matching transformers and an RF splitter. These are neither particularly expensive nor difficult to find. If your video dealer doesn't have them, the nearest Radio Shack should. Read the owner's manual carefully and follow the instructions step-by-step. You can also be consoled that there isn't much you can do, short of plugging your antenna leads in the AC outlet, that can do much damage to your equipment. And perhaps this is the perfect time to invite that video-freak friend over for the afternoon.

Once you've accomplished a successful installation, we remind you to forego your victory celebration until you've tested *all* of the features on your VCR. If anything is malfunctioning, this is the time to get the unit replaced.

But now that your VCR is hooked up and working, you may find you'd like to move it next to your easy chair where it's more convenient to operate. Or perhaps you want to run more than one TV set from you VCR. In either case, you'll need more in the way of cables and connectors than the typical manufacturer provides.

If you are lucky enough to be on a cable network, you may discover that the cable tuner prevents your VCR from recording one program while you watch another. You can solve this problem yourself, too, but you'll need something called an "up-converter." It isn't standard issue.

Perhaps your dream is to watch commercial television without the commercials. Here, too, your VCR will help. But you'll need another black box, one that can sense when a TV commercial is coming on the air and put your VCR on pause until the commercial break has ended.

If any, or all, of these objectives tickle your fancy, read on. As you'll quickly discover, you can do a whole lot more with your VCR than just keep up with the "6 O'clock News."

● **Cables.** There are 3 basic types of cables involved in home video installations. If you own a component stereo system you are already familiar with the first of these—the audio cable. And if you have an exterior antenna on your house, you will certainly recognize the second type of cable, the distinctive flat insulated wire that leads from the roof to your TV set. This one's called a "300-ohm cable." (An ohm, of course, is a Tibetan chant, or mantra. It is also a measure of "impedance," or the electrical resistance of a wire. If you think of wire as a "pipe" for electrons you'll get the idea. The greater the impedance, or resistance, the harder it is for the electrons to squeeze through the pipe.) The last type of wire is known as a 75-ohm cable. (You guessed it, less resistance.) The 75-ohm cable is also "shielded" to prevent any stray signals from entering it, which earns it the designation "coaxial" cable.

● **Connectors.** Each of these cables has its own unique style of connector. The audio cable generally uses what is known as the RCA plug. 300-ohm cable has a simple spade lug that fits easily under a screw head such as the typical antenna terminals on the back of older model TV sets. The 75-ohm cable uses a so-

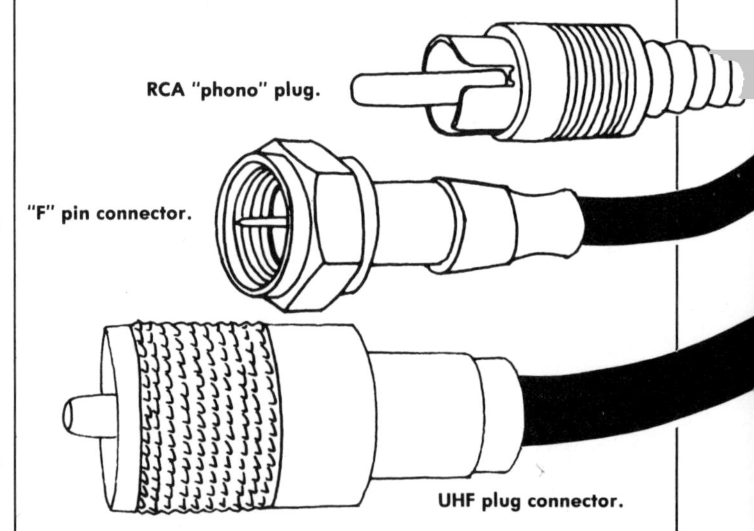

called "F-type" connector—a rather delicate affair with female threads. Most recent model TVs have a male version, called an F-type jack, to which a 75-ohm cable can be directly connected. If you don't have an F-type jack on your TV set you probably need to include a 75 to 300-ohm "matching transformer" between the 75-ohm cable from your VCR and the 300-ohm antenna terminal on your TV.

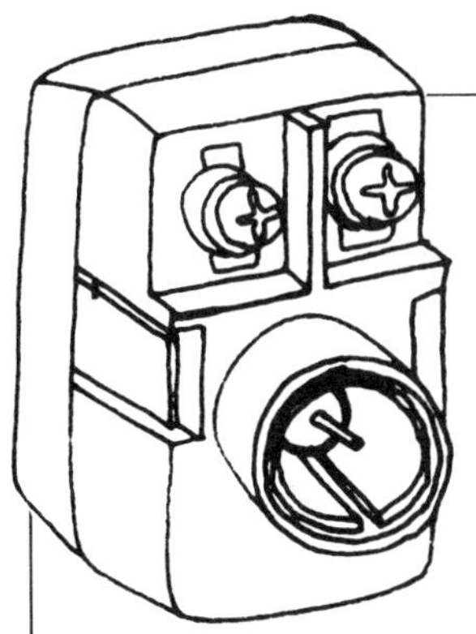

Matching transformers are usually, but not always, supplied with your new VCR. They can also be obtained at most Radio Shack stores.

● **Matching transformers.** The little gizmo we just mentioned merely *transforms* the impedance of the audio and video signals passing through it from 75 to 300 ohms. If you have an outside antenna on your house you may have to include another matching transformer between your antenna lead and VCR, this one to reduce the impedance of the signals from 300 to 75 ohms.

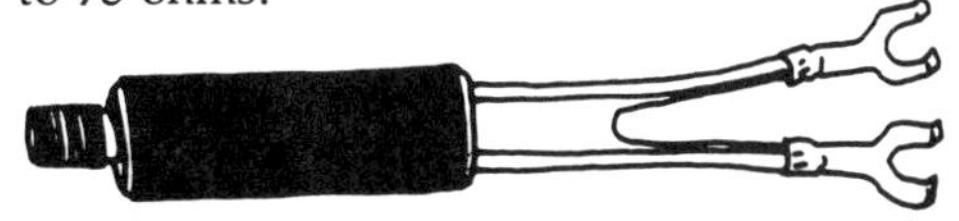

● **Extention cables.** Having successfully installed your VCR, you'll see what a snap it is to fashion yourself a set of extention cables so you can operate your VCR from beside your favorite TV chair. Just don't forget you'll need 2 cables, one to bring the antenna signals (either from the rabbit ears on your set, from your cable tuner box, or from your outside antenna) to your VCR, and a second to feed the signal from the VCR to the TV. As for the 75-ohm coaxial cable you're using, this stuff is about as sturdy as Steuben crystal, so don't run it anywhere it'll get stepped upon, twisted, kinked, or crimped (such as under rugs in major traffic areas or near window or door sills).

● **2 or more TVs from one VCR.** You'll need a widget called an RF splitter (RF stands for Radio Frequency—all the audio and video signals we're talking about are actually classified as being on the RF band). This will cost you a couple of bucks at Radio Shack. If you don't have a F-type jack on the back of your TV you'll also need one of those 75 to 300-ohm matching transformers. Just splice the RF splitter into the 75-ohm cable that runs from your VCR to the TV set. Then run the new cable from the splitter to the second TV, including the matching transformer if necessary. If you see a definite loss of image quality you can rectify the situation with a trip back to Radio Shack for an "RF amplifier" to boost the signal.

● **Up-converter.** This little black box will *up-convert* all your VHF, "midband" and "superband" cable TV frequencies to UHF channels. Loosely translated, this means your VHF channels 2 to 13, your midband cable channels A to I, and your superband cable channels J to R will show up on your videocassette recorder's UHF tuner, usually on UHF channels 36 to 71. If you found that you couldn't program your VCR for more than one channel or watch one show while recording another, the up-converter should solve your problem. The only drawback? The up-conversion process itself has a tendency to create interference. If you install an up-converter, keep the cables short and as far away from each other and the AC cord as possible.

● **Dubbing cables.** Some folks find the cheapest way to make dubs from one VCR to another is to use a short (under 2 feet) pair of ordinary audio cables. But to get the best possible copies with a minimum of signal loss or interference we suggest you purchase, or make up, a pair of dubbing cables. Many video dealers sell dubbing cable kits, as do companies listed in "The Buyer's Guide to Accessories" such as Cableworks and Vidicraft. Before you go to buy a pair of dubbing cables check the type of connectors on both VCRs. If you use the "direct video" method, your cables will be going from the video and audio outputs on one VCR to the video and audio inputs on the second. Most videocassette and audio recorders use RCA jacks for their *video* inputs and outputs. *Audio* jacks are not quite so standardized, with some machines using RCA-type jacks while others are equipped with MINI jacks. Hence, while your video cable will probably require RCA *plugs* at each end, your audio cable may need an RCA plug and a MINI plug, a pair of MINI plugs, or a pair of RCA plugs. If you aren't certain what type of connectors are on your friend's VCR, we suggest you buy a pair of dubbing cables with RCA plugs as well as a couple of RCA to MINI plug adapters. These adapters will cost a few dollars extra, but chances are excellent they will prove useful sooner or later.

Vidicraft's Guard Stabilizer will defeat the Coypguard process used on most prerecorded videocassettes to prevent illegal duplication.

THE DETAILER II

● **Video enhancer.** If you are doing a great deal of dubbing, but aren't satisfied with the quality, consider a video enhancer. Much like the fine-tuning control on your TV, the video enhancer can increase the sharpness of a video signal, thus improving the appearance of a dubbed copy. For best results you should use a video enhancer for every step: recording a master tape, dubbing the master, and viewing the dub. The video enhancer is not without its tradeoffs, however. If your master copy contains too much video noise, you may find your enhanced copy actually looks worse, since this black box also tends to "enhance" dropouts, imperfections, and video noise. The more expensive models, such as Vidicraft's Detailer II and the Image Enhancer from Video Interface, are equipped with "coring" circuits that help reduce the video noise brought out by sharpness enhancement. If you plan to buy a projection TV someday, you may want to invest in a video enhancer now. This way you can build a library of enhanced recordings for future use on the big-screen TV, where sharpness is essential.

● **Commercial-free TV.** There are 2 devices available that purport to automatically eliminate commercials from your video recordings. For black and white programs The Killer, marketed by Video Services, claims to be 95% effective in eliminating ads. The Killer continually scans your video signal for the presence of a "color burst"—a special pulse designed to let your TV set know that material being broadcast is in color. When a color burst is detected—such as at the start of a station break or advertisement—The Killer puts your VCR on pause until the color signal has ended. It is highly effective—as long as the TV station isn't careless enough to leave their color burst generator turned on when they return to your b&w movie. If you're a real classic film afficionado, The Killer, at $95, may be the greatest thing to happen to you since *The Jazz Singer* (original 1927 version, of course).

Exterminating ads from color shows, unfortunately, is not quite so easy. One black box that makes the attempt is Shelton Video's The Editor (it can also be used on b&w shows). When a scene fades to black The Editor goes on alert. As soon as the audio track also fades The Editor trips the pause button on your VCR until it senses that 35 seconds of continous programming has elapsed. There are 2 problems. First, The Editor doesn't always sense a commercial is coming on the air. Local TV stations, for instance, frequently don't bother to fade the audio before cutting to a commercial break, so your Editor never gets its cue to axe the ads. The other problem is with the timer. Occasionally The Editor not only eliminates the stuff you don't want, it also edits out the first half minute of the show you do want.

Opinions on the efficiency of The Editor vary from user to user, with many owners putting its success rate at about 70%. Still, at only 70%, your savings in time and tape can be dramatic. If you use The Editor to "de-commercialize" 1½ hours of TV programming a day, you'll be missing over 90 hours (that's almost 4 full days) of Madison Avenue's finest hype each year. If you record a lot of broadcast TV, The Editor at $249 may save you time and money in the long run. Although, think of all those great ads you may be missing. One note of caution: if your VCR is equipped with electronic transport controls some extensive rewiring of the pause mechanism may be necessary to accommodate either The Killer or The Editor. We suggest you contact Vidicraft or Shelton to see if these units are compatible with your VCR.

● **RF switchers.** Once you start adding accessories, you may discover yourself spending more time plugging and unplugging cables than watching those good shows you want to see. The solution? You guessed it—yet another black box. Beta Video's Distrivid is a good bet if you find yourself with a healthy appetite for video accessories.

Not exactly inexpensive at $199.95, the Distrivid does have 2 advantages. First, the number of connections you can make involving 2 VCRs, 2 TVs, cable TV, subscription-TV decoders, an external antenna, and the many accessories such as stabilizers, enhancers, and commercial eliminators is virtually limitless. Second, we can vouch for the fact that the interlocking pushbuttons are well-shielded to prevent interference. For less-complicated hookups the same company, Beta Video, also offers a 4-input, 1-output RF switcher so that you can choose between a VCR, a video-game or home computers, cable-TV or subscription-TV decoder, and one other program source—say a videodisc player—without having to connect or disconnet any cables. The price is $59.95. Another RF switcher, the VM-600 Videomate from Total Video Supply, allows you 6 inputs and 1 output for $99.95. The TRJ Corporation offers a device called The Channelizer that even eliminates having to push buttons. The Channelizer lets you combine any Channel 3 (or Channel 4) input source such as a VCR, subscription decoder box, cable tuner, or videodisc players for "switchless' viewing. Price is $79.95.

● **Video faders.** There's another black box out there if you have the yen to give your dubs that professional look that comes with being able to fade in or out. Twist the knob from black to full video to start your programs or edited scenes, then turn it back when the action has ended to fade to black. The most economical model, Vidicraft's Video Fader costs $149.95. If you own a video camera, you may find you already have a video fader of sorts. Some deluxe cameras have an "auto fade" switch that you can also use for adding fade to black segments to your off air recordings.

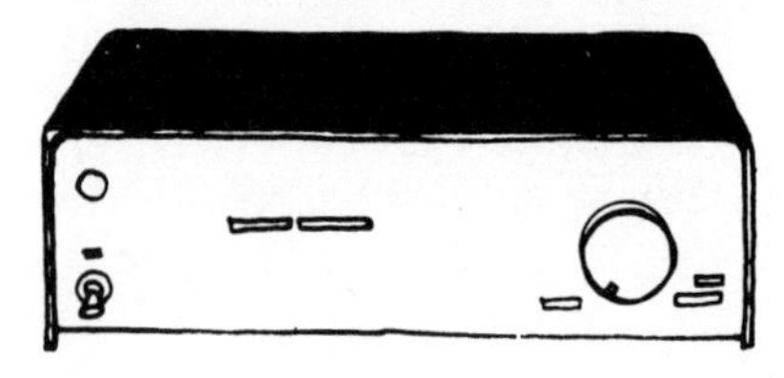

VIDEO FADER

THE BUYER'S GUIDE TO VIDEO ACCESSORIES

Cables, connectors, matching transformers, RF splitters, RF amplifiers, dubbing cable kits:

Cableworks, 4228 Santa Anna Street, Southgate, CA 90208, 213-563-2710.

Comprehensive Video Supply Company, 148 Veterans Drive, Northvale, NJ 07647, 201-767-7990.

RMS Electronics, 50 Antin Place, Bronx, NY 10462, 212-892-6700.

Vidicraft, Inc., 3357 SE 22nd Street, Portland, OR 97202, 503-231-4884.

Other suppliers closer to your home may include Radio Shack, Lafayette Electronics, most video outlets, and hobby-oriented electronics stores.

Up-converters:

CABLE CONVERTER ($29.95) Lindsay Products, 50 Mary Street W, Lidsay, Ontario, Canada K9V4S7, 705 324-2196.

CB-1 CABLE BYPASS KIT ($34.95) Cableworks, see address above.

MX 44C CABLE CONVERTER ($39.95) Magnavox CCTV Division, 183 West Seneca Street, Manlius, NY 13104, 315-682-9105.

VCR-522 VCR CHANNEL CONVERTER ($69.95) Marshall Electronics, Box 2027, Culver City, CA 90230, 213-836-4228.

VIDCOR 2000 SIGNAL CONVERTER ($79.95) Vidcor, Inc., 200 Park Avenue South, Suite 1441, New York, NY 10003.

Copyguard stabilizers:

VIDEO STABILIZER ($85) Vancouver Video Center, 4611 NE 122 Avenue, Vancouver, WA 98662, 206-892-6851.

GUARD STABILIZER ($98) Vidicraft, Inc., see address above.

MOD BOX ($99.95) Video Mods, Box 2591, Sepulveda, CA 91341, 213-361-4694.

Video enhancers:

DETAILER I ($149) Vidicraft, Inc., see address above.

VIDEO ENHANCER ($229.95) Video Interface Products, 19310 Ecorse, Allen Park, MI 48101.

DETAILER II ($295) Vidicraft, Inc., see address above.

Commercial-free TV:

THE KILLER ($95) Video Services, Inc., 80 Rock Ridge Road, Fairfield, CT 06430, 203-259-6899.

THE EDITOR MODEL 100 ($250) Shelton Video Editors, Box 860, Vashon, WA 98070, 206-463-3778.

RF Switchers:

IC-08 DISC SWITCH ($59.95) Beta Video, 9612 F Lurline Avenue, Chatsworth, CA 91311.

THE CHANNELIZER ($79.95) TRJ Corporation, 9200 Sunset Boulevard, Suite 915, Los Angeles, CA 90069.

ESS EDITING SOURCE SELECTOR ($95) Smith-Mattingly Productions, 515 Kirby Hill Road, Oxon Hill, MD 20022.

VM-600 VIDEOMATE RF SWITCHING SYSTEM ($99.95) Total Video Supply, 9060 Clairemont Mesa Boulevard, San Diego, CA 92123, 714-560-5616.

VDA-104 VIDEO DISTRIBUTION AMPLIFIER ($100) Metro Systems, 3834 Catalina Street, Los Alamitos, CA 90720.

IC-28 DISTRIVID ($199.95) Beta Video, see address above.

Video fader:

VIDEO FADER ($149.95) Vidicraft, Inc., see address above.

VS-200 COLOR EDIT FADER ($295) Columbia Tele-Com, 2241 West Burnside, Portland, OR 97210.

VIDEO FURNITURE

Once you've bought that shiny new VCR, videodisc player, or home computer—the next problem is where to put it. Because electronic equipment requires both a tangle of connecting cables and a dust-free, moisture-free environment, your home video unit may place more demands on the room that houses it—and the furniture that holds it—than you might realize. Fortunately, there's help in the form of a number of companies that build furniture specifically for video equipment.

Obviously, a stuffed laundry hamper is not the proper base for a videocassette recorder—and neither is a delicate glass-shelved etagere. If you have solid-state console TV, you may find the best place for your VCR is right up on top. Or, if the area where you plan to do your viewing is relatively dry and dust free, there's no good reason you can't use an existing wall shelf or bookcase—providing there's adequate support and space for access, ventilation, and hookup cables. You may even find your present TV stand or cart has a lower shelf that works just fine—although this is just about the dustiest place in the house.

As for home computers and keyboard video games, you'll probably want to place them on a small table or desk near your TV set. But fear not, the furniture builders haven't overlooked you either. You can buy a set-up designed specifically for home computers, known as the Compu-table, for $130,00.

How much space do you devote to your video equipment? If you live in studio apartment the answer is obvious—not much. If you've got a rambling Tudor mansion, the sky's the limit. Perhaps you'll want your own viewing room, centered around a projection TV. Or maybe an island in the midst of an informal living room. Wherever you put it, just remember your video equipment will create its own traffic flow.

In video furniture, as in love, there's no accounting for taste. If your home or apartment is appointed with French Provencial antiques, finding the right piece of furniture is going to take some time and energy, not to mention money. If you're a fanatic about natural hardwoods, you, too, will have to devote some time searching—or else commission a cabinetmaker to build to your specifications. For those lucky enough to be handy and own a good set of woodworking tools the answer may be to build it yourself. If, on the other hand, you're willing to settle for factory-built furniture—usually Formica or vinyl-clad simulated-wood finish—you'll discover a selection on the following pages, of models designed specifically for video equipment.

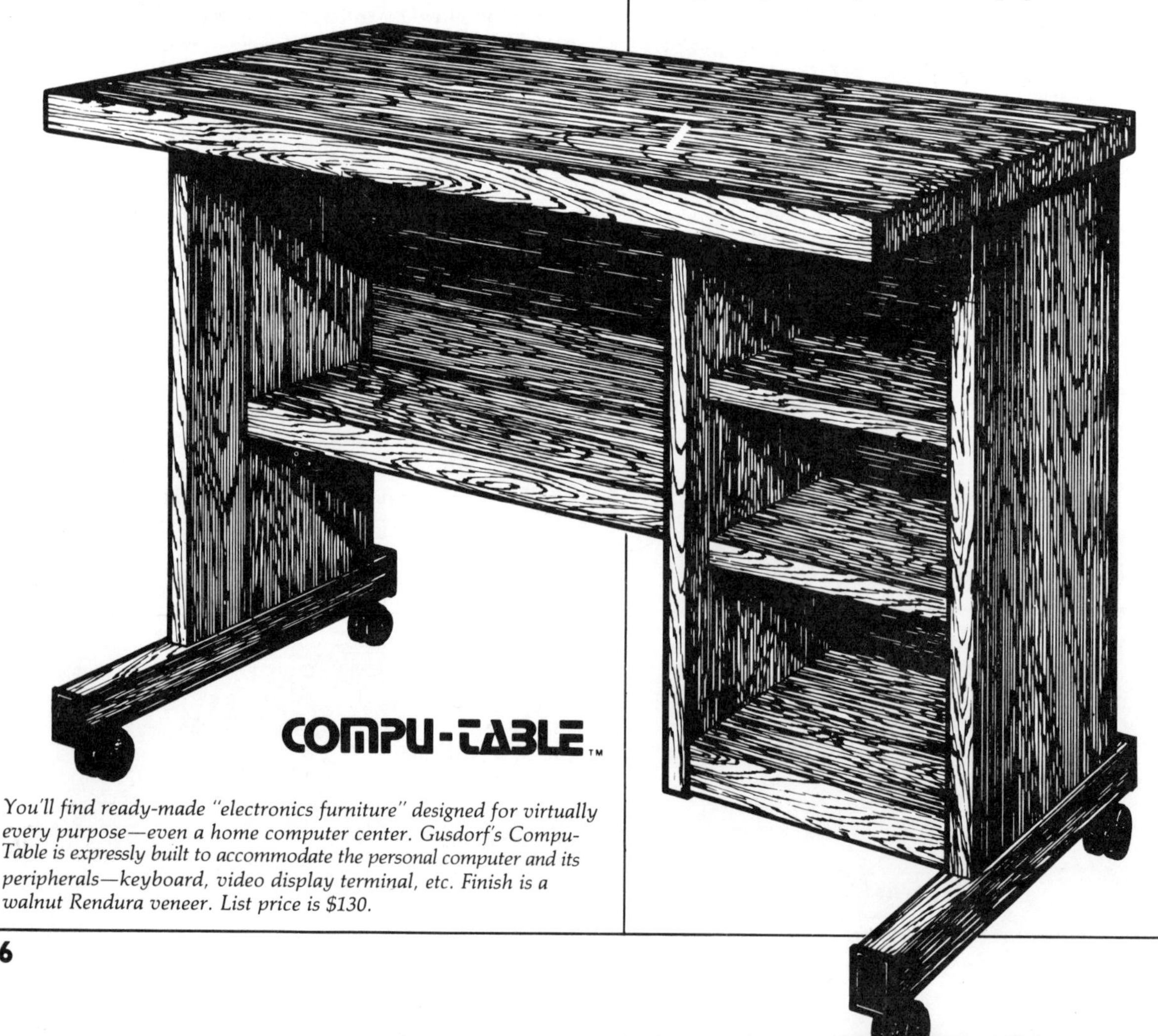

You'll find ready-made "electronics furniture" designed for virtually every purpose—even a home computer center. Gusdorf's Compu-Table is expressly built to accommodate the personal computer and its peripherals—keyboard, video display terminal, etc. Finish is a walnut Rendura veneer. List price is $130.

Photo Courtesy Epic Video

Epic Video's Tape Cabinet stores up to 100 Beta or VHS cassettes. Suggested retail is $290.

Regardless of the path you follow to the perfect asylum for your home video equipment, here are some pointers to consider.

Build yourself a model. Put the equipment in front of you on the floor, and play with it! In so doing, you can:

- Find the ideal viewing height and placement for the TV set (use that stuffed laundry basket if you must). Be sure to avoid strong backlight or glare.
- Place your video hardware so that it is comfortable to operate from your viewing position. Don't put your VCR so low that you'll be continually bending over to load cassettes and operate the controls.
- Decide if you want the audio equipment nearby. If you prefer to route the sound through your stereo speakers, you'll probably want to place the speakers near the TV set. Beware, however, of storing tapes near a turntable or stereo amplifier and their magnetic fields.
- No matter how many tapes or discs you currently have, you could easily double that amount within a year. Block out sufficient storage space and allow extra space for the future must-haves.

Following are some of the special requirements to keep in mind when shopping for video furniture.

1 *Video equipment produces heat.* Ventilation is essential to prevent the equipment from over-heating, which can shorten the lifespan of certain components. A video cabinet should be designed to allow air to circulate through and around the equipment. If the VCR is enclosed behind glass, or within a cabinet, ventilation holes are a must.

2 *Most VCRs and disc players are top-loading.* Make sure that the shelf you select for your VCR has enough room above it to allow space for easy insertion and the ejection of cassettes, as well as sufficient clearance to adjust the controls. A slip-out shelf, such as provided on some ready-made furniture, can work well.

3 *A TV set without a cable or roof antenna requires rabbit-ears.* Allow enough space for rabbit-ear antenna on top of the TV set. (Even with cable, allow space in case the cable conks out.) This doesn't mean the TV must go on top of the cabinet. A slide-out shelf can have adequate clearance for most standard televisions. But if there's doubt, check the manufacturer's specifications, or measure your set to make sure it will fit.

4 *Heavy cords and cables interconnect the video equipment.* A video cabinet should have an open back to increase ventilation and make necessary connections without having cords and cables snaking in and out of the front. If you have a lovely piece of furniture (you finally found just the right French Provincial piece!) that meets all your requirements, but has a completely enclosed back, you may cut holes where necessary...if cutting holes won't sabatoge the structure's integrity. Sandpaper the edges smoooth so wires can't chafe.

5 *The combined weight of a TV, VCR, disc player and other components can add up to the weight of a full-grown human.* Also, the components are not equally weighted although their base sizes may be similar: if you put a 19-inch TV set beside a VCR, the TV will weigh more. The furniture should be sturdy *and* stable. You don't want to spend up to a hundred dollars a pound on video hardware only to have everything crash to the floor because you bought a flimsy—or unbalanced—piece of furniture to hold it.

6 *There is a right location for each piece of equipment.* Before buying video furniture, measure *height* and *width,* and *depth* of the space where it will sit. Be sure there are sufficient electrical outlets, and that nearby doors won't bang into it when opened. Will the TV set be too high or too low for normal viewing comfort? Will the screen be blocked by another piece of furniture? (Perhaps you'll need a swivel-base for the TV shelf.) If you have an entire wall available, you can get a larger piece of furniture than if the area is limited—and a unit housing the TV and VCR side-by-side offers opportunities for more storage space underneath than a vertical configuration with the equipment stacked on top of each other.

7 *Dust and moisture kill VCRs and tapes.* Be sure the space you've decided on is a safe distance from windows, radiators, air conditioners, ovens and other sources of major temperature fluctuation. Don't put the video equipment where water seeps in, or beneath a potentially leaky pipe. Beware of waterbeds. Extremes of heat and cold and be harmful to both video hardware and software. *Avoid magnetic fields* from hi-fi/stereo audio systems. If you want to store videodiscs in a cabinet together with your disc player, remember that, while laser-optical discs can be stored horizontally, capacitance discs must be stored vertically (like audio records) to *avoid warping.*

Now you're ready to get that furniture. Make a sketch of your room—with exact dimension—and go out to shop for furniture that will enhance it. You may not find *exactly* what you want but you'll be surprised by the variety that is available.

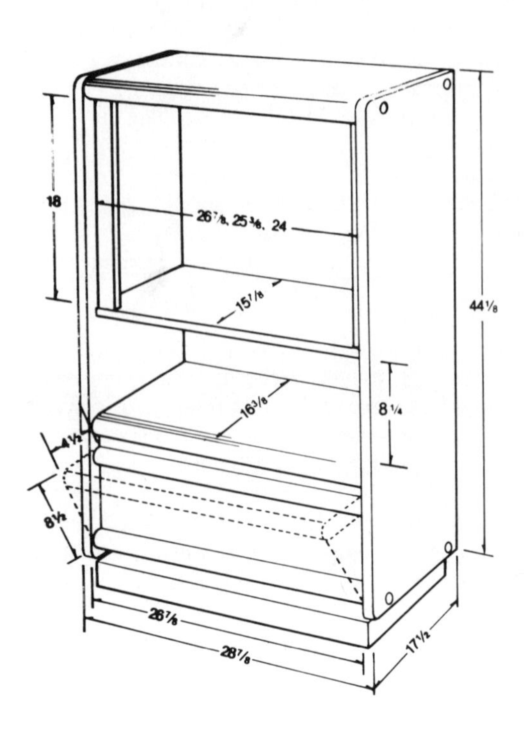

This "Swivel VCR/TV Cabinet" is one of the best buys we've seen in video furniture. The finish is a honey elm veneer in a round rail design. The key feature, however, is the swivel base that allows comfortable viewing from any direction. There is also a concealed tape storage shelf. Approximate price is $140.

Gusdorf's 6-foot Status Pro 1990 electronics tower accommodates both your audio and video systems. Audio equipment is protected behind bronze-toned glass doors while the VCR is mounted on a retractable shelf within a concealed 2-door cabinet. Suggested retail price is $380.

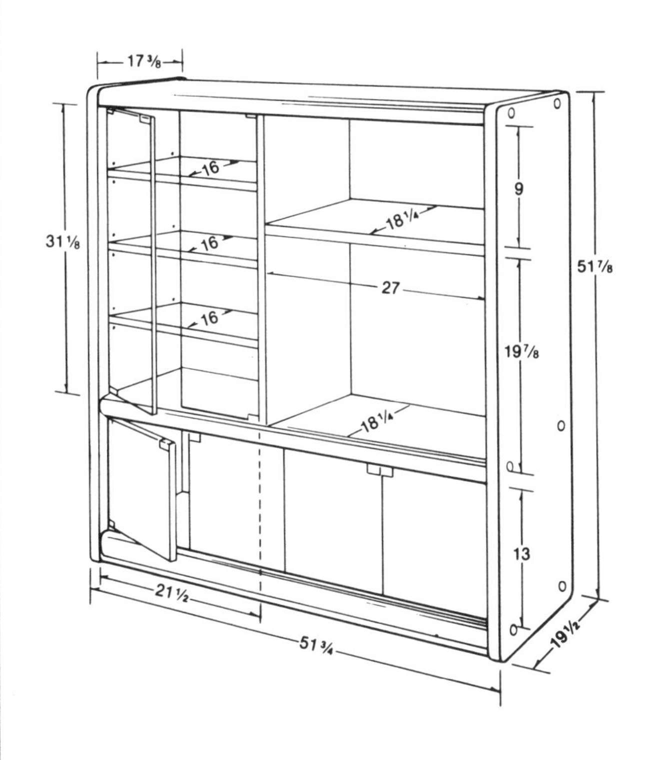

The O'Sullivan DC930 "Audio/Video Center" offers side-by-side display of stereo and home video gear. Audio equipment sits behind solar bronze glass doors. A sliding shelf above the TV accommodates the VCR. One advantage of this unit: You can operate the VCR controls from a standing position. List price is approximately $340. Another bonus—twin storage cabinets for records and tapes.

The Status Pro 1920 is a dedicated video unit with a slide-out VCR shelf and videocassette storage. Gusdorf's suggested list is $235. Available in walnut or pecan Rendura veneer.

For additional information contract:
Epic Video
7 West 51st Street
New York, NY 10019
212-765-7190

Gusdorf Corporation
6900 Manchester Avenue
St. Louis, MO 63143
314-647-1207

O'Sullivan Industries
19th & Gulf Streets
Lamar, MO 64759
417-682-3322

SATELLITE RECEIVERS

In 1977 Ted Turner perpetuated the longest winning streak in the history of sport—over 30 years without a defeat—by successfully defending the America's Cup on Block Island Sound, near the summer resort of Newport, RI. And while Turner's victory over the Australian challenger was based on classical yacht racing strategy, the electronic paraphernalia that accompanied the Turner entourage to Newport was right out of a science-fiction flick that was also sailing into the record books that summer—*Star Wars.*

The biggest giveaway that Turner was in town was not the arrival of his yacht at Bannister's Warf, but the appearance on the front lawn of Connally Hall, Turner's Newport residence that year, of a 16-foot parabolic dish antenna, one of three main components of a device known as a "downlink receiver"—or satellite earth station.

Why a frontyard satellite station in Newport? Mainly so that when Turner wasn't sailing, he could keep up with the ballpark progress of his home team, the Atlanta Braves. And if anyone in Newport thought *that* a trifle eccentric, nobody said so. After all, Turner did happen to own the Atlanta Braves, the Atlanta superstation that relayed their games via satellite, *and* the yacht that was defending the America's Cup.

In the summer of '77 an earth station like Turner's cost upwards of $80,000. Half a decade later, that price has deflated by 1,000%—to less than $8,000. Prices are still fa 'ing. By 1990 some electronics experts predict you'll be able to own a backyard satellite receiver for under $500.

Which is great. Except, you may ask, what's a satellite receiver, and what good is it anyway? Fair question. Let's take the second part first.

On page 162 is a complete list of programs that are being relayed to earth right now via Satcom I, one of 10 synchronous orbit satellites (synchronous because they remain at a fixed point in the sky, they are also called geo-stationary) that are used to relay television signals over North America. Among your options today—Nickelodeon, The Movie Channel, Ted Turner's WTBS, New York City's WOR, ESPN, CBN, C-Span, USA Network, Black Entertainment Network, Calliope, The English Channel, Showtime, American Educational Television Network, Window on Wall Street, Galavision, Cinemax, Home Theater Network, Modern Satellite Network, Home Box Office, and Turner's Cable News Network. And these are just for starters. You've still got 9 more North American satellites and, depending on where you live, you may be able to train your antenna on such off-the-wall programs as live talk shows from Siberia (in Russian), South American Soccer games (in Spanish and Portuguese), and even jolly old BBC (in British). With a satellite receiver on the lawn, it's fair to say that your TV does, indeed, become a window on the world.

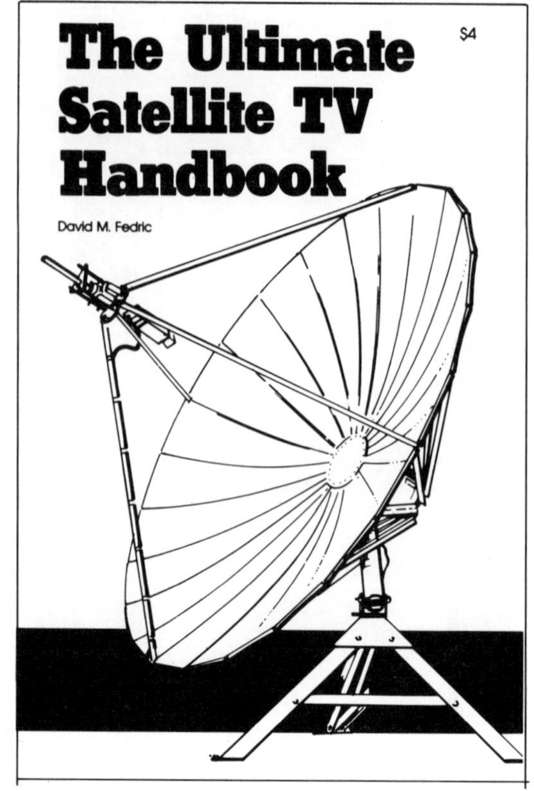

This 24-page booklet explains the basics of setting up a home satellite station. See page 163 for details

Who can own a satellite receiver? Just about anyone with $8,000 in their pocket (some models go for as low as $3,000) and a place to put the antenna. In October 1979 the FCC de-regulated its mandatory (and expensive) licensing of earth-station receivers. You can now legally own your own downlink, although the question of what you can *legally* watch is still up in the air, so to speak.

Also up in the air, about 22,300 miles up, are the geostationary satellites—the Comstars, Westars, Aniks, and Satcoms. These satellites receive signals from powerful earth ("uplink") transmitters, then re-broadcast the transmissions with a mere 5 watts of power (about the same as your legal CB radio). Programming that once could barely travel a hundred miles in the best of conditions with a conventional line-of-sight TV transmitter can now blanket an entire continent using satellite microwave relays. The only hitch—it takes some very sophisticated equipment to collect and amplify these faint microwave transmissions.

The drawing on page 162 shows the "footprint" of Satcom I, the weakest of the North American satellites. As you can see from the EIRP numbers (Effective Isotropic Radiating Power—the higher the better), the best reception for Satcom I is in New England and the plains states. The worst Satcom I reception is in Florida and Southern California.

The most troublesome component of the home satellite station is without a doubt the antenna. In most areas of the United States a 12-foot antenna is required to collect and concentrate the satellite signal. Usually constructed of fiberglass and aluminum, the typical downlink antenna on its pedestal can be taller than a single story house and is definitely *not* recommended for apartment dwellers.

Mounted at the prime focus of your antenna is a device known as the LNA (Linear Noise Amplifier), which boosts the satellite signal approximately 100,000 times without adding any unwanted electronic noise. From the LNA the amplified signal runs via shielded cable to your house and the satellite receiver. This is the device that lets you switch from "transponder to tranponder," which for our purposes is essentially the same thing as switching channels. Finally, the signal is run from your receiver to a "dedicated" TV monitor, or through your VCR to the family television screen (video buffs will realize that your are using your VCR to "modulate" the signal). The end result, if all is well, is up to 70 satellite channels—all of them just as crisp and vibrant as the highest-quality broadcast signal.

Sound irresistible? Before you call any of the earth-station suppliers listed on page 163, read a bit further. Here are some additional factors you should consider before spending money on a home satellite receiver.

● **Terrestrial interference.** About 2% of rural locations, and as much as 10% of all urban areas, are unsuitable for downlink stations because of microwave contamination—usually caused by neighboring telephone relay systems. Your earth-station installer should test for this *before* you order your unit or pour any concrete footings.

● **Physical obstructions.** Something else only your installer *should* survey for you. It's often difficult, if not impossible, to find one antenna location that will let you beam in on all the available satellite signals. Satcom I, in particular, is rather low on the horizon in many areas, and is sometimes obscured by tall trees, buildings, or neighboring hills.

● **Zoning laws.** Chances are you'll have to get zoning clearances, on-site building inspections, and all the rest. This is your responsibility, not the installers.

● **Antenna hazards.** A 12-foot antenna weighs nearly 250 pounds, and becomes a highly effective sail in high winds. Unless properly anchored, it can rip your roof off or sashay through your living room wall.

● **Antenna security.** There's nothing more likely to attract the attention of every kid on the block than a new backyard satellite station. You may find a chainlink fence a worthwhile investment to protect your downlink antenna.

● **Satellite switching.** You *can* receive up to 70 channels, but you will have to re-aim your antenna every time you want to switch from one satellite to another. This can be done either with a hand crank or an electric motor. Either way, you have to stand next to the antenna *with a TV set* to zero in on the exact spot in the heavens from which your satellite is transmitting. You can use pre-marked guidelines to sight the general vicinity, but even so, until the antenna is almost on the money, you won't get any signal from the satellite.

The Crowded Heavens— A Guide To Synchronous Orbit Satellites In North America

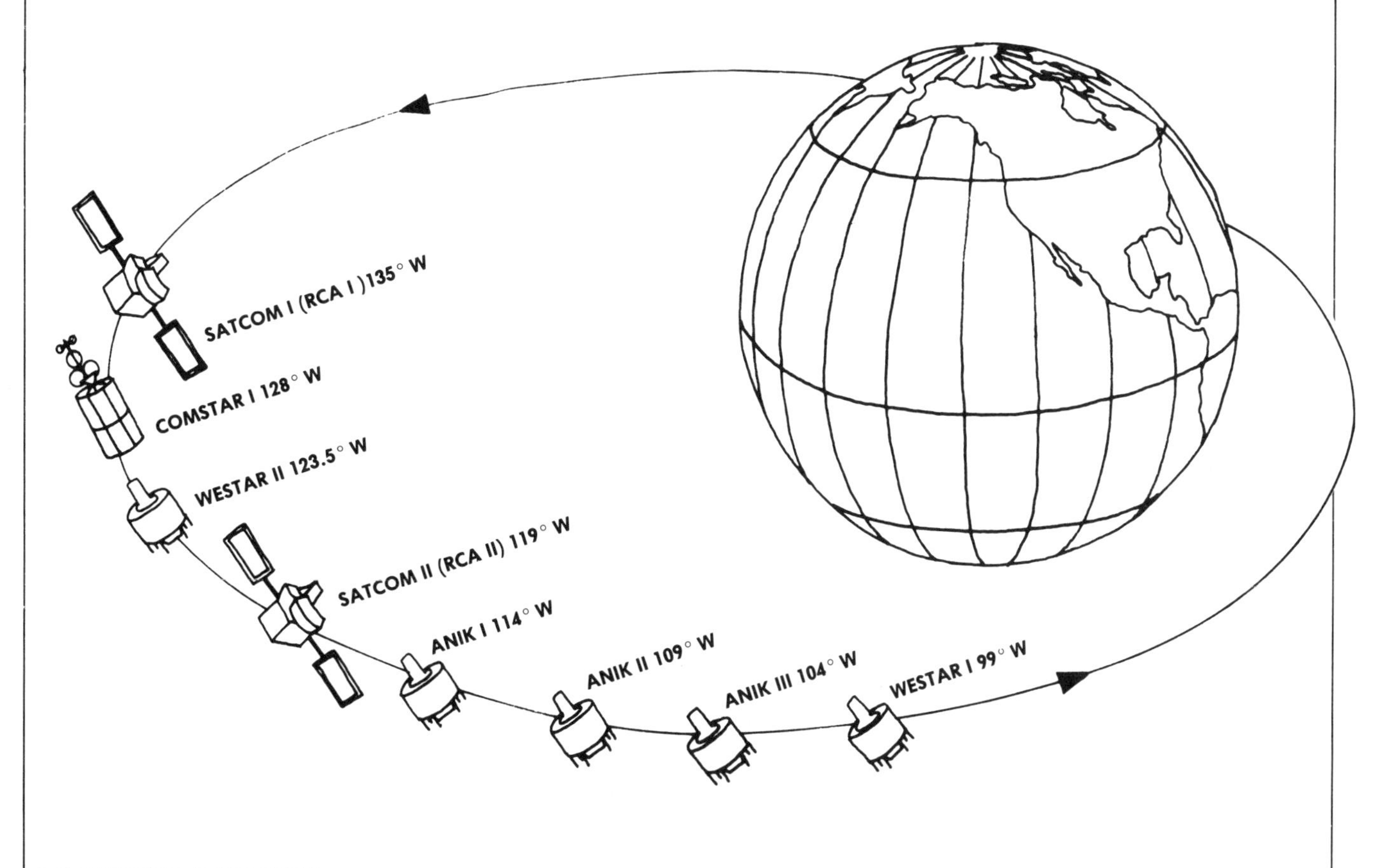

©1982 THE HOME VIDEO SOURCEBOOK

Satcom I "Footprint"

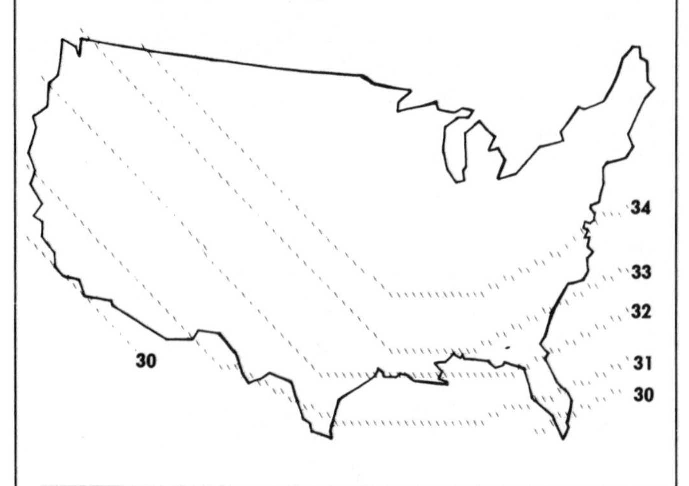

The higher the "EIRP" number, the better the Satcom I reception.

©1982 THE HOME VIDEO SOURCEBOOK

PROGRAMMING AVAILABLE ON SATCOM I (RCA I BASIC CABLE SATELLITE)

Transponder Number	Program Description
1.	**Nickelodeon**—Children's shows.
2.	**PTL** (People That Love)—Round-the-clock Christian programming.
3.	**WGN**—Chicago's independent TV station. Chicago Cubs baseball, Bulls basketball, and one of the world's largest classic movie libraries.
5.	**The Movie Channel**—24 hours a day, uncut and uninterrupted.
6.	**WTBS**—Turner Communications family-oriented programming with Braves baseball, Hawks basketball and lots of golden oldie movie classics and not-so classics.
7.	**ESPN**—24 hours a day of sports. Over 1,400 NCAA events.
8.	**CBN** (Christian Broadcasting Network)—Nonstop Christian programming from over 60 sources. Music, news, sports, and family entertainment shows. Free phone-in prayer and counseling services.
9.	**USA Network**—Madison Square Garden sports events—375 of them each year, including NHL hockey, NBA basketball, and major league baseball. Also broadcasts Calliope for children and The English Channel.
9.	**BET** (Black Entertainment Network)—films, sports, musicals, and other black oriented programming.
9.	**C-SPAN**—live, gavel to gavel coverage of Congress.
10.	**Showtime**—movies, Broadway and off-Broadway shows, night club acts, specials, and original programming.
10.	**AETN** (American Educational Television Network)—continuing education programming to professionals required to meet mandated licensing requirements.
14.	**CNN** (Cable News Network)—another Ted Turner enterprise. Round-the-clock news, sports, special reports, interviews, and national weather forecasts.
16.	**ASCN** (Appalachian Community Service Network)—public service programming, including college and continuing education programs offered for credit.
16.	**WWS** (Window On Wall Street)—financial news and trends.
17.	**WOR**—New York's independent superstation. Mets baseball, classic movies, and vintage TV shows.
18.	**Galavision**—Spanish programming with sports and movies from Mexico and South America.
20.	**Cinemax** (East)—All movies, no commercials.
21.	**HTN** (Home Theater Network)—family movie channel, G and PG only.
22.	**MSN** (Modern Satellite Network)—consumer oriented programs.
22.	**HBO** (Home Box Office)—First-run movies, sports, and entertainment. **HBO** was the original satellite movie channel.

SATELLITE RECEIVER GUIDE

Fortunately, once you are aimed at a satellite like Satcom I you will be able to switch from transponder to transponder (Satcom has 24 transponders, most carry a different program) as easily as changing channels on broadcast TV.

● **Program information.** TV Guide is not a big help here. Neither is your local newspaper. You can at least find out what's coming up on Satcom I by subscribing to *SatGuide*, published by Commtek Publishing Co., Box 1569, Hailey, ID 83333. At $36 a year it isn't exactly cheap, but then neither is a home earth station.

● **Licenses.** You don't need one, but it is not a bad idea. Otherwise you're likely to wake up one morning and discover Ma Bell has just installed a microwave relay across the street, rendering your $8,000 downlink installation useless.

● **Obsolescence.** Yes, as soon as more powerful communications satellites are launched, your 12-foot monster may look a mite antiquated. By the end of the decade it's expected a 2-foot antenna is all you'll really need.

● **Legal issues.** There's no problem owning an earth station. But using it is something else. The downlink dealers and installers are assuming it's perfectly legal to tune in on HBO, Showtime, Cinemax, Escapade, The Movie Channel, and others. The cable services disagree. The issue has yet to be resolved in court.

● **Scrambled signals.** Because of the proliferation of backyard satellite receivers, networks like HBO are seriously considering the idea of "scrambling" their uplink signals. Invest in a satellite receiver now, and there's a very real possibility that by next year all those premium services you thought you'd be getting free will only be so much static on your screen. But then the movie studios came up with Copyguard to prevent videocassettes duping—and we know how effective that was.

ACCESS TO INFORMATION

SATELLITE STUDY PACKAGE
A study kit that gives you a full description of what home satellite reception is and how it works. The "Home Satellite TV Reception Handbook" is written in comprehensible English prose and explains each component of the downlink system and what it does. Also included is a color poster positioning over 30 satellites for you. $15 from STT, Box 2476, Napa, CA 94558. An excellent investment if you're serious about downlink.

THE ULTIMATE SATELLITE TV HANDBOOK
This 24-page booklet gives you a concise and informed picture of the basics of home satellite TV. A little overpriced at $4, but worthwhile nevertheless. National Microtech, Inc., Box 417, Grenada, MS 38901.

ANTENNAS

ADM MODEL 11
This 11-footer is the best-selling satellite antenna in the world. It promises to give studio-quality pictures in most areas of the United States. Conversion panels are available to expand it to 13 feet. Price $2,500. ADM/Antenna Dev. & Mfg. Co., Box 1178, Poplar Bluff, MO 63901.

LOW-NOISE AMPLIFIERS

AMPLICA
This company produces a complete line of LNAs Their products are widely considered among the best in the industry. Amplica, Inc., 950 Lawrence Drive, Newbury Park, CA 91320.

AVANTEK
The Model AWC-4215 is the best selling consumer amplifier in the country. Our info is that this is an excellent product for the money. You will find it offered by many of the complete downlink system suppliers. Avantek, Inc., 3175 Bowers Avenue, Santa Clara, CA 95051.

SATELLITE RECEIVERS

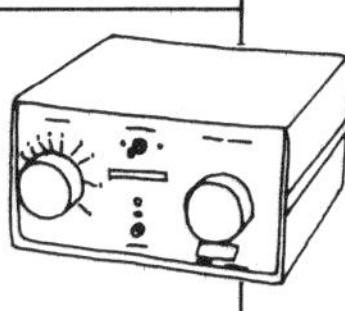

AVCOM
Model PSR-3 has 24-channel receivers with scan tune circuitry that automatically sweeps the satellite band. The company itself is well known within the home satellite industry for integrity. $3,295 from AVCOM of Virginia, Inc., 500 Research Road, Richmond, VA 23235.

KLM
The Sky Eye I is the world's best-selling receiver. You'll find it offered in many downlink packages. KLM Industries, Inc. 17025 Laurel Road, Morgan Hill, CA 95037.

COMPLETE DOWNLINK SYSTEMS

DOWNLINK, INC.
At $2,995 (plus installation) the Downlink system with its 12-foot Skyview spherical antenna is among the most economical complete home satellite systems available. Downlink offers a demonstration videocassette that includes samples of satellite programming. $29.95 for Beta, $34.95 for VHS. Black and White Enterprises, Ltd., Putnam, CT 06260.

NATIONAL MICROTECH, INC.
The widely advertised company offers three packages ranging from $6,800 to $8,800. They also carry a variety of quality components for putting your own system together. National Microtech, Inc., Box 417, Grenada, MS 38901.

CABLE & PAY-TV

There's turmoil in Tinseltown, and the main reason for it is spelled C-A-B-L-E T-V.

In 1980 a strike by the Screen Actor's Guild closed the major studios for months, delayed the production of hundreds of movies, wiped out the fall TV season, and torpedoed dozens of special projects—including a 25th reunion of the original Mouseketeers (Cheryl, Bobbie, Annette, Karen, Chubby, Darlene, Sharon, Tommy and Doreen) for which Disney Productions had invested over $2 million in lavish sets. The following year it was the Screen Writers turn to walk off the job.

Why such discontent among entertainment professionals, some of whom are paid in excess of $15,000 a week for their services? The answer, of course, is money. More specifically, a slice of the profits from videocassettes, videodiscs, and, most important, from cable and pay-TV.

With over 20,000,000 subscribers already on cable hookups—and millions more joining each year—cable and pay-TV are changing the way the entertainment industry does business.

Because of cable's well-financed appetite for feature films, Hollywood is finding it can turn a profit even on such box-office duds as Steven Spielberg's *1941.* Meanwhile, the networks are learning it is difficult to go head-to-head against pay-TV's uncut, uninterrupted feature films. The result? Fewer "movie nights" on prime time. In addition, both ABC and CBS are rushing to establish their own cable TV satellite services. Even the venerable—but financially troubled—institution of Public Television is feeling the heat. In a serious blow to PBS, a new culture-cable network called The Entertainment Channel, won the rights of first refusal to all BBC programming offered in the next decade. There is a new power to be reckoned with in television. The power of cable.

All cable TV is pay-TV, but not all pay-TV is actually cable. If this sounds confusing, it's not. Cable-TV is a way of distributing programs by means of coaxial cables (shielded wires) rather than over the airways. As a cable subscriber you'll pay a fixed monthly fee. You'll usually be able to add pay-TV channels, like HBO or Showtime, for an additional monthly charge. But not all pay-TV arrives via cable. Wometco Home Theater, for instance, broadcasts its programs like conventional TV stations. But you have to rent a "decoder" that is installed on your TV to unscramble the sound and picture.

Cable has been with us since the late 1940s when the television industry was just emerging. In the beginning cable brought TV programs to rural areas that were beyond the range of commercial TV signals. The cable

©1982 THE HOME VIDEO SOURCEBOOK

companies, also called community antenna television (CATV), built powerful receiving antennas and then re-transmitted the commercial TV signals over the coaxial cable system. The result dramatically improved reception, and added channels–including automated channels such as time, weather, wire-service news, and stock quotations.

A quarter of a century later, urban viewers are catching on to the joy of cable. And with communications satellites like Satcom I capable of beaming programming to local cable companies from coast to coast at low cost, the cable explosion has merely begun. You can expect to see new "basic" (included in your monthly cable fee) and pay-TV (you pay extra for it) services cropping up almost monthly as transponders on newly launched domestic becomes available. Rockefeller Center and RCA's The Entertainment Channel, ABC's Arts Network, and CBS Cable are among the new services to be on the lookout for.

The big breakthrough for cable came in 1976—the same year VCRs were beginning to appear in a few homes around the country. It was Time, Inc., that pioneered the idea of pay-TV by distributing the fledgling "premium" (meaning you have to pay extra for it) movie service, Home Box Office, via satellite. HBO went SRO, and the rest is history.

HBO's success spawned a number of satellite-distributed imitators, including Showtime, The Movie Channel, and Cinemax—the HBO clone—as well as regional pay-TV movie services like L.A.'s Z Channel, On-TV, and the Select-TV. Today there are over 4,300 local cable systems in the United States, over half of which offer at least one pay-TV service. In fact, the latest word in the cable TV lexicon is "tiering." A term devised to describe the situation when a cable subscriber buys a combination of pay-TV services, like HBO and Showtime.

If you're lucky enough to have cable in the first place (alas, many exurban dwellers aren't—too close to the big city to get on a rural system, too far away to get on an urban one), you'll find there are a number of great programming alternatives available to you that are included in your "basic" monthly fee. ARTS, Black Entertainment Television, Calliope, Cinemerica, C-Span, ESPN (Entertainment & Sports Programming Network), Nickelodeon, USA Network, and Cable News Network are among the channels you may find on your cable system. Right now the buzzword in cable programming is "narrowcasting." The idea is to develop an audience with shared interests—sports fans, working women, the elderly, blacks, culture buffs, etc. Such audiences have big appeal to advertisers who can reach their target markets efficently and inexpensively. You also may be able to use your cable system to tune in one of the regional superstation—like Ted Turner's WTBS, which offers round-the-clock nostalgia and sports.

As with HBO and Showtime, these stations relay their programming via satellite to your cable TV service.

Ironically, many of the early cable-TV systems can handle only 12 channels (10 years ago who'd have thought you could have more than 12 channels to choose from?). But new equipment has been developed, allowing you to tune in 52—or more—different channels. One catch for VCR owners: With a multi-channel cable box that contains its own tuner, you may lose the ability to record one TV show while you watch another.

If you also have subscription TV, there's another problem to contend with. The subscription TV decoder also works as your TV speaker. You'll be able to record the video image, but no audio. Fortunately, all these problems can be solved with a little assistance from your cable company or local video dealer, who will be delighted to see you buy the proper accessories.

What's in store for the future of cable TV? No one knows for sure, but one thing is certain—Cable TV will play a bigger and bigger role in peoples lives. One service, called Viewtron, is letting subscribers in Coral Gables, FL, shop from their living rooms via a two-way—or interactive—TV system. Interactive cable is finding other applications in business, government, and education.

Paying for cable is also getting easier. You can now charge your monthly HBO bill to Mastercard or Visa. And what happens if you don't pay up? Well, in Hampton, VA, one Warner Amex channel broadcasts the photos of wanted criminals. So far, viewers have helped police nab two felons.

As a new cable subscriber you'll pay a modest installation fee–$16 to $22 seems to be the typical range, and perhaps a permanent deposit of $25 to $50 for your cable tuner box. You will be billed monthly for your basic cable service, plus and expanded basic tiers or pay-TV channels you chose.

In most areas of the country the basic fee ranges from as low as $6 to as high as $12 or $13. Like the basic charges, the price of additional pay-TV channels can also vary by as much 100% from franchise to franchise, with a service like Showtime selling for as little as $5 to $6 in some areas to over $12 in others. The recommended price for HBO is $9 per subscriber.

Exactly how much you pay depends, to some extent, on how you live—with people living in single-family houses (which cost more to hookup to cable) often paying higher monthly fees than apartment dwellers. The number of channels offered, the size of your cable company, and the percentage of the take that goes to your local city or municipal government (or government officials) are all factors in determining how much you pay. As for your cable operator—he gets some services like Cinemericá and Christian Broadcasting Network for nothing, while others like Nickelodeon, USA, and ESPN cost any where from 2¢ to as much as 20¢ (for Cable News Network) per subscriber per month.

While services like Nickelodeon, Cabletext, the UPI Newswire, CNN, and ESPN are nice to have, the area of greatest interest to most VCR owners is movies—preferably uncut and uninterrupted. The 3 leading pay-TV movie services—with almost 90% of all pay-TV subscribers—are HBO, Showtime, and The Movie Channel. The remaining 10% of pay-TV subscribers have been divided among 8 services: Cinemax, Spotlight, Prism, Home Theater Network, Bravo, Escapade, Private Screenings, and GalaVision.

The majority of cable operators offer at least one pay-TV option—with HBO currently leading the pack with well over 2,000 affiliates and 6 million subscribers. Showtime is number two, with over 1,100 cable systems carrying the channel to over 2 million subscribers. Some cable operators will offer you a tiered pay-TV service, with a choice of two or more pay-TV channels (sometimes including the new Playcable channel) at reduced rates. In Manhattan, the center of the nation's largest cable TV market, a serious TV junkie can subscribe to Teleprompter's basic service for $11.95, but up his monthly bill to $44 by adding pay-TV channels like HBO, Showtime, and The Movie Channel. If he then decides to also include a separate subscription pay-TV service like Wometco Home Theater at $19, his monthly viewing bill can exceed $60.

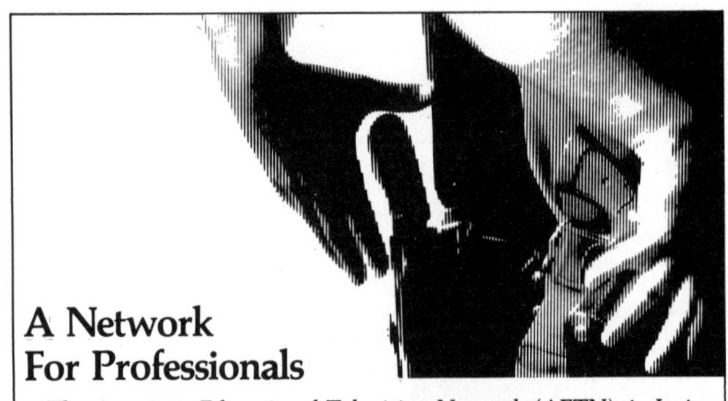

A Network For Professionals

The American Educational Television Network (AETN), in Irvine, California, is the first adult continuing education TV network operating on a nationwide basis through cable and broadcast TV distribution systems.

AETN was developed to reach the nation's 30 million professionals. It provides national delivery—for the first time—of educational programming required by professionals, from physicians to firemen, to keep abreast changing developments in their fields. AETN offers specialized classes in such career areas as law enforcement, insurance, accounting, emergency medical services, dentistry, public administration, nursing and medicine.

What To Expect From Your Cable Franchise

Virtually each of the nation's 4,300 cable-TV services offers its viewers a slightly different package of "basic" and pay-TV options. But the fact remains that most cable programming is nationally distributed by networks like ESPN and USA via microwave satellite transmission. Whether you live among the scotch pines on Presque Isle, Maine, or adjacent to the back lots of the Disney Studios in Burbank, California, your cable TV service will probably have 3 fundamental components.

- Local and regional TV stations. These are the commercial and public TV stations you get free anyway, but your reception may be much improved over cable. Also, you may find you are getting a number of channels from nearby cities that you couldn't receive before.
- Basic satellite-distributed programs. With the number of TV households on cable hookups approaching 1 out of every 4, new cable networks are springing up almost overnight. The biggest limiting factor is the availability of satellite transponder space. Your cable franchise can now select from well over 20 basic satellite distributed services that range from cultural programming like ABC's ARTS or CBS CABLE to ESPN's 24-hour sports broadcasts or Ted Turner's Atlanta superstation WTBS. A selection of these channels is almost always included in your basic monthly fee. A few operators offer an "expanded basic tier" for which you will have to pay more to add more channels.
- Pay-TV services. Like the basic services, these are nationally distributed by satellite. But pay-TV concentrates on just one thing—movies. The pay-TV tier will cost you more, usually a lot more.

How Does Your Cable Service Rate?

Here's a rundown of the nation's most popular basic cable and pay-TV services. Very few cable subscribers get all these services. But if your cable franchise is up to snuff, you should be receiving a generous sampling of the basic services, along with at least one or two of the pay-TV channels.

Pay-TV

HBO was the first, and is still the biggest. We wish we could differentiate between HBO, Showtime, The Movie Channel, Wometco Home Theater, and some of the others. But the bottom line is that all of these services show similar, if not sometimes identical, selections of recent box-office hits. There's no advertising. You'll get to see big smash hits like *Alien, Close Encounters,* and *The Godfather Parts I & II.* There will be some sports, some concerts, some specials, and an occasionally amusing short feature. You'll also find a lot of films you've already seen on the network movie nights, but at least they'll be unedited and easier to tape.

Showtime is a distant second to HBO in terms of subscribers and affiliates, but this only means they are trying harder to catch up. It used to be that all the pay-TV services had to do was talk the *cable franchises* into taking them on. But with tiering coming on strong, there's increasing competition to win over the viewer, not just the cable company manager. This means more and better films. But if you look at the titles, Showtime's movies are about the same as everybody else's. Showtime is also airing some original programming including a consumer affairs show with Ralph Nader, an off-beat magazine called "What's Up America," and a raunchy comedy series called "Bizarre."

Prism. A relative newcomer to the pay-TV scene, this Philadelphia-based service offers Hollywood features and regional sports. So far only a few hundred affiliates carry Prism, but as demand for "second tier" services increases, so will Prism's market penetration. One drawback is Prism's high cost, over $5 per subscriber per month to the cable operator.

Home Theater Network was started by a local Portland, Maine, cable operation and recently sold to Westinghouse. HTN is aimed squarely at the family market, and offers only G and PG rated films. Cost to the cable operator is $2 per subscriber per month, making HTN a popular option to offer in a second tier package. HTN is carried by several hundred affiliates across the country.

Escapade. This R-rated movie service focuses in on hot stuff for adults. It's satellite transponder, however, is located on Comstar D2, not RCA's Satcom I like most other basic and pay-TV services. Since very few CATV systems have downlink stations for Comstar, Escapade's penetration is limited to less than 100 affiliates.

Private Screenings. Another R-rated, sex-oriented movie channel. But don't be surprised if PS isn't on your cable system's pay-TV plan. Private Screenings is also on an offbeat satellite, Westar 3, and so far only reaches a handful of affiliates.

GalaVision. With under 100,000 subscribers, this is the smallest of all the pay-TV services. Since the programming is designed for a hispanic audience, it's mostly films from Mexico and Spain.

The Movie Channel. Like their ads say, "just movies, nothing but movies." As with HBO and Showtime, you can expect at least 5 or 6 real box-office biggies to premiere each month, plus lots and lots of warm leftovers. The Movie Channel and Showtime both cost your cable operator about $4 a month, but how much he charges you is another question. If you're a compulsive movie collector, it's arguable that The Movie Channel's all-movie policy will give you more to choose from, but like all pay-TV services, the rerun rate is atrocious. Over 1,000 cable franchises offer The Movie Channel to their subscribers. The Movie Channel was also the first pay-TV movie service to offer 24-hour-a-day programming. You can credit The Movie Channel with forcing HBO, Showtime, and Prism to expand their schedules to 24 hours.

Cinemax. HBO cloned itself and the result was Cinemax. This second Time Life-owned movie channel was the child of a marketing study that showed TV executives that there were a lot of viewers willing to shell out for more than one pay-TV movie service. Cinemax tries to avoid showing the same movies in the same month as HBO, and sometimes it's the first of the pair to air a new release. Presently Cinemax carries only theatrical films, with slightly more emphasis on children's movies, foreign films, classics, and adventure films than HBO. Several hundred cable franchises around the country now offer Cinemax, and the number is growing rapidly. It's a little cheaper for the cable operator than Showtime or The Movie Channel, but once again, whether the savings get passed on to you, the subscriber, is a different story.

Spotlight. The movie channel is available only to Times Mirror Cable Television Systems. It carries movies, specials, sports and reaches about 200,000 subscribers.

Advertisement Courtesy PlayCable

Subscription TV (STV)

Since subscription-TV is broadcast over the public airwaves on the UHF band, you may be able to receive an STV channel like WHT even if there is no cable service in your area.

Unfortunately, STV is about the most expensive way of watching movies short of going to the Triplex. This is partly because you have to rent a UHF antenna and STV "decoder box," all to be able to tune in one channel that may only be on the air a few hours a day, and even then is airing mostly reruns. STV prices run from a low of $13 or $14 to over $20 a month, plus a deposit and installation fee. For the same price you can usually get a 15 or 20-channel cable service *plus* a pay-TV channel like Showtime or HBO that will have the same movies you'll find on STV. So why bother with STV? Like we said, if cable isn't available, and you want uncut, uninterrupted movies, you don't really have much choice.

Another annoying aspect of STV is the Blounder-Tongue decoder used by WHT and several other services. This decoder has a built-in speaker which delivers poorer sound than just about all but the shoddiest bargain-basement black & white portables. Yet, there is no audio output to run the signal to your stereo should you happen to object to watching soundtrack-oriented films like *One Trick Pony* or *The Conversation* accompanied by an audio quality that a $9.95 transistor radio would be ashamed of.

On-TV is the oldest and most successful of the STV stations with movies, special features, and sports. Home base is L.A., but it is moving into Detroit and other urban markets.

©1982 THE HOME VIDEO SOURCEBOOK

Wometco Home Theater (WHT) broadcasts to New York, Long Island and New Jersey from transmitters in Newark, New Jersey, atop the World Trade Center in Manhattan, and in Smithtown on Long Island. The movies are standard fare with 5 or 6 top-quality premieres each month, a few classy classics, and lots and lots of real losers you'll never have heard of before or will ever want to hear of again. Biggest drawbacks are price: one of the most expensive around at $19 a month, and limited broadcast hours, usually less than 8 hours a weekday, of which all but 2 or 3 hours are devoted to reruns.

Z-The Movie Channel reaches many noncable areas of Los Angeles County, including most of San Fernando and San Gabriel Valleys, the South Bay, Long Beach, and Whittier. About the same movies as everybody else, but a longer broadcast day than most.

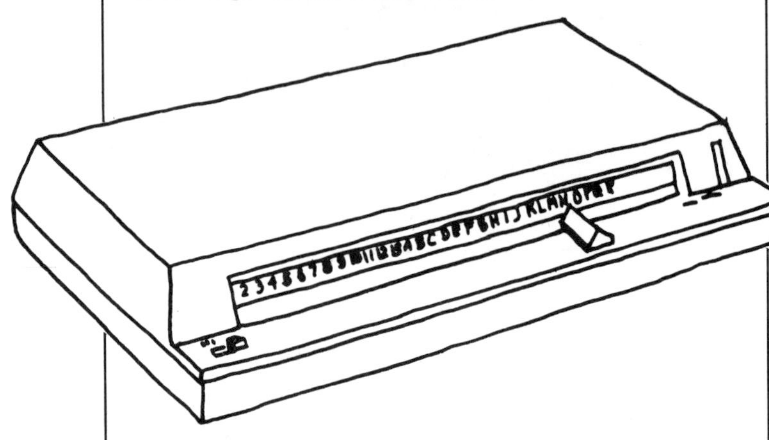

Basic Cable Services

CBN (Christian Broadcasting Network) is the oldest and the largest of the basic cable services. CBN, which advertises itself as "the fourth network" (after ABC, CBS, and NBC), has over 3,000 affiliates and reaches more than 10 million subscribers (CBN puts the figure at 20 million people). In a recent attempt to expand its appeal, CBN is de-emphasizing their religious programs and is including more and more family-oriented entertainment shows.

ESPN (Entertainment and Sports Programming Network). This recent addition to the cable line-up is a sports addict's delight, with 24 hours a day of college and pro events not featured on network TV. ESPN has skyrocketed into second place among the basic services, with almost as many subscribers as CBN. To find out what's on ESPN today call toll-free, 800-243-0000.

USA Network has also been growing at a prodigious rate, and now ranks among the most widely available cable services. Programming is heavily sports-oriented, but also features The English Channel with quality documentaries and classic films, Calliope, "YOU Fashion Magazine for Woman," "Night Flight" rock films and concerts, and "Alive and Well" (a 5-day-a-week health magazine produced under a $40 million contract with Bristol-Myers).

C-Span (Cable Satellite Public Affairs Network). We suspect it has more to do with politics than consumer demand, but this network—which is distributed free to cable operators—is now among the top 5 mostly widely broadcast cable TV channels in the country. What's on C-Span? Why those enlightened proceedings of the U.S. House of Representatives, that's what. If you find network programming like "Love Boat" or "The Dukes of Hazzard" an insult to your intelligence, try spending an hour or two observing Congress in action.

BET (Black Entertainment Network) also ranks among the top 5 most widely distributed cable channels, and is growing rapidly as more urban areas are wired for cable. Programming consists of black-oriented films, music specials featuring black entertainers, and college sports.

CNN (Cable News Network) is the fastest growing of all cable services. In the 6 months between November 1980 and May 1981, CNN doubled its subscriber base from 3 to 6 million, and is continuing to add new affiliates and viewers at an incredible pace. CNN offers 24-hour-a-day news coverage, with an emphasis on in-depth reporting, sports coverage, and national weather. CNN is now available on over half of the nation's cable systems.

Nickelodeon is also adding new affiliates and subscribers at an explosive rate. Well over 1,000 cable franchises now carry this Warner-Amex children's network, which features live performances, music, nature films, animated shorts, and on-going kids shows like "Livewire" and "Pinwheel." Also being carried on the Nickelodeon transponder is ABC's culture channel, ARTS, which is devoted to the visual and performing arts with specials on jazz, dance, opera, painting, and the like.

MSN (Modern Satellite Network) distributes films produced primarily by business and government agencies for educational purposes.

SIN (National Spanish Television Network). Features Spanish-language movies, programs, and sports.

SPN (Satellite Program Network). A mixed bag of programs with a general focus on women's issues.

Cinemerica is a network designed for men and women 45 and older. Programming ranges from consumer education to entertainment.

UPI Newstime and **Cabletext** both provide teletext wire service reports of national and international news events.

Superstations

The 3 most widely distributed independent TV stations are WTBS, WOR, WGN. Not only are these stations broadcast in their respective cities—Atlanta, New York, and Chicago—but they are also relayed to thousands of CATV affiliates by satellite, much like pay-TV and the basic cable service channels like ESPN.

WTBS, Ted Turner's Atlantic superstation, was the first independent TV station to go-cable in the 1970s. WTBS stresses its 40 movies a week, and sports programs emphasizing Turner-owned teams like the Atlanta Braves.

WOR originates in New York City and provides sports coverage of the NY Rangers, Islanders, Knicks, and Cosmos, as well as many golden-oldie film classics.

WGN broadcasts Chicago's professional sports teams, including both White Sox and Cubs as well as claiming one of the nation's top libraries of classic films.

THE RENTAL ALTERNATIVE

Most of us are willing to invest in a videocassette recorder for 3 reasons —we want to watch what we *want, where* we want, *when* we want. That usually means good movies—at home.

But when it comes to spending $40 to $90 to get an uncut, prerecorded film that we'll probably only view a couple of times, we draw the line. Why spend big bucks on a flick that's only going to gather dust on the bookshelf? Especially when you can rent the same film at Fotomat or from your video specialty shop for as little as 1/10th the price.

So, if your personal prerecorded film collection doesn't look like the window display at Video To Go, don't despair. You are not alone. In one telephone survey the A.C. Nielson Company found fewer than 1 out of 3 VCR households owned any prerecorded tapes at all.

Cover Courtesy VIDEO SWAPPER Magazine

The Local Rental Scene

We'll venture to predict the day is not far off when videocassette (and perhaps videodisc) rental outlets will be as common as the corner record store. If you live in a major metropolitan area—L.A., New York, Chicago, Philadelphia, etc.—you should already have at least 2 or 3 rental sources within a short drive, with more on the way. Witness the growth of Video Station, a Los Angeles-based franchise operation founded in 1977. Today Video Station claims to have several hundred stores in business, with over 20 new shops opening every month.

The big daddy of videocassette rentals—with over 3,800 outlets leasing more than 1,200 tapes every day—is Fotomat. (See "Fotomat" on page 51.) Having started their national program on May 25, 1979, Fotomat can take credit for being a principal trailblazer in the brave new world of videocassette rentals.

The current Fotomat catalog offers over 300 titles—including many blockbuster releases from the Paramount, Columbia, EMI, RKO, and Disney inventories. Rental fees range from $5.95 to $11.95, depending on the film. You keep the cassette for 5 days.

To order a film from Fotomat, just dial 800-325-1111. In theory, you should be able to pick up your choice within a day or two at the nearest Fotomat booth. For sheer convenience Fotomat's "Drive Thru Movies" are hard to beat, particularly if one of the booths happens to be located along your regular commute.

Ironically, at the same time that a *photographic* service operation—Fotomat—was locking up the videocassette rental market from coast to coast, local *video* dealers were under increasing pressure from their ultimate suppliers—the big Hollywood studios—to keep out of the rental business.

If people like us could rent, instead of having to buy, the studio heads figured they'd lose out on some very generous royalties—sometimes as much as $30 per cassette. Dealers who wanted to sell films were forced to sign contracts prohibiting them from renting. The result was a booming—but clandestine—"under-the-counter" rental industry.

The turning point came in September, 1980, when a Highland Park, Illinois, video dealer named Gene Kahn defied the studios. Kahn announced he was going to do openly what almost everybody was already doing secretly—rent videotapes. If his suppliers didn't like it, he'd buy his tapes elsewhere.

Kahn's timing couldn't have been better. With the studio bigwigs losing millions of dollars a week as the result of the Screen Actor's Guild Strike—then entering its 3rd month—the rebellion of a local mid-west video dealer went almost unnoticed. Only MCA, the distributor for Universal, withheld tapes from Kahn.

The Highland Park incident, however, resulted in a bonanza for movie-hungry VCR owners. Overnight, extensive rental libraries appeared on the shelves in thousands of video specialty shops.

Today, almost any video store manager worth his signal-to-noise ratios stocks an inventory of 300 to 800 or more rental tapes. Video Station reports rentals outnumber sales in its 300-plus affiliate stores by 15 to 1. Video Connection, a small chain with stores in the affluent suburbs of Long Island, say 90% of its prerecorded videotape business is in rentals. Even at Philadelphia's Movies Unlimited—where sales, not rentals, are stressed—4 tapes are being rented for each that is sold. In Chicago the City Fathers have found a way to get a piece of the action—a new law socks a 6% tax on videocassette rentals.

In most stores you select the film you want, usually from a mimeographed list, and leave an imprinted credit-card slip, or a $50 to $75 cash deposit, as an indication of your intention to return the tape. Most of the libraries we've seen contain 100% legally licensed tapes. A few don't. Bootleg tapes and illegal dubs are cheaper (no royalties to pay)—and pirate tapes don't seem to bother the average customer, as long as the picture quality is up to par.

Most of the store-operated rental libraries are heavy on "major motion pictures," i.e., commercial Hollywood fare. These are the films the studios spend big money to promote, and for which most people are willing to spend money to rent. But mixed among yesterday's fading box-office hits, you're likely to uncover at least a handful of the more obscure, but delightful film classics. If you don't like your dealer's current selection, check back in a few months. Most video shops are adding to their rental libraries at the rate of one new title a day. You'll also find many shops maintain separate catalogs—one for their family movies and one for films of the more erotic vein.

How much can you expect to pay for a videotape rental? First, if the film you want is a big-box office smash that's just been released on videotape, the chances are you'll be lucky to rent it at all, since you and every other VCR owner in town is asking for it. Also, until a film has proved its rental staying power, most small dealers are reluctant to invest heavily in multiple rental copies. When you can get them, the hot new hits will bring premium prices—usually $7 to $12 for a minimum rental, with a dollar or two a day tacked on if you keep it past an agreed date. The adult flicks also fall into this premium catagory at most shops.

For the vast majority of films, however, you should expect to pay between $5 to $8 for a short rental. But temporary price wars are common, so shop around. L.A. and Chicago have both seen rentals advertised for a low as $1.

Video Exchanges—Home Grown To Commercial

If you enjoy savoring your films in repeated showings served up over a couple of weeks, or if perhaps you just get high on the idea of actually *owning* a film—albeit only temporarily—the rental scene may not be for you. One alternative is the "video exchange."

Video exchanges come in more flavors than Baskin Robbins. You can exchange tapes directly with other VCR owners, or through an intermediary such as the Video Guild, or you can join a "time-sharing" club like the American Video Library. Tape swapping is enormous fun, and it can be considerably less expensive than renting. It also requires a little video savvy—just ask the guy who swapped a legal $70 copy of *Alien* for a presumably equivalent videocassette of *Patton.* Except this *Patton* happened to be a 2nd-generation-homemade-dub-of-the-edited-for-television version, complete with commercials, recorded on a Hong Kong bargain-basement cassette. Don't get us wrong, most tape collectors and swappers are scrupulously honest hobbyists whose goal, like yours, is to watch good movies at reasonable prices. But there are always a couple of exceptions.

Almost anywhere you find VCRs, an informal video exchange network is sure to follow. Check the "Collector's Bulletin Board," or equivalent, at your nearest video specialty shop. Here you should find lists of programs that other VCR owners in your area want to swap. Tape trading, incidentally, isn't a bad way to get plugged into your local video network. You may even discover a well-organized club with a regular newsletter that catalogs entire video collections that are up for trade. And it's not likely that someone you'll be running into from time to time around town will be dealing in defective merchandise.

For swapping on a national level we suggest you write *Video Swapper,* Box 684, Sterling Heights, Michigan 48077, and enclose $2 for a sample copy. This video trading magazine aimed at VCR owners and tape collectors printed their first issue in January 1981 and went from bi-monthly to monthly editions later the same year. Send them your "swap" or "want" list typed on a sheet of 8½" by 11" white paper, and *Video Swapper* will reprint it for their 10,000 subscribers. The charge for this service is $12 per sheet. $12 is also the cost of an annual subscription. About 70% of the *Video Swapper's* space is devoted to the sale of used tapes, 30% to exchanges or trades.

If you're nervous about trading videocassettes with someone you don't know, the Video Guild at 500 Davis Center, Evanston, Illinois 60201, has a unique service for its members. Their monthly newsletter—"The Cassette Gazette"—runs a list of used films and programs that Video Guild members want to sell, and all of those cassettes have been pre-screened to check for defects.

The Video Guild, which was founded in 1980, also maintains one of the largest mail-order rental collections in the country—over 800 titles available for 1-week rentals for between $9.95 and $13.95. Their toll-free telephone number—800-323-4227—is a bonus should you have any problems. The service is prompt and reliable, and the annual membership fee of $18 is not unreasonable.

Perhaps the most interesting of the video exchange permutations is a "time sharing" video club—the American Video Tape Library (AVTL)—started in September 1978 by Nancy Payne in the Denver suburb of Littleton, Colorado. To join you pay a one-time fee of $89.95, plus monthly dues of $5. There are no other costs. As a member of the AVTL you can borrow *as many videocassettes as you want* (one at a time) and keep each cassette for up to 30 days. You pay only return postage—about $1.50 for VHS tapes, less for Beta.

The AVTL has over 600 titles in its "member-owned" collection, including many of the movies you'll find reviewed in our "Best Of Video" section. A full-time staff of 6 runs the AVTL computer, answers the toll-free phone—800-525-8998—and keeps track of the more than 4,000 videocassettes.

Normally we're skeptical of any operation that asks for your money in return for future services. Two former video exchange clubs come immediately to mind—Dial-A-Movie and the Free Video Exchange Club. Both promised unbeatable deals after you paid a healthly membership fee, and both are no longer in business. But we like the quality of the AVTL collection, the simplicity of their rules, their unlimited check-out potential, their reasonable dues, and the fact that they've survived when many similar clubs haven't. If you foresee yourself screening 2 or 3 prerecorded films a month and don't mind an occasional trip to the post office, we think AVTL is a good bet. The address is: American Video Tape Library, 6650 South Broadway, Littleton, Colorado 80121.

For sheer longevity we can recommend Discotronics, now located at 713 North Military Trail, West Palm Beach, Florida 33406. This company originated the commercial videocassette exchange in 1977, the same year Magnetic Video began releasing prerecorded movies. Short of renting or joining a "time sharing" club like AVTL, the Discotronics' one-for-one swap is among the most inexpensive ways we know for obtaining prerecorded films.

True, you'll need a legally licensed videocassette to start. And it has to be a title listed in the Discotronics catalog. But with well over 400 family features and 200 adult films on the list, the odds are definitely in your favor.

What we particularly like is that Discotronics will accept your tape for a credit of 50% of the recommended list price *even if you purchased it somewhere else.* It merely has to be a legal copy in good condition.

So, if you've finally grown tired of watching Christopher Reeve snatch Margot Kidder from the jaws of death, you can mail in your $69.95 copy of *Superman* for a $35 credit. If you want you can apply this credit toward another new film. In which case a new $69.95 film—say *All That Jazz*—will cost $35 plus your trade-in. No big savings here.

You can, however, apply your credit towards a *used* tape (Discotronics has to do something with its trade-ins, right). And here's where you can really save some money, since the used tapes are sold at about a 30% discount. The bottom line? It'll cost you $15 plus postage to swap your legal *Superman* for any other $69.95 film in the Discotronics catalog. If the film you want isn't available used, Discotronics will put your name on their reserve list and send you one of "back-up" choices in the meantime.

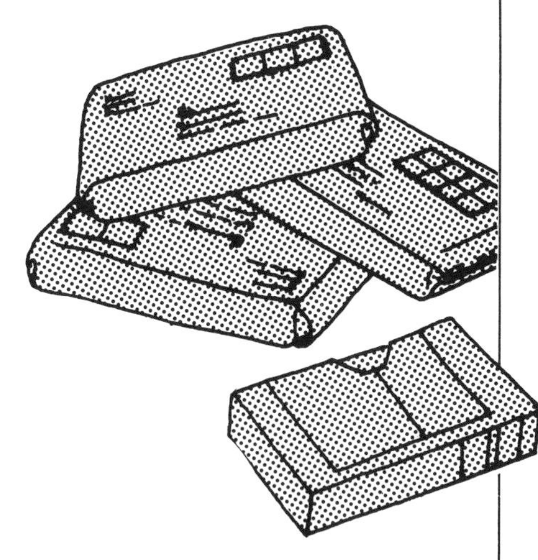

DISCOUNT SHOPPING

At least once in your life you have probably experienced the sinking feeling that comes from discovering the gadget for which you just shelled out $499 being advertised across town for $249.95. We assume you are not in the habit of repeating such errors—and just to make sure, here's a list of 17 toll-free telephone numbers.

Each of these numbers will connect you with a discount outlet that claims to carry a large inventory of home video equipment. If you know the brand and model number of the item you want, it shouldn't take more than a couple of minutes to get 2 or 3 price quotes. And it won't cost you a dime.

Please understand that we are not necessarily recommending you buy from any of these establishments. But we are strongly advising that before you buy anywhere, you take a couple of minutes to find the rock-bottom lowest price that you can get away with paying.

Actually, you may find some of these quotes very enticing, especially if you've already done some looking around at your neighborhood appliance dealer or big-name department store. If you have the luck of the Irish with electronics equipment—nothing you buy ever needs servicing or repair—then you may want to go ahead and give one of those voices on the other end of the phone your MasterCard number. Should the machine you order happen to be DOA (Dead On Arrival), you can always ship it back by return mail and wait for a replacement. But with most mail-order firms, after a week or 10 days has elapsed, you are on your own. After that, any service problems are strictly between you and the manufacturer.

That usually means sending your unit—in its original shipping carton—to an authorized service center. That usually means you won't see your beloved widget for anything between the next 3 weeks and 6 months. And this can happen even if the problem is a minor adjustment that could have been corrected by a knowledgeable video technician in a couple of minutes.

What is the alternative? One answer is to buy over-the-counter from a discounter with his own service center. Tape City in Manhattan is one. The Video Service Center in Sacramento, California, is another. Check the Yellow Pages under "Television & Radio Dealers." Many regional discount store chains will have a central service center to which your machine is forwarded for repairs. Crazy Eddie's in the New York metropolitan area is a 10-store chain with their own service and repair center.

The Discount Shopper's Guide To Toll-Free Numbers*

Attic, Inc.	800-221-0118	Forest Hills, NY
Buescher Enterprises	800-841-4443	Valdosta, GA
Electronics Distributors	800-327-3376	Miami, FL
Electronics Playground	800-228-4097	Lincoln, NB
47th Street Photo	800-223-5661	NYC
Hardside (Computers)	800-258-1790	Milford, NH
International Wholesalers	800-327-0596	Miami, FL
Inovision	800-527-0264	Dallas, TX
J&R Music World	800-221-8180	NYC
Mike Brody's Camerama	800-645-6323	Roslyn, NY
S&S Sound City	800-223-0360	NYC
Southwest Video Wholesalers	214-669-2331	Richardson, TX (Call collect)
Tape City	800-223-1586	NYC
Video Service Center	800-824-7875	Sacramento, CA
Videotime Corporation	800-645-2317	Westbury, NY
Video Wholesalers, Inc.	800-327-0337	Miami, FL
Warren Processing	800-221-0246	Brooklyn, NY

*If you live in the same state check with information for a local telephone number.

The Do's And Don'ts Of Discount Shopping

● Do decide on the specific brand and model you want, along with an acceptable alternative.
● Do get the manufacturer's specifications before you buy (either from a local dealer or by mail from the manufacturer).
● Do use "The Discount Shopper's Guide to Toll-Free Numbers" to find the lowest possible mail-order price.
● Do find the nearest video dealer who can really service what he sells.
● Do ask about extended contracts.
● Do compare prices between the mail-order discount houses and your local dealer.
● Don't buy mail-order unless you just can't find a comparable deal (we think an extra 5 to 10% for personal service and a well-equipped repair shop on the premises is a comparable deal) closer to home.
● Don't buy anywhere until you have compared prices.
● Don't ever throw away your warranty card, sales receipt, original shipping carton, or owner's manual.
● Don't forget to make sure *all* the features work when testing out a new piece of video equipment.
● Don't fail to return the item *immediately* if something doesn't work.

Crazy Eddie's also sells extended service contracts. Buy a VCR from them and you will be given the option to purchase an extended 2-year service plan for $130 that will cover parts and labor on any repairs. If you use your equipment a lot, an extended service contract is not a bad idea.

If you don't have a discount outlet in your area with its own service center, we suggest you try the Yellow Pages again, this time under "Video Dealers," to find the nearest store that *specializes* in video equipment. See if they stock the model you want, or can offer an acceptable alternative. Also find out if they maintain a service department. If the answer is yes, pay them a visit. Talk to the salespeople and, if possible, the repair technicians. How many full-time technicians work there? Is the facility neat and well-organized? What kind of repairs can be made on the premises, and do *they* ever have to return the unit you want to the manufacturer for repairs? Now get the dealer's best price, and don't be afraid to haggle. If it's within 5 to 10% of the discount prices you got over the phone, our advice is go for it.

Retail electronics is a highly competitive business right now. We've been pleasantly surprised to discover our storefront video dealer pricing his brands (JVC, Sony, RCA, Quasar) at just 5 to 6% over the lowest discount prices available in New York City. In our opinion the peace of mind that comes with good personal service is well worth paying an extra $50, or even $100, on a $1,000 purchase.

As you may have guessed, we are not particularly sanguine about buying from appliance or department stores. Some can service the video equipment they sell, many can't. But if the price looks right, and you think the service is the best you'll find in your area—why not?

VIDEOSPEAK

A

AC (Alternating Current). Direct Current (DC) electricity, such as from a battery, always flows in one direction along a wire. AC electricity, however, first flows in one direction, then in the other, reversing itself with a certain frequency, usually 50 or 60 times (or cycles) each second. Household current in the US is 120 volts AC. All video equipment sold in the US is designed to run on this AC current unless it is battery operated. For battery-powered units, like portable VCRs, an "AC Adapter" is usually available.
Amplifier. An electronic circuit that strengthens any signal sent through it. If you use "splitters" to divide a signal between several TVs or VCRs, it may be necessary to add a "booster amp" in the line to prevent signal loss.
Aperture. Still photographers will recognize this term as referring to the size of a lens opening created by the iris—usually given in f-stops. Most home video cameras have automatic irises that electronically adjust the aperture according to the prevailing light conditions.
Aperture Grille. A metallic screen on the inside of the Sony Trinitron TV screen which helps to increase the apparent sharpness of the video image.
Aspect Ratio. The ratio of height to width. The video frame has an aspect ratio of 3:4, while conventional movie film has an aspect ratio of 2:3. Hence, some part of the frame must always be cropped when transferring to video, or vice versa.
Audio Dubbing. A feature on virtually all VCRs that lets you record on the audio (or soundtrack) without erasing your video image.
Audio Inputs/Outputs. These are the jacks on most VCRs, videodisc players, projection TVs, and cameras that allow you to route your soundtrack. You will need the audio jacks when dubbing between two VCRs, when making an audio dub, or when playing back your soundtrack through a stereo amplifier.
Audio Head. The magnetic device within your VCR that records and plays the soundtrack.
AGC (Automatic Gain Control). This is a very handy circuit in VCRs and cameras that automatically limits the audio and video signal intensity going onto the video tape, hence insuring properly recorded sound and video tracks.
Available Light. A term borrowed from photography referring to the light that is naturally present in a scene, without the addition of flood lamps or other artificial light sources. Because of their tremendous sensitivity, video cameras can shoot in almost any available light situation—although your results will be far better if you do use flood lamps where light levels are exceedingly low.

B

Back Light. Another photography term. It refers to a light source which is located *behind* your principal subject. Strong back light usually confuses an automatic iris, resulting in an underexposed image. Some cameras are equipped with a back-light control switch (BLC) that will compensate for this situation.
BASIC. Beginner's All-purpose Symbolic Instruction Code. BASIC is the simplest of the computer programming languages to learn.

©1982 THE HOME VIDEO SOURCEBOOK

Beta. The VCR format originated by Sony and licensed to Sanyo, Toshiba, and Zenith. The Beta format videocassette is somewhat smaller and lighter than the VHS format cassette. It also holds less magnetic tape and has a somewhat shorter maximum recording time.
BNC. A professional type of bayonet cable connector found on some video products.
Boom Mike. On video cameras this is a microphone that is extended from the camera on a short boom. A boom mike helps to eliminate camera and background noises from your soundtrack.
Broadcast TV. Any station that arrives at your home via the public airwaves—this includes all UHF, VHF, and many subscription TV channels like WHT and LA's Z-Channel.
Burn. A lingering after-image that occurs when the light-sensitive tube in a video camera is exposed to a bright light source.

C

Cable TV. Also called CATV for Community Antenna Television. The cable TV service collects programming from various local, regional, and national sources and relays it to you via special cables. In addition to the basic channels covered by their monthly fee, most cable TV services also offer pay-TV channels like HBO, Showtime, or Cinemax for which you will pay extra. If you have the opportunity to get on a cable service, do it.
CED (Capacitance Electronic Disc). A 12-inch, phonograph-like disc developed by RCA for the playback of audio and video signals. The CED videodisc player uses a diamond-tipped stylus to pick up electronic signals from the shallow disc grooves, about 38 of which will fit into one groove on an ordinary audio record.
Capstan. The motor-driven roller in a VCR that determines the exact speed of your tape.
CRT (Cathode Ray Tube). The electron-gun-type picture tube found in most TVs.
Channel. A specific assigned frequency. There are, for instance, 12 channels on the television VHF band and 68 channels on UHF.
C-Mount. A mounting device with a threaded collar which allows the interchangeability of lenses on a video camera.
Color Burst. A set of TV transmissions that indicate the color phase of each horizontal line on a video screen.
Color Phase. The timing of a color signal. Incorrect color phasing will affect the hues of a video image.
Color Temperature. A measurement of the exact color of a light source expressed in Degrees Kelvin. Bright sunlight is about 6000°K, while incandescent lighting is about 3000°K. Most color video cameras have a switch that allows you to adjust for the color temperature of various shooting conditions.
Contrast. Technically referred to as "contrast ratio." It is the ratio between the brightest and darkest areas on a TV screen. A picture with well adjusted contrast will give you a maximum of middle tones. When contrast is poor the image will appear either washed out or too dark.
Control Track. A portion of a video tape that contains information not used in the video or soundtracks. Most control tracks indicate the tape speed at which the recording was made.
Coaxial Cable. A cable with one ground and one conductor that can simultaneously carry a number of different frequencies with little loss of signal strength. 75-ohm coaxial cables with "F" type connectors are used in many home video applications.
Crosstalk. The unwanted interference that can occur between adjoining video or audio tracks.

D

Decibel (dB). Decibels are used to measure the volume of sound, or to express a ratio between two amounts of electrical or accoustical power. In evaluating video equipment, we are particularly interested in the ratio of the audio or video signal to the unwanted noise that accompanies that signal. This signal-to-noise ratio, as it is called, is measured in terms of decibels—and for our purposes, the higher the number of dBs, the "cleaner" the audio or video image will be.
Definition. The amount of sharpness or detail in a TV picture.
Dew Indicator. A moisture sensitive circuit that shuts down your VCR if there is a potentially damaging level of humidity on the video heads. Video tape will stick to a moist head, destroying the tape and possibly the head as well.
Disc. There are currently 3 disc formats on the market. These are LV (LaserVision), CED (Capacitance Electronic Disc), and VHD (Video High Density). While these playback-only discs are used in videodisc players, a fourth type of disc, known as the "floppy disc" is used with many home computers to play back and store information.
Dropout. A temporary loss of picture that appears on the TV screen as a white streak. Usually caused by impurities or imperfections in the magnetic video tape.
Dubbing. The duplication of a videotape, or the addition of a new soundtrack to an existing tape. The duplicate (or next generation) tape is called a "dub," or occasionally a "dupe."

E

Electron Gun. The device at the back of a CRT that produces the beam of scanning electrons.
Electronic Viewfinder. A tiny black & white TV set built into the back of a video camera. An electronic viewfinder allows you to see the video image as you record it, or to play back a tape in the field without having to carry along a bulky monitor. It is definitely the best viewfinder for location shooting, but adds at least $100 to the cost of a video camera.
Extended Length Tapes. Tapes that are longer (and hence give you more playing time) because they are thinner (and hence more susceptible to damage). We like them for storing films you won't be viewing too frequently, but for an everyday workhorse we recommend the Beta L-750 or VHS T-120.
EP (Extended Play). The "6-hour" playing speed on VHS format recorders. Called SLP, for Super Long Play, on most brands. But if you can have 3 ways to say the same thing, it's bound to confuse 6 times as many people.

F

Fade To Black. A screenwriter's device used to indicate a pause in the action. Often, but not always, used before a commercial interruption in TV serials or made-for-TV movies.
First Generation. The first tape made from a film or live event.
Flagging. A term used to describe the situation in which the top portion of your picture bends to one side of the screen. This rarely happens with first generation tapes you've recorded, or legal prerecorded tapes.

Footcandle. A measurement of light intensity. For our purposes footcandles are used to compare the sensitivity of video cameras in low light. A 60-watt lightbulb produces about 400 footcandles. Many video cameras will record a recognizable image in light as low as 10 footcandles, but don't expect the greatest quality.
Format. In video, format denotes a technologically compatible system. The two basic VCR formats are called Beta and VHS. In videodisc players there are 3 formats—CED, LV, and VHD.
Frame. A frame (or complete TV picture) is composed of 525 horizontal lines which are scanned by the electron gun 30 times per second.
Freeze Frame. A special effects feature in which a single frame (there ar 30 frames per second in video, 24 in film) is stopped—or "frozen"—on the screen, usually by pushing the pause button.
Frequency. The number of times a signal vibrates each second. Usually expressed in Hertz (Hz).
FM (Frequency Modulated). Any signal that has been impressed onto a radio wave. UHF and VHF are both FM signals.
Frequency Modulator. An electrical circuit that produces a radio wave on which an audio and video signal can be impressed. Most VCRs have a frequency modulator built into them.
Frequency Response. This is the range of frequencies to which an electronic circuit is receptive. The human ear can usually detect signals between 30 and 15,000 Hz. Video signals range from 0 to 4 *million* Hz. Hence, the enormous demands place upon video tape and VCR circuitry.
F-Stop. A measurement of the amount of light passing through the aperture of a lens.
F Connector. See "coaxial cable."

G

Gain. Signal amplification. Turning up the gain increases the strength of a signal, turning down the gain does the reverse.
Generation. Each generation is one dub further away from the original, and one step lower in audio and video picture quality.
Glitch. Picture distortion.

Guardbands. The separation between tracks on a video tape.

H

Hardware. A word borrowed from computer lingo. It's used to describe any piece of electronic equipment—VCR, videodisc player, projection TV, home computer, video game console, satellite receiver. Software, on the other hand, is the videocassettes, discs, game programs, or programming instructions used with this equipment.
Head. The electromagnetic component of a VCR that paints the audio or video signals onto the magnetic video tape.
Head Alignment. The all-important angle of the video head. A poorly aligned head will result in signal loss, or no signal at all.
Head Clogging. Happens when the VCR head gap gets filled with dirt, dust, grease, food particles, oxide, or any other foreign material. The result is a loss in image quality or a loss of image altogether. The sure cure is a head cleaning. Do it yourself with a cleaning kit or head-cleaning videocassette. Or you can take it to your nearest video dealer.
Helical Scanning. The method used for getting the electromagnetic information onto the video tape in most consumer VCRs. The tape follows a partial helical (or spiral) path around a slightly slanted video head.
Hertz (Hz). Cycles per second. Audio signals range from 20 to 15,000 Hz. Video signals are from 0 to 4 million Hz. Household electrical current is 60 Hz.
High Density. Refers to video tape that achieves a maximum of magnetic particles per square inch.
Horizontal Resolution. The number of lines that can be counted on a TV screen. The more lines of horizontal resolution, the sharper the picture. The maximum horizontal resolution possible under American broadcast standards is 525. Most VCRs produce an image with a horizontal resolution in the vicinity of 240 to 340 lines.

I

Impedance. The resistance of a wire or electronic component to the flow of a signal toward it.
Interface. To interconnect equipment, particularly computer hardware.
IPS. Inches per second. Used to describe the speed at which video tape passes the VCR head.
Iris. The adjustable diaphragm in a camera lens that controls the amount of light reaching the light-sensitive material. In the case of video cameras, many models have automatically controlled irises. See: aperture, back light.

J

Jack. Male cable connector or plug.
Joystick. A lever-like device allows a user to control the position of an object on a video screen. Used especially with video games. Don't expect the same kind of precision from a home video game joystick as you find in the amusement arcade.

K

Kelvin. After Lord Kelvin, the scientist. A unit of measurement used to describe the color temperature of light. See: color temperature.

L

Lag. Image retention, usually a ghost-like image that remains on the screen momentarily after the camera angle has changed. Most consumer video cameras can be expected to produce a certain amount of lag, particularly in the highlight areas of a picture.
Lens Speed. The ability of a lens to transmit light, usually measured in f-stops. The smaller the f-stop number (ie. f.4 is smaller than f2.8) the faster the lens.
Line In. Audio input.
Line Scanning. The path of the electron beam as it moves across the CRT.
LCD (Liquid-Crystal Display). A highly accurate digital display now used on many VCRs, cameras, and other electronic devices. Usually dark numbers on a light background.
LED (Light-Emitting Diode). Another widely used type of digital display, usually bright red or green numbers on a dark background.
LP (Long Play). The 4-hour recording mode on VHS format VCRs.
Low-Light Level. For videocameras this is usually anything under about 50 footcandles. Recording in low-light will usually produce an image with excessive video snow and low contrast.
Luminance. The brightness of an object, or the brightness of a video signal.

M

Macro Lens. A close-up lens capable of focusing within inches of a subject, often combined with a zoom lens.
Matching Transformer. A device for equalizing the impedance of 2 different inputs. In home video installations 300-ohm to 75-ohm, and 75-ohm to 300-ohm matching transformers are frequently required. They cost $2 to $3 a piece and can be purchased from most video dealers of Radio Shack outlets.
Microwave. High-frequency radio waves—1,000 Megahertz and up—that are used to relay television programming, especially by satellite.
Mini Plug. A small jack used for audio cables.
M-Load. The loading mechanism used in the VHS format in which the video tape makes a M-shaped pattern around the audio and video heads.
Modem. A device for translating a computer's digital information into audio signals for telephone transmission.

N

Noise. Random interference—either audio or video—a certain amount of which is found in most audio and video components. Excessive noise will create an audible speaker hiss or show up on the screen as video snow. The signal-to-noise ratio is a measure of how much video noise is created by the components of a VCR or videocamera. The higher the number in the ratio, as expressed in decibels, the better. See decibel.
NTSC (National Television System Committee). The Federal Communications Commission committee that determines standards for U.S. television broadcast systems. All VCRs, video cameras, videodisc players, and TV sets sold in the U.S. are designed to the NTSC Color Standard.

O

Ohm. A measure of electrical resistance within a component or wire.
Omni-Directional Mike. A microphone sensitive to sound coming from any direction. Built-in omni-directional mikes are standard in most inexpensive video cameras.
Optical Viewfinder. A viewfinder that uses lenses rather than electronics. Through-The-Lens (TTL) viewfinders allow you to see exactly what the camera sees. There are still a few very basic video cameras on the market that have optical viewfinders which are not TTL, but merely a retangular guide used to frame your scene. We recommend avoiding the non-TTL optical viewfinders.
Oxide. The magnetic particles that are bound to a polyester ribbon to form video tape. It is the oxide that stores the audio and video signals created by the VCR heads.

P

Pause. The VCR transport control that momentarily stops the tape without unloading it from the heads. Pausing is very convenient since you do not have to wait for the video tape to become loaded and unloaded. However, it placed enormous strain on video tape and heads. Over-pausing can damage both.
Phase. Relative timing of one signal to another.
Playback Head. Audio or video head on a VCR that reads information that has been previously written on the video tape. In most helical scan VCRs one video head serves for both the recording and playback functions.
Phono Plug. A type of jack commonly used for audio, and some video, cable connections. Also called an RCA plug.
Power Zoom. A motorized zoom lens available on many top-of-the-line video cameras. Power zoom often comes with a manual override and multiple-speed settings.
Printer. A computer-driven typewriter that can provide you with a "hard copy" version of a CRT display.
Projection TV. A large-screen TV in which the image of 1 to 3 video tubes is magnified by means of an optical lens system.
Programmable VCR. A videocassette recorder with a microprocessor-assisted tuner that can record multiple events off a number of different channels over an extended period of time—up to as many as 8 events over 14 days.
Public Domain (PD). Legal status of works not protected by current copyright laws. All PD programs can be legally duplicated and sold by anyone.

R

Radio Frequency (RF). The range of frequencies used for the transmission of electric waves.
RF Amplifier. An amplifier used to boost radio frequency signals.
RF Converter. A device that transforms audio and video signals into RF signals for TV playback. Most VCRs have a built-in RF converter.
RF Splitter. An inexpensive device for dividing an RF signal among several outputs. Used for multiple TV or VCR installations. Installations using RF splitters sometimes require an RF amplifier to boost the signal.
Rapid Scan. An extremely useful special feature with which the tape speed can be accelerated without losing the video image. Very handy for locating program segments.
Raster. The pattern created by an electron beam across the face of a CRT.
RAM (Random Access Memory). This is the computer memory you can store, locate, and remove.
ROM (Read Only Memory). The computer memory you cannot change.
Resolution. The amount of detail visible in a video picture.
Roll. Loss of verticle sync causing the picture to move up or down across the screen.
RPM. Revolutions per minute.

S

Scanning. Horizontal and vertical movement of an electron beam across CRT or vidicon tube.
Satellite Receiver (Also, Satellite Station, Earth Station, or Downlink Receiver). A set of electronic components including an antenna and tuner/receiver that is capable of collecting, tuning, and amplifying satellite microwave transmissions.
Search. Memory search is a VCR feature that automatically stops the fast forward or reverse mode when an electric impulse is sensed on the tape. Also called memory cue. It is a nonvisual method for program location. In most VCRs the electronic memory search cue is placed on the tape when the record button is pushed. Speed search is a visual method for locating program segments by scanning a tape at a faster than normal speed. See speed scan.
Shadow Mask. A metal sheet with 500,000 small holes that is located between the electron guns and the CRT screen.
Signal. Information that has been translated into an electrical impluse or wave.
Signal-To-Noise Ratio. See noise and decibel.
Skew. Tape tension.
Smear. A blurring or bleeding of the video picture, usually caused by insufficient lighting. Also called streaking.
Software. Cassettes, discs, game cartridges, or other programs or programming information used in home video equipment. See hardware.
Snow. A partial obliteration of the video image by "snow-like" white dots or streaks on the picture screen caused by video noise or other interference. The most common and easiest to cure cause of video snow is head clogging.
Solenoid. An electromagnetic circuit control system that can be operated by a "soft touch" electronic switch instead of a piano key type mechanical linkage.
Solid State. Electronic circuitry—usually transistors and integrated circuits—that does not require vacuum tubes.
Special Effects. A range of recently developed VCR playback features including slow motion, freeze frame, frame-by-frame advance, and speed scan.
Speed Scan. A special effects feature useful for locating program segments, replaying certain scenes, and reducing the time spent viewing commercials. In speed scan the tape speed is accelerated—usually from 2 to 40 times—without losing the video image. Also called speed search, rapid scan, cue and review, visual scan, Betascan, and super scan.
Switcher (RF Switcher). A black-box accessory for combining the outputs from a number of different sources such as cable TV, video games, home computers, videodisc players, and VCRs.

T

Teletext. A system for transfering written or graphic information via standard TV signals.
TTL (Through The Lens) Viewfinder. The type of optical viewfinder used on most of the less expensive models of video cameras. The TTL viewfinder uses a mirror or prism to let you see the same image that is reaching the vidicon tube. The major disadvantage to the TTL viewfinder is that, unlike the electronic viewfinder, it cannot be used as a TV monitor to view your recordings in the field.
Transport. The mechanical components of a VCR that load, move, and align the video tape. Play, record, fast forward, rewind, and pause are the basic transport functions and are operated by buttons called the "transport controls."
Tuner/Timer. The tuner is a component of a VCR or TV set that "tunes in" the specific RF frequency of each individual UHF and VHF channel. The timer is a digital clock that allows a VCR to be turned on and off automatically. All console VCRs have some sort of built-in tuner/timer. Portable VCRs require a separate tuner/timer unit. Most electronic tuner/timers are programmable.

U

UHF (Ultra High Frequency). A term describing the general frequency band of RF signals used to carry Channels 14 to 83.
U-Loading. The loading technique used by Beta format VCRs in which the video tape follows a U-shaped pattern around the audio and video heads.

V

VHF (Very High Frequency). The frequency band of RF signals used to carry Channels 2 through 13.
Vertical Interval. The unused band of horizontal lines that becomes visible when a video image rolls.
Video. A catch-all term that is used to describe almost anything related to the transmission or reception of television signals. The invisible electrical waves that carry TV signals from the transmitter to your TV set or VCR are video signals. The picture on your TV screen is a video image, the screen itself is a video display, the camera that recorded the video image is a video camera, the magnetic tape on which it was recorded is video tape, and the machine which plays it back is a videocassette recorder. It can all be traced back to a Latin verb, *videre,* to see. So the next time your VCR goes on the blink, blame Caesar.
Videodisc. See disc.
VHS (Video Home System). The VCR format developed by Matsushita Electric and JVC. VHS format recorders now outnumber Beta VCRs by about 3 to 1.
Vidicon Tube. The light-sensitive electron gun tube in a video camera that translates the optical image created by the lens into electrical impulses.

W

White Balance. An automatic circuit in most video cameras that maintains proper white tones. From time to time it may be necessary to adjust the white balance setting for optimum results.

X

X-1, X-2, X-3. The tape speeds designed for Beta format VCRs. In *THE SOURCEBOOK* we generally refer to these speeds as Beta I, Beta II, and Beta III. The X-1—or Beta 1—speed gets just 1 hour of recording time from an L-500 cassette and is now generally obsolete. A few Sony VCRs include Beta I, but for playback only.

Z

Zoom Lens. A lens that can be adjusted from wide angle to telephoto. In video cameras we recommend buying a camera with a zoom lens, preferably one that is attached to the camera with a C-mount.